A Course in Regression and Smoothing Methods

This book provides a concise account of four components of regression and smoothing methods: linear regression, generalized linear models, spline and kernel methods, and generalized linear mixed models. By bringing together parametric regression and non-parametric smoothing methods, the book emphasizes connections across methods, enabling readers to recognize common structures and to adapt techniques to new problems.

While standard texts often focus on the application of statistical methods from a user's perspective, this book covers the foregoing topics from a developer's perspective, with systematic attention to the mathematical, statistical, and computational ideas and results that underlie the methods. The distinction is analogous to that between a user's manual and a developer's manual for software: the goal is not only to demonstrate how to apply the methods, but also how they are derived and implemented.

Assuming a basic knowledge of undergraduate statistics, the book is intended primarily as a graduate textbook for the teaching and studying regression and smoothing methods. It serves as a useful resource for students and researchers in Statistics, Data Science, and related fields who wish to move beyond routine application and study modern regression and smoothing methods at a more advanced level.

Key Features:

- Focuses on core and representative topics in regression and smoothing while addressing important methodological issues often omitted at the introductory level.
- Presents regression and smoothing methods in a coherent, interconnected manner that highlights their common structures and relationships.
- Explains and demonstrates numerical algorithms in a self-contained way, with R code that implements the methods directly rather than solely relying on existing packages.
- Reinforces learning through not only end-of-chapter exercises but also questions and exercises integrated into the main text.

Zhiqiang Tan is a Distinguished Professor in the Department of Statistics, Rutgers University. His research and teaching interests include Monte Carlo methods, causal inference, statistical learning, and related areas. He is a Fellow of the American Statistical Association, a Fellow of the Institute of Mathematical Statistics, and an Elected Member of the International Statistical Institute.

CHAPMAN & HALL/CRC

Texts in Statistical Science Series

Recently Published Titles

A Course in the Large Sample Theory of Statistical Inference
W. Jackson Hall and David Oakes

Statistical Inference, Second Edition
George Casella and Roger Berger

Nonparametric Statistical Methods Using R, Second Edition
John Kloke and Joseph McKean

Generalized Linear Mixed Models
Modern Concepts, Methods and Applications, Second Edition
Walter W. Stroup, Marina Ptukhina, and Julie Garai

Analysis of Categorical Data with R, Second Edition
Christopher R. Bilder, Thomas M. Loughin

Applied Nonparametric Statistical Methods, Fifth Edition
Nigel C. Smeeton, Neil H. Spencer, and Peter Sprent

Linear Models with R, Third Edition
Julian J. Faraway

Statistics in Survey Sampling
Jae Kwang Kim

Time Series: A Data Analysis Approach Using R
Robert H. Shumway and David S. Stoffer

Bayesian Statistical Methods: With Applications to Machine Learning
Brian J. Reich, Sujit K. Ghosh

Experimental Design for Data Science and Engineering
V. Roshan Joseph

Survival Analysis: Principles and Applications in Clinical Trials and Beyond
Song Yang

Foundations of Bayesian Statistics for Data Scientists: With R and Python
Alan Agresti, Maria Kateri, Ranjini Grove, and Antonietta Mira

A Course in Regression and Smoothing Methods
Zhiqiang Tan

For more information about this series, please visit: https://www.routledge.com/Chapman--HallCRC-Texts-in-Statistical-Science/book-series/CHTEXSTASCI

A Course in Regression and Smoothing Methods

Zhiqiang Tan

CRC Press
Taylor & Francis Group
Boca Raton London New York

CRC Press is an imprint of the
Taylor & Francis Group, an **informa** business
A CHAPMAN & HALL BOOK

Designed cover image: Zhiqiang Tan

First edition published 2026
by CRC Press
2385 NW Executive Center Drive, Suite 320, Boca Raton FL 33431

and by CRC Press
4 Park Square, Milton Park, Abingdon, Oxon, OX14 4RN

CRC Press is an imprint of Taylor & Francis Group, LLC

ISBN: 978-1-041-15277-4 (hbk)
ISBN: 978-1-041-15686-4 (pbk)
ISBN: 978-1-003-68072-7 (ebk)

DOI: 10.1201/9781003680727

Typeset in CMR10 font
by KnowledgeWorks Global Ltd.

Publisher's note: This book has been prepared from camera-ready copy provided by the author.

Contents

Preface

The subject of this book consists of four components, as indicated by the four chapters of the book:

(i) Linear regression,

(ii) Generalized linear models,

(iii) Spline and kernel methods for smoothing, and

(iv) Generalized linear mixed models.

These topics are fundamental and standard in Statistics, being taught in various undergraduate and graduate courses. In fact, each of the four topics is often covered by a separate semester-long course, with its focus on important sub-topics for the *application* of the relevant methods in practical data analysis. In other words, the courses are usually designed for students to learn the methods from a *user's* perspective. While such courses are warranted at undergraduate and Master's levels, there is also a need to teach (or present) the foregoing topics to more advanced students (or readers), for example, PhD students, from a *developer's* perspective, including mathematical, statistical, and computational ideas and results which underlie the *development* of the methods. This distinction is similar to that between a user's manual and a developer's manual for a software system such as R (R Core Team, 2020). The advanced (or developer's) material tends to be buried and scattered among specialized research papers and books and hence not readily accessible to students and readers who want to study beyond the basic (or user's) material. To bridge different levels, it is desirable to present the advanced material in an accessible, compact, and integrated manner.

The book aims to provide an accessible and concise account of both parametric regression and nonparametric smoothing methods at an advanced level, while assuming basic knowledge of Statistics at the undergraduate level. These goals are pursued through the following choices.

(i) The book focuses on main representative topics in regression and smoothing methods, and devotes careful attention to various issues which are important but ignored at introductory levels, including random- vs fixed-design regression and model misspecification.

(ii) The book covers regression and smoothing methods in a coherent and inter-connected manner, for example, residual-on-residual regression related to doubly robust estimation, ridge regression related to reproducing kernel Hilbert space regression and mixed-effect regression, and Lasso regression related to total variation splines.

(iii) The book explains and demonstrates numerical algorithms with R code for all statistical methods studied (for example, Lasso estimation, Fisher scoring, smoothing splines), in a direct and self-contained manner instead of calling existing R packages.

(iv) The book not only provides exercises at the end of each chapter, but also includes questions and exercises alongside regular discussions.

As facilitated by the four features, the book is mainly intended as a graduate textbook (used for teaching and studying representative topics), instead of as a research book (used for documenting and disseminating research). The primary audience for the book is graduate students and researchers in Statistics, Data Science, and related fields. A secondary audience may include anyone who is interested in studying the book material, i.e., both parametric regression and nonparametric smoothing methods.

All coding and computation demonstrated in the book are performed using R Statistical Software (R Core Team, 2020). Putting a method (or algorithm) in computer code is not only necessary for numerically applying the method, but also instrumental to conceptually understanding the method. Accompanying R code and data files are included as supplementary material on the publisher's book webpage and on my book webpage:

`https://statweb.rutgers.edu/ztan/books.html`.

Errata and updates will also be posted on my book webpage.

This book can be studied together with a sister book, "A Course in Large-sample and High-dimensional Theory" (Tan, 2025). The theory book aims to present both classical large-sample theory and modern high-dimensional theory at a more advanced level than, for example, Casella and Berger (2001), but in an accessible and concise manner. I hope that the two books may provide an up-to-date account of statistical theory and methods, moving from elementary, conventional material to advanced, modern material.

This book is based on lecture notes used for several years in a first-year PhD course, titled "Advanced Applied Statistics I", at the Department of Statistics, Rutgers University. From my experience, the book provides sufficient material including numerical demonstration to be covered in a semester of 14 weeks with 2 classes each of 80 minutes per week. The theory book mentioned above is based on my teaching of another first-year PhD course, titled "Advanced Modern Statistical Inference II". I am grateful to the Department for the opportunity to design and teach these courses. I am also grateful to students in my classes over the years for their questions and comments. Our discussions have helped to shape the two books.

I thank Peter McCullagh for detailed and constructive comments. I also thank my editor, David Grubbs, for encouragement and support.

Zhiqiang Tan, December 2025

1

Linear regression

In this chapter, we study linear regression through the following topics, while assuming knowledge of basic material on linear regression. First, we discuss estimation and properties in fixed-design and random-design settings with correct model misspecification and then model-robust evaluation and inference allowing for model misspecification. Second, we discuss residual on residual regression and, in a closely related manner, doubly robust estimation in partially linear regression. Third, we discuss ridge and Lasso penalized linear regression, with particular focus on numerical algorithms for Lasso regression. We provide numerical demonstration in R, using both existing packages and self-contained code for implementation and application of main methods.

Linear regression is one of the basic statistical methods, widely used and heavily studied. Useful books on linear regression include Draper and Smith (1998), Searle (1971), and Weisberg (2014) among others. We assume understanding of basic material on linear regression, and study several topics which may not have been discussed before in an explicit or accessible manner. All asymptotic properties considered are in the classical setting where the number of parameters is fixed and the sample size tends to infinity. To help understanding, we provide direct and complete proofs as much as possible.

1.1 Data and model

1.1.1 Data

Let $\{(X_i, Y_i) : i = 1, \ldots, n\}$ be a collection of n *observations*, where X_i is a *covariate vector* (also called a regressor vector, or a predictor vector) and Y_i is a scalar *response variable* (also called an outcome variable). We write $X_i = (X_{i1}, \ldots, X_{ip})^{\mathrm{T}}$ as a column vector, with X_{ij} denoting the jth component of the ith covariate vector X_i.

DOI: 10.1201/9781003680727-1

1.1.2 Model

A linear regression model postulates that

$$Y_i = X_i^{\mathrm{T}}\beta^* + \varepsilon_i, \quad i = 1, \ldots, n, \tag{1.1}$$

where $\beta^* = (\beta_1^*, \ldots, \beta_p^*)^{\mathrm{T}}$ is an unknown parameter vector, and $\varepsilon_i = Y_i - X_i^{\mathrm{T}}\beta^*$ is a noise variable. To complete the model definition, statistical assumptions are required in addition to (1.1). It is important to distinguish two settings (or designs):

- *Fixed design.* The covariate vectors $X_1, \ldots, X_n$ are fixed, and the responses $Y_1, \ldots, Y_n$ are random. A fixed design is plausible in the context of designed experiments (after conditioning on any randomization). For example, the covariates may record the levels of temperatures and pressures applied in an industrial experiment. The covariate levels may be either deterministically picked or randomly picked but then conditioned on in subsequent analysis. The following assumptions are often made:

$$Y_1, \ldots, Y_n \text{ are independent with } X_1, \ldots, X_n \text{ fixed}, \tag{1.2a}$$
$$E(Y_i) = X_i^{\mathrm{T}}\beta^*, \quad i = 1, \ldots, n. \tag{1.2b}$$

  Assumption (1.2b) can be equivalently stated as $E(\varepsilon_i) = 0$ for $i = 1, \ldots, n$. The independence assumption (1.2a) is satisfied when the individuals are independently sampled from sub-populations with fixed covariate levels. Conceptually, randomization of covariate levels, if any, can be considered to be done prior to sampling.

- *Random design.* The pairs $(X_1, Y_1), \ldots, (X_n, Y_n)$ are random. A random design is plausible in observational studies. For example, the covariates may record the heights and weights of patients in a medical study. The following assumptions are often made:

$$(X_1, Y_1), \ldots, (X_n, Y_n) \text{ are IID as } (X, Y), \tag{1.3a}$$
$$E(Y|X) = X^{\mathrm{T}}\beta^*, \tag{1.3b}$$

  where IID refers to "independent and identically distributed", and (X, Y) is an identically distributed pair as (X_i, Y_i). The marginal distribution of X is left unspecified. Assumption (1.3b) can be equivalently stated as $E(\varepsilon|X) = 0$ with $\varepsilon = Y - X^{\mathrm{T}}\beta^*$. The IID assumption is justified when the individuals are considered a simple random sample with replacement, or approximately justified for a simple random sample without replacement from a large population.

For a random design with IID pairs (X_i, Y_i), it is convenient to introduce (X, Y) representing the underlying population and state the conditional mean restriction compactly as (1.3b), which otherwise can be stated as $E(Y_i|X_i) = X_i^{\mathrm{T}}\beta^*$ for $i = 1, \ldots, n$. Given a random design, a fixed design can also be considered to be derived by conditioning on (or fixing) $(X_1, \ldots, X_n)$ for the purpose of statistical inference.

In the fixed or random design, the noise variables $\varepsilon_1, \ldots, \varepsilon_n$ are independent of each other by assumption (1.2a) or (1.3a), although ε_i may not be independent of X_i in the random design. Assumption (1.2b) or (1.3b) only restricts that the means or conditional means of the response variables depend linearly on the covariates. Additional assumptions can also be made, beyond the mean restrictions. For example, *homoscedastic* linear regression further postulates that in the fixed design

$$\text{var}(Y_i) = \text{var}(\varepsilon_i) = \sigma^{*2}, \quad i = 1, \ldots, n, \tag{1.4}$$

or in the random design

$$\text{var}(Y|X) = \text{var}(\varepsilon|X) \equiv \sigma^{*2}, \tag{1.5}$$

where σ^* is an unknown parameter. Moreover, distributional assumptions can be introduced on the noises such as marginally in the fixed design,

$$Y_i \sim \text{N}(X_i^{\text{T}}\beta^*, \sigma^{*2}) \iff \varepsilon_i \sim \text{N}(0, \sigma^{*2}), \quad i = 1, \ldots, n, \tag{1.6}$$

or conditionally on X in the random design,

$$Y|X \sim \text{N}(X^{\text{T}}\beta^*, \sigma^{*2}) \iff \varepsilon|X \sim \text{N}(0, \sigma^{*2}). \tag{1.7}$$

[Exercise: Show that $\text{var}(\varepsilon) = \sigma^{*2}$ marginally under (1.5) in the random design. In general, $\text{var}(\varepsilon)$ remains an unknown constant, but $\text{var}(\varepsilon|X)$ is a function of X. Show that ε is independent of X, and hence $\varepsilon \sim \text{N}(0, \sigma^{*2})$, under (1.7) in the random design.]

For clarity, we distinguish two types of statistical assumptions stated above. Assumption (1.2a) or (1.3a) represents a (nonparametric) *sampling assumption* about the basic probabilistic structure of the observed data, such as randomness or independence. Assumption (1.2b), (1.3b), or any of (1.4)–(1.7) represents a (parametric) *modeling assumption* about a specific aspect of the probability distributions of the observed data. From this perspective, a fixed design or random design amounts to a choice of the sampling assumptions. [In a broader sense, modeling assumptions may also be said to include sampling assumptions. But then some other terminology is needed to indicate the two distinct types of assumptions.]

1.2 Fixed-design linear regression

Consider linear regression (1.1) in the fixed design, with assumptions (1.2a)–(1.2b). In other words, both sampling assumption (1.2a) and modeling assumption (1.2b) are correctly specified. Hence statistical properties and inference discussed below are said to be *model-based.*

1.2.1 Estimation

The least squares (LS) estimator of β^* is defined as

$$\hat{\beta} = \text{argmin}_\beta \frac{1}{n} \sum_{i=1}^{n} (Y_i - X_i^{\mathrm{T}} \beta)^2. \tag{1.8}$$

By solving the stationary condition that the gradient of the objective is 0, a closed-form expression for $\hat{\beta}$ is

$$\hat{\beta} = \left(\sum_{i=1}^{n} X_i X_i^{\mathrm{T}} \right)^{-1} \left(\sum_{i=1}^{n} X_i Y_i \right). \tag{1.9}$$

provided that the Hessian of $L_n(\beta) = n^{-1} \sum_{i=1}^{n} (Y_i - X_i^{\mathrm{T}} \beta)^2$,

$$\hat{H} = \frac{1}{n} \sum_{i=1}^{n} X_i X_i^{\mathrm{T}},$$

is nonsingular, which is assumed unless otherwise stated. [See Exercises 1.1 and 1.2 for some properties of multivariate differentiation.] We introduce the following two sets of notations, which are useful in different ways. First, we denote by $\tilde{E}(\cdot)$ a sample average over $(X_1, Y_1), \ldots, (X_n, Y_n)$, which are treated as a sample of (X, Y) in a loose sense. Then the displays (1.8) and (1.9) can be written as

$$\begin{aligned} \hat{\beta} &= \text{argmin}_\beta \, \tilde{E} \left\{ (Y - X^{\mathrm{T}} \beta)^2 \right\} \\ &= \tilde{E}^{-1}(X X^{\mathrm{T}}) \tilde{E}(XY). \end{aligned}$$

Alternatively, we use the matrix notation, by defining

$$\mathbf{X} = \begin{pmatrix} X_1^{\mathrm{T}} \\ \vdots \\ X_n^{\mathrm{T}} \end{pmatrix}, \quad \mathbf{Y} = \begin{pmatrix} Y_1 \\ \vdots \\ Y_n \end{pmatrix},$$

which is an $n \times p$ matrix or an $n \times 1$ vector respectively. Then the displays (1.8) and (1.9) can also be written as

$$\begin{aligned} \hat{\beta} &= \text{argmin}_\beta \, \|\mathbf{Y} - \mathbf{X}\beta\|_2^2 \\ &= (\mathbf{X}^{\mathrm{T}} \mathbf{X})^{-1} (\mathbf{X}^{\mathrm{T}} \mathbf{Y}), \end{aligned}$$

where $\|u\|_2 = (\sum_{i=1}^{k} u_i^2)^{1/2}$ is the L_2 norm for a vector $u = (u_1, \ldots, u_k)^{\mathrm{T}}$. The matrix $\hat{H} = \mathbf{X}^{\mathrm{T}} \mathbf{X}/n$ is also called the (scaled) Gram matrix, whose entries are the scaled inner products from the p columns of $\mathbf{X}$.

Under homoscedasticity assumption (1.4), a common estimator of σ^{*2} is

$$\hat{\sigma}^2 = \frac{1}{n-p} \sum_{i=1}^{n} (Y_i - X_i^{\mathrm{T}} \hat{\beta})^2. \tag{1.10}$$

Under the normality assumption (1.6), the maximum likelihood (ML) estimation leads to the same estimator $\hat{\beta}$ for β^*, but a slightly different estimator for σ^{*2}:

$$\hat{\sigma}^2_{\text{ML}} = \frac{1}{n}\sum_{i=1}^{n}(Y_i - X_i^{\text{T}}\hat{\beta})^2.$$

The use of the denominator $n-p$ instead of n for $\hat{\sigma}^2$ is an adjustment of degrees of freedom such that the estimator $\hat{\sigma}^2$, unlike $\hat{\sigma}^2_{\text{ML}}$, is unbiased for σ^{*2} as indicated by Proposition 1.1 below.

1.2.2 Properties

Given point estimators, it is important to study their sampling properties, which not only indicate the goodness of the estimators but also facilitate the derivation of inference procedures (confidence intervals or hypothesis tests). First, the following result is standard about the sampling properties of $\hat{\beta}$ and $\hat{\sigma}^2$ under homoscedastic linear regression. As seen from the proof, the unbiasedness of $\hat{\beta}$ holds regardless of homoscedasticity assumption (1.4).

Proposition 1.1 (Linear regression with homoscedastic noises in fixed design)**.** *Suppose that linear regression (1.1) is correctly specified in the fixed design, with assumptions (1.2a)–(1.2b). Then $\hat{\beta}$ is unbiased for β^*: $E(\hat{\beta}) = \beta^*$. Under homoscedasticity assumption (1.4),* $\text{var}(\hat{\beta}) = (n\hat{H})^{-1}\sigma^{*2}$*, and $\hat{\sigma}^2$ is unbiased for σ^{*2}: $E(\hat{\sigma}^2) = \sigma^{*2}$.*

Proof. We use the $\tilde{E}(\cdot)$ notation. [Exercise: Complete the proof using the matrix notation.] First, we show that $E(\hat{\beta}) = \beta^*$:

$$\begin{aligned} E(\hat{\beta}) &= \hat{H}^{-1}E\left\{\tilde{E}(XY)\right\} \\ &= \hat{H}^{-1}\frac{1}{n}\sum_{i=1}^{n}X_iE(Y_i) = \hat{H}^{-1}\frac{1}{n}X_iX_i^{\text{T}}\beta^* = \beta^*. \end{aligned}$$

The first two equalities use the linearity of expectations, and the third equality uses $E(Y_i) = X_i^{\text{T}}\beta^*$ by (1.2b). Second, we show that $\text{var}(\hat{\beta}) = (n\hat{H})^{-1}\sigma^{*2}$:

$$\begin{aligned} \text{var}(\hat{\beta}) &= \hat{H}^{-1}\text{var}\left\{\tilde{E}(XY)\right\}\hat{H}^{-1} \\ &= \hat{H}^{-1}\left\{\frac{1}{n^2}\sum_{i=1}^{n}X_i\text{var}(Y_i)X_i^{\text{T}}\right\}\hat{H}^{-1} \qquad (1.11) \\ &= \hat{H}^{-1}\left\{\frac{1}{n^2}\sigma^{*2}\sum_{i=1}^{n}X_iX_i^{\text{T}}\right\}\hat{H}^{-1} = \frac{1}{n}\hat{H}^{-1}\sigma^{*2}. \end{aligned}$$

The second equality uses the independence of $(Y_1,\ldots,Y_n)$ by (1.2a). The third equality uses $\text{var}(Y_i) = \sigma^{*2}$ by (1.4). It is left as an exercise to show that $E(\hat{\sigma}^2) = \sigma^{*2}$. □

The sampling properties in Proposition 1.1 hold in finite samples, but only regarding the moments of $\hat{\beta}$ and $\hat{\sigma}^2$. For distributional properties (which are needed for deriving confidence intervals), we would expect that in large samples, $\hat{\beta}$ is approximately $\mathrm{N}(\beta^*, (n\hat{H})^{-1}\sigma^{*2})$. Under suitable conditions as $n \to \infty$ and p is fixed, it can be shown that (Section 1.10.2)

$$\sqrt{n}\hat{H}^{1/2}(\hat{\beta} - \beta^*) \underset{\mathcal{D}}{\to} \mathrm{N}(0, \sigma^{*2} I_p), \tag{1.12a}$$

$$\hat{\sigma}^2 \underset{\mathcal{P}}{\to} \sigma^{*2}, \tag{1.12b}$$

where $\underset{\mathcal{D}}{\to}$ denotes convergence in distribution, $\underset{\mathcal{P}}{\to}$ denotes convergence in probability, and I_p is the $p \times p$ identity matrix. For a symmetric nonnegative-definite matrix A, the square root $A^{1/2}$ is defined as a symmetric matrix B such that $B^2 = A$. If A is positive-definite, then $A^{1/2}$ is unique. [Question: Is it valid to state

$$\sqrt{n}(\hat{\beta} - \beta^*) \underset{\mathcal{D}}{\to} \mathrm{N}(0, \sigma^{*2}\hat{H}^{-1}) \text{ ?]}$$

From these results, a Wald confidence interval for β_j^*, the jth component of β^*, with asymptotic size $1 - c$ is

$$\hat{\beta}_j \pm z_{c/2}\hat{\sigma}\sqrt{\frac{(\hat{H}^{-1})_{jj}}{n}}, \tag{1.13}$$

where $\hat{\beta}_j$ is the jth component of $\hat{\beta}$, $(\hat{H}^{-1})_{jj}$ is the (j,j)th element of $\hat{H}^{-1}$, $z_{c/2}$ is the $(1 - c/2)$-quantile of $\mathrm{N}(0,1)$. [Question: From (1.12a), how can you show $\sqrt{n}\{(\hat{H}^{-1})_{jj}\}^{-1/2}(\hat{\beta}_j - \beta_j^*) \underset{\mathcal{D}}{\to} \mathrm{N}(0, \sigma^{*2})$?] The estimated standard deviation, also called the standard error (SE), for $\hat{\beta}_j$ is

$$\mathrm{SE}(\hat{\beta}_j) = \hat{\sigma}\sqrt{\frac{(\hat{H}^{-1})_{jj}}{n}}. \tag{1.14}$$

Hence the Wald confidence interval is of the form $\hat{\beta}_j \pm z_{c/2}\mathrm{SE}(\hat{\beta}_j)$. See Section 1.10.1 for a quick explanation of Wald confidence intervals.

Under the additional normality assumption (1.6), the following distributional results can be shown [Exercise]:

$$\hat{\beta} \sim \mathrm{N}(\beta^*, (n\hat{H})^{-1}\sigma^{*2}), \quad (n-p)\frac{\hat{\sigma}^2}{\sigma^{*2}} \sim \chi^2_{n-p},$$

independently, where χ^2_k denotes the χ^2 distribution with k degrees of freedom. From these results, confidence intervals for β_j^* can be derived, similarly to (1.13) but using t-distributions, to achieve exact (finite-sample) coverage sizes.

To accommodate *heteroscedastic* noises, let

$$\mathrm{var}(Y_i) = \mathrm{var}(\varepsilon_i) = \sigma_i^{*2}, \quad i = 1, \ldots, n, \tag{1.15}$$

where $(\sigma_1^*, \ldots, \sigma_n^*)$ are unknown parameters. The following result can be directly deduced from the proof of Proposition 1.1, in particular, Equation (1.11).

Proposition 1.2 (Linear regression with heteroscedastic noises in fixed design). *Suppose that linear regression (1.1) is correctly specified in the fixed design, with assumptions (1.2a)–(1.2b). Then* $E(\hat{\beta}) = \beta^*$ *and*

$$\mathrm{var}(\hat{\beta}) = \hat{H}^{-1} \left\{ \frac{1}{n^2} \sum_{i=1}^n \sigma_i^{*2} X_i X_i^{\mathrm{T}} \right\} \hat{H}^{-1}, \tag{1.16}$$

where $(\sigma_1^*, \ldots, \sigma_n^*)$ *are defined in (1.15), allowing heteroscedastic noises. Under homoscedasticity (1.4),* $\sigma_1^* = \cdots = \sigma_n^* = \sigma^*$*, the preceding expression of* $\mathrm{var}(\hat{\beta})$ *reduces to* $(n\hat{H})^{-1}\sigma^{*2}$ *as in Proposition 1.1.*

Apparently, each individual parameter σ_i^{*2} cannot be consistently estimated, without introducing further restrictions, for example, on how σ_i^{*2} may be related to X_i. Nevertheless, $\mathrm{var}(\hat{\beta})$ can be consistently estimated without requiring each individual σ_i^{*2} to be consistently estimated. Let

$$\hat{\Sigma}(\beta) = \hat{H}^{-1}\hat{G}(\beta)\hat{H}^{-1}, \tag{1.17a}$$

$$\hat{H} = \frac{1}{n}\sum_{i=1}^n X_i X_i^{\mathrm{T}}, \quad \hat{G}(\beta) = \frac{1}{n}\sum_{i=1}^n (Y_i - X_i^{\mathrm{T}}\beta)^2 X_i X_i^{\mathrm{T}}. \tag{1.17b}$$

A natural estimator for $\mathrm{var}(\hat{\beta})$ is $n^{-1}\hat{\Sigma}(\hat{\beta}) = n^{-1}\hat{H}^{-1}\hat{G}(\hat{\beta})\hat{H}^{-1}$, obtained by replacing σ_i^{*2} with $(Y_i - X_i^{\mathrm{T}}\hat{\beta})^2$ in (1.16). Either $n^{-1}\hat{\Sigma}(\hat{\beta})$ or $\hat{\Sigma}(\hat{\beta})$ is called a *sandwich variance estimator*. Under suitable conditions as $n \to \infty$ and p is fixed, it can be shown that (Section 1.10.2)

$$\sqrt{n}\{\hat{\Sigma}(\beta^*)\}^{-1/2}(\hat{\beta} - \beta^*) \underset{\mathcal{D}}{\rightarrow} \mathrm{N}(0, I_p), \tag{1.18a}$$

$$\|\hat{\Sigma}(\hat{\beta}) - \hat{\Sigma}(\beta^*)\| \underset{\mathcal{P}}{\rightarrow} 0, \tag{1.18b}$$

where $\|\cdot\|$ is a matrix norm, and $\{\hat{\Sigma}(\beta^*)\}^{1/2} = \hat{H}^{-1}\{\hat{G}(\beta^*)\}^{1/2}$ and hence $\{\hat{\Sigma}(\beta^*)\}^{-1/2} = \{\hat{G}(\beta^*)\}^{-1/2}\hat{H}$. [Question: Is it valid to state $\hat{\Sigma}(\hat{\beta}) \underset{\mathcal{P}}{\rightarrow} \hat{\Sigma}(\beta^*)$?] From these results, a Wald confidence interval for β_j^* with asymptotic size $1 - c$ is

$$\hat{\beta}_j \pm z_{c/2}\sqrt{\frac{\{\hat{\Sigma}(\hat{\beta})\}_{jj}}{n}}, \tag{1.19}$$

where $\{\hat{\Sigma}(\hat{\beta})\}_{jj}$ is the (j,j)th element of $\hat{\Sigma}(\hat{\beta})$. The associated standard error for $\hat{\beta}_j$ is

$$\mathrm{SE}(\hat{\beta}_j) = \sqrt{\frac{\{\hat{\Sigma}(\hat{\beta})\}_{jj}}{n}}. \tag{1.20}$$

Hence the Wald confidence interval can be written as $\hat{\beta}_j \pm z_{c/2}\mathrm{SE}(\hat{\beta}_j)$.

Returning to the initial idea of estimating $\mathrm{var}(\hat\beta)$ by $n^{-1}\hat\Sigma(\hat\beta)$, it can also be shown that as $n \to \infty$ and p is fixed,

$$\|\hat\Sigma(\hat\beta) - n\,\mathrm{var}(\hat\beta)\| \underset{\mathcal{P}}{\to} 0. \tag{1.21}$$

Combining (1.18b) and (1.21) gives $\|\hat\Sigma(\beta^*) - n\,\mathrm{var}(\hat\beta)\| \underset{\mathcal{P}}{\to} 0$. Hence the limit of the rescaled finite-sample variance $n\,\mathrm{var}(\hat\beta)$ and the asymptotic variance $\hat\Sigma(\beta^*)$ for $\sqrt{n}(\hat\beta - \beta^*)$ are the same. In general, the limiting finite-sample variance and asymptotic variance are two distinct concepts. See Casella and Berger (2001), Example 10.1.8. For convenience, we may ignore such a distinction in heuristic discussion.

By the matrix notation, the sandwich variance estimator $\hat\Sigma(\hat\beta)$ can be written as

$$\hat\Sigma(\hat\beta) = \hat H^{-1}\hat G(\hat\beta)\hat H^{-1},$$
$$\hat H = \mathbf{X}^{\mathrm{T}}\mathbf{X}/n, \quad \hat G(\hat\beta) = \mathbf{X}^{\mathrm{T}}\hat D(\hat\beta)\mathbf{X}/n,$$

where $\hat D(\hat\beta) = \mathrm{diag}(\hat r_1^2, \ldots, \hat r_n^2)$ and $\hat r_i = Y_i - X_i^{\mathrm{T}}\hat\beta$, the ith residual. The matrix expressions are convenient for numerical implementation.

1.3 Random-design linear regression

Consider linear regression (1.1) in the random design, with assumptions (1.3a)–(1.3b). In other words, both sampling assumption (1.3a) and modeling assumption (1.3b) are correctly specified. As in Section 1.2, statistical properties and inference discussed below are said to be *model-based.*

1.3.1 Estimation

Despite the change to a random design, the LS estimator $\hat\beta$ for β^* and, under homoscedastic noises, the usual estimator $\hat\sigma^2$ for σ^{*2}, are defined the same as in Section 1.2. Under the normality assumption (1.7), the ML estimators of (β^*, σ^{*2}) also remain the same in Section 1.2, although the likelihood function involved is a conditional likelihood derived from the conditional densities of Y_i given X_i for $i = 1, \ldots, n$.

1.3.2 Properties

The change from a fixed design to a random design brings subtle changes to the study of sampling properties. The finite-sample properties in Propositions 1.1 and 1.2 can be directly extended if interpreted conditionally on $(X_1, \ldots, X_n)$.

Proposition 1.3 (Conditional properties in random design). *Suppose that linear regression (1.1) is correctly specified in the random design, with assumptions (1.3a)–(1.3b). Then*

$$E(\hat{\beta}|X_1,\ldots,X_n) = \beta^*,$$
$$\mathrm{var}(\hat{\beta}|X_1,\ldots,X_n) = \hat{H}^{-1}\left\{\frac{1}{n^2}\sum_{i=1}^n \sigma^{*2}(X_i)X_iX_i^{\mathrm{T}}\right\}\hat{H}^{-1},$$

where $\sigma^{*2}(x) = \mathrm{var}(Y|X = x)$, *allowing heteroscedastic noises. Under homoscedasticity (1.5),* $\sigma^*(x) \equiv \sigma^*$, *the preceding expression of* $\mathrm{var}(\hat{\beta}|X_1,\ldots,X_n)$ *reduces to* $(n\hat{H})^{-1}\sigma^{*2}$, *and* $\mathrm{var}(\hat{\sigma}^2|X_1,\ldots,X_n) = \sigma^{*2}$.

Conducting inference about β^* conditionally on $(X_1,\ldots,X_n)$ in the random design proceeds in the same way as in the fixed design discussed in Section 1.2. Alternatively, it is informative to study unconditional properties of $\hat{\beta}$ and then unconditional inference about β^*, which are technically simpler than those in the fixed design; compare Section 1.10.2 and the proof of Proposition 1.4 below. Further differences between the fixed and random designs will be discussed in Section 1.4, where linear regression may be misspecified.

By the conditional unbiasedness, $\hat{\beta}$ remains unconditionally unbiased for β^*: $E(\hat{\beta}) = \beta^*$. Similarly, under the homoscedasticity assumption (1.5), $\hat{\sigma}^2$ remains unconditionally unbiased for σ^{*2}: $E(\hat{\sigma}^2) = \sigma^{*2}$. [Exercise: Prove these statements.] The unconditional variance of $\hat{\beta}$ can determined as $\mathrm{var}(\hat{\beta}) = E\{\mathrm{var}(\hat{\beta}|X_1,\ldots,X_n)\}$, for example, $\mathrm{var}(\hat{\beta}) = n^{-1}\sigma^{*2}E(\hat{H}^{-1})$ under homoscedasticity (1.5). [Exercise: Prove this statement.] But no further simplification is feasible, without additional information about the marginal distribution of X_i. Nevertheless, the asymptotic normality of $\hat{\beta}$ can be established in the random design, which turns out to be more straightforward than in the fixed-design, (1.12a) or (1.18a), due to the IID structure by assumption (1.3a). The proof below uses the law of large numbers and central limit theorem for IID variables, instead of triangular-array central limit theorems as discussed in Section 1.10.2. Let

$$\Sigma(\beta) = H^{-1}G(\beta)H^{-1}, \tag{1.22a}$$
$$H = E(XX^{\mathrm{T}}), \quad G(\beta) = E\left\{(Y - X^{\mathrm{T}}\beta)^2XX^{\mathrm{T}}\right\}, \tag{1.22b}$$

where (X,Y) is an identically distributed pair as (X_i,Y_i).

Proposition 1.4 (Unconditional properties in random design). *Suppose that linear regression (1.1) is correctly specified in the random design, with Assumptions (1.3a)–(1.3b). Assume that* H *is nonsingular and the diagonal elements of* $G(\beta^*)$ *are finite. As* $n \to \infty$ *and* p *is fixed,*

$$\sqrt{n}(\hat{\beta} - \beta^*) \underset{\mathcal{D}}{\rightarrow} \mathrm{N}(0, \Sigma(\beta^*)), \tag{1.23a}$$
$$\hat{\Sigma}(\hat{\beta}) \underset{\mathcal{P}}{\rightarrow} \Sigma(\beta^*), \tag{1.23b}$$

where $\hat{\Sigma}(\beta)$ *is defined as in (1.17a). Under homoscedasticity (1.5),* $\mathrm{var}(Y|X) \equiv \sigma^{*2}$, *(1.23a) reduces to* $\sqrt{n}(\hat{\beta} - \beta^*) \xrightarrow[\mathcal{D}]{} \mathrm{N}(0, \sigma^{*2}H^{-1})$, *and* $\hat{\sigma}^2\hat{H}^{-1} \xrightarrow[\mathcal{P}]{} \sigma^{*2}H^{-1}$.

Proof. We use the $\tilde{E}(\cdot)$ notation. By the definition of $\hat{\beta}$ and simple algebra,

$$\begin{aligned}\hat{\beta} - \beta^* &= \tilde{E}^{-1}(XX^{\mathrm{T}})\tilde{E}(XY) - \hat{\beta} \\ &= \tilde{E}^{-1}(XX^{\mathrm{T}})\tilde{E}\{X(Y - X^{\mathrm{T}}\beta^*)\}, \end{aligned} \tag{1.24}$$

and hence

$$\sqrt{n}(\hat{\beta} - \beta^*) = \tilde{E}^{-1}(XX^{\mathrm{T}}) \cdot \sqrt{n}\tilde{E}\{X(Y - X^{\mathrm{T}}\beta^*)\}. \tag{1.25}$$

By the central limit theorem for IID variables, $\sqrt{n}\tilde{E}\{X(Y - X^{\mathrm{T}}\beta^*)\} \xrightarrow[\mathcal{D}]{} \mathrm{N}(0, G(\beta^*))$. By the law of large numbers for IID variables, $\tilde{E}(XX^{\mathrm{T}}) \xrightarrow[\mathcal{P}]{} E(XX^{\mathrm{T}}) = H$ and hence $\tilde{E}^{-1}(XX^{\mathrm{T}}) \to H^{-1}$. With these properties, applying Slutsky's theorem in (1.25) yields (1.23a). The foregoing argument can also be seen to establish the asymptotic expansion

$$\sqrt{n}(\hat{\beta} - \beta^*) = H^{-1}\sqrt{n}\tilde{E}\{X(Y - X^{\mathrm{T}}\beta^*)\} + o_p(1),$$

which leads to asymptotic normality (1.23a) by the central limit theorem and Slutsky's theorem. The proof of the remaining conclusions is left as an exercise.
□

By Slutsky's theorem with (1.23a)–(1.23b), the Wald confidence interval (1.19) for β_j^* achieves asymptotic size $1-c$ as in the fixed design. Similarly, under homoscedasticity (1.5), the confidence interval (1.13) for β_j^* also achieves asymptotic size $1 - c$. Therefore, these confidence intervals remain valid in both the fixed and random designs, although the theoretical justifications are distinct from each other.

1.4 Model-robust evaluation and inference

The material in the preceding sections is developed in a *model-based* (frequentist) approach, which can be summarized with two components as follows (from a frequentist perspective):

(i) A linear regression model is introduced including the sampling and modeling assumptions (as described in Section 1.1), and point estimators for relevant parameters are constructed by exploiting implications of the model through moments or likelihoods.

(ii) Sampling properties of the estimators are studied and inferences (confidence intervals or hypothesis tests) are derived *provided* that the model is correctly specified.

The two components can be referred to as (i) construction (of a model and point estimators) and (ii) evaluation (of sampling properties of point estimators) and inferences (or uncertainty quantification). For inferences, whenever possible, we focus on *Wald confidence intervals and hypothesis tests*, which are plainly determined once asymptotic variances and consistent variance estimators are derived (see Section 1.10.1). Hence the main task in component (ii) is studying sampling properties. There are other methods for conducting inferences based on point estimators, for example, bootstrap or resampling-based methods. But from a frequentist perspective, the derivation and justification of inferences, implicitly or explicitly, depend on understanding of sampling properties of the associated point estimators.

The two components, *after* introducing a model in (i), are related to the concepts of constructing estimators and evaluating estimators popularized by Casella and Berger (2001). A subtle difference, however, is that both point estimators and interval estimators (i.e., confidence intervals) are included as two types of estimators in Casella and Berger (2001), whereas only point estimators are constructed in component (i) and then confidence intervals are derived in (ii) *after* studying sampling properties of the point estimators.

In this section, we discuss a *model-robust* approach for linear regression. We retain component (i) as above but proceed differently in component (ii): we consider the linear regression model and the LS estimator as in Sections 1.1–1.3, but study sampling properties and inferences under the standard sampling assumptions (fixed or random design) while allowing that the modeling assumption in linear regression may be misspecified (to be formalized below). In this approach, a more substantial difference emerges between the fixed design and the random design than in the model-based approach.

On a general note, a focal point of the model-robust approach is to treat a (point) estimator as a data-based algorithm (taking the sample data as input and producing an estimate as output). The estimator is usually motivated by some modeling assumptions in addition to (more basic) sampling assumptions, but its sampling properties can be studied, including the probability limit and asymptotic variance for the estimator, under only the sampling assumptions but possible misspecification of the modeling assumptions. Based on the sampling properties, statistical inference (or uncertainty quantification) can be conducted, without assuming that the modeling assumptions are correctly specified. See Figure 1.1.

1.4.1 Fixed-design linear regression

In the fixed design with assumption (1.2a), the linear regression model (1.2b) is said to be *misspecified* if Assumption (1.2b) does not hold, i.e., there exists no parameter value β^* such that $E(Y_i) = X_i^{\mathrm{T}}\beta^*$ for $i = 1, \ldots, n$. With possible model misspecification, Proposition 1.2 can be extended, with the true value

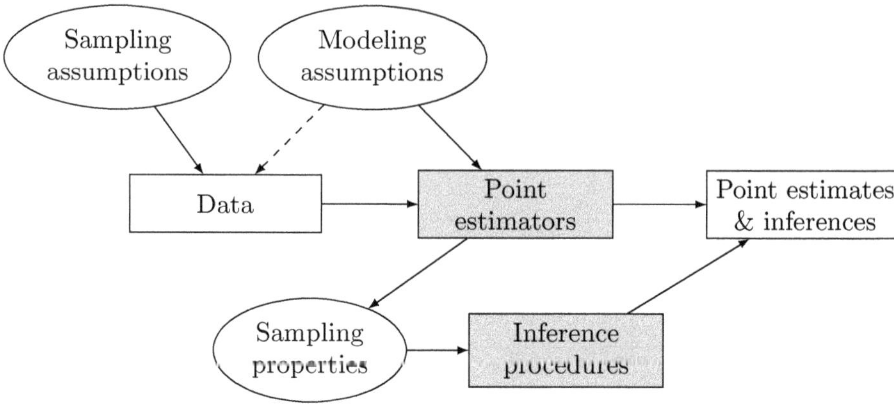

FIGURE 1.1
Flow chart in studying statistical methods: point estimators and inference procedures are data-based algorithms (shaded), taking data as input and producing point estimates and inferences as output. Both the sampling and modeling assumptions are required to be valid in model-based evaluation (of sampling properties) and inference, whereas the modeling assumptions are used to motivate point estimators but allowed to be misspecified, as signalized by a dashed arrow, in model-robust evaluation and inference.

β^* replaced by

$$\begin{aligned}\bar{\beta} &= \text{argmin}_\beta \sum_{i=1}^n E\{(Y_i - X_i^{\mathrm{T}}\beta)^2\} \\ &= \left(\sum_{i=1}^n X_i X_i^{\mathrm{T}}\right)^{-1} \left(\sum_{i=1}^n X_i E(Y_i)\right), \end{aligned} \tag{1.26}$$

which is called the target value for $\hat{\beta}$ (or limit value as mentioned below). [Exercise: Prove the second equality, with the first equality being a definition.] If linear regression (1.2b) is correctly specified, then $\bar{\beta}$ is identical to the true value β^*:

$$\bar{\beta} = \left(\sum_{i=1}^n X_i X_i^{\mathrm{T}}\right)^{-1} \left(\sum_{i=1}^n X_i (X^{\mathrm{T}}\beta^*)\right) = \beta^*.$$

Otherwise, $\bar{\beta}$ remains well-defined, but β^* does not exist. In this case, $\varepsilon_i = Y_i - X_i^{\mathrm{T}}\beta^*$ does not exist either. Hence σ^{*2} in (1.4) and σ_i^{*2} in (1.16) can only be defined in terms of Y_i, not ε_i.

Proposition 1.5 (Linear regression in fixed design with model misspecification). *Suppose that linear regression (1.2b) may be misspecified in the* *fixed design, with Assumption (1.2a). Then* $E(\hat\beta) = \bar\beta$ *and the expression of* $\text{var}(\hat\beta)$ *in (1.16) remains valid, where* $\sigma_i^{*2} = \text{var}(Y_i)$*. Under homoscedasticity,* $\sigma_1^* = \cdots = \sigma_n^* = \sigma^*$*, the expression of* $\text{var}(\hat\beta)$ *in (1.16) reduces to* $(n\hat H)^{-1}\sigma^{*2}$ *as in Proposition 1.1.*

The estimator $\hat\beta$ can be shown to be consistent for the target value $\bar\beta$: $\hat\beta \xrightarrow{\mathcal{P}} \bar\beta$ as $n \to \infty$ and p is fixed. Hence $\bar\beta$ is the limit value in probability for $\hat\beta$. However, a fundamental change from Section 1.2 is that $\text{var}(\hat\beta)$, after rescaled by n, in general cannot be consistently estimated. The sandwich variance estimator

$$\hat\Sigma(\hat\beta) = \hat H^{-1} \left\{ \frac{1}{n} \sum_{i=1}^n (Y_i - X_i^{\mathrm T}\hat\beta)^2 X_i X_i^{\mathrm T} \right\} \hat H^{-1}$$

tends to be inconsistent for the rescaled variance

$$n\,\text{var}(\hat\beta) = \hat H^{-1} \left\{ \frac{1}{n} \sum_{i=1}^n \sigma_i^{*2} X_i X_i^{\mathrm T} \right\} \hat H^{-1},$$

even if the latter converges to a limit as $n \to \infty$. In fact, although $\hat\beta \xrightarrow{\mathcal{P}} \bar\beta$, the individual term $(Y_i - X_i^{\mathrm T}\bar\beta)^2$ may no longer provide unbiased information about σ_i^{*2}, i.e., $E\{(Y_i - X_i^{\mathrm T}\bar\beta)^2\} \neq \sigma_i^{*2} = \text{var}(Y_i)$ with $X_i^{\mathrm T}\bar\beta \neq E(Y_i)$ under model misspecification. Furthermore, the estimator $\hat\beta$ may not satisfy the extension of asymptotic normality (1.18a) with β^* replaced by $\bar\beta$. From Section 1.10.2, the basic identity (1.73) underlying (1.18a) can be extended as

$$\hat\beta - \bar\beta = \hat H^{-1} \frac{1}{n} \sum_{i=1}^n X_i (Y_i - X_i^{\mathrm T}\bar\beta).$$

However, the term $Y_i - X_i\bar\beta$ may not be mean-zero, $E(Y_i) \neq X_i^{\mathrm T}\bar\beta$, and hence central limit theorems are not applicable to deduce asymptotic normality for $\hat\beta$. Without the extension of (1.18a), the confidence interval (1.19) does not achieve desired converge about $\bar\beta_j$ (the jth element of $\bar\beta$), i.e., it does not achieve asymptotic size $1-c$ in covering $\bar\beta$. [Question: How does the discussion hold under homoscedasticity, $\sigma_1^* = \cdots = \sigma_n^* = \sigma^*$?]

1.4.2 Random-design linear regression

In the random design with assumption (1.3a), the linear regression model (1.3b) is said to be misspecified if assumption (1.3b) does not hold, i.e., there exists no parameter value β^* such that $E(Y|X) = X^{\mathrm T}\beta^*$. With possible model misspecification, in contrast with the fixed design, the asymptotic normality of $\hat\beta$ in Proposition 1.4 can be extended by seemingly replacing the non-existent true value β^* with the target value for $\hat\beta$, defined as

$$\begin{aligned} \bar\beta &= \text{argmin}_\beta\, E\{(Y - X^{\mathrm T}\beta)^2\} \\ &= E^{-1}(XX^{\mathrm T})E(XY), \end{aligned} \tag{1.27}$$

[Exercise: Prove the second equality, with the first equality being a definition.] By Proposition 1.6 below, $\bar\beta$ is the limit value for $\hat\beta$: $\hat\beta \underset{\mathcal{P}}{\longrightarrow} \bar\beta$. If linear regression (1.3b) is correctly specified, then $\bar\beta$ is identical to the true value β^* provided that $H = E(XX^{\mathrm{T}})$ is nonsingular:

$$\begin{aligned}\bar\beta &= E^{-1}(XX^{\mathrm{T}})E(XE(Y|X)),\\ &= E^{-1}(XX^{\mathrm{T}})E(XX^{\mathrm{T}}\beta^*) = \beta^*.\end{aligned}$$

Otherwise, $\bar\beta$ remains well-defined, but β^* does not exist. In this case, $\varepsilon = Y - X^{\mathrm{T}}\beta^*$ does not exist either. Hence homoscedasticity in (1.5) can only be defined as $\mathrm{var}(Y|X) \equiv \sigma^{*2}$.

Proposition 1.6 (Linear regression in random design with model misspecification). *Suppose that linear regression (1.3b) may be misspecified in the random design, with assumption (1.3a). Assume that H is nonsingular and the diagonal elements of $G(\bar\beta)$ are finite. As $n \to \infty$ and p is fixed,*

$$\sqrt{n}(\hat\beta - \bar\beta) \underset{\mathcal{D}}{\longrightarrow} \mathrm{N}(0, \Sigma(\bar\beta)), \tag{1.28a}$$

$$\hat\Sigma(\hat\beta) \underset{\mathcal{P}}{\longrightarrow} \Sigma(\bar\beta), \tag{1.28b}$$

where $\hat\Sigma(\beta)$ and $\Sigma(\beta)$ are defined as in (1.17a) and (1.22a).
[Exercise: Under homoscedasticity, $\mathrm{var}(Y|X) \equiv \sigma^{*2}$, *explain that the convergence (1.28a) does not in general reduce to* $\sqrt{n}(\hat\beta - \bar\beta) \underset{\mathcal{D}}{\longrightarrow} \mathrm{N}(0, \sigma^{*2}H^{-1})$*.]*

Proof. The result follows from the proof of Proposition 1.4 with β^* replaced by $\bar\beta$ throughout. The asymptotic expansion of $\hat\beta$ about $\bar\beta$ is

$$\sqrt{n}(\hat\beta - \bar\beta) = H^{-1}\sqrt{n}\tilde{E}\{X(Y - X^{\mathrm{T}}\bar\beta)\} + o_p(1). \tag{1.29}$$

The central limit theorem for IID variables can be applied to deduce that $\sqrt{n}\tilde{E}\{X(Y - X^{\mathrm{T}}\bar\beta)\} \underset{\mathcal{D}}{\longrightarrow} \mathrm{N}(0, G(\bar\beta))$ because $E\{X(Y - X^{\mathrm{T}}\bar\beta)\} = 0$ by the definition of $\bar\beta$. □

From the two properties (1.28a)–(1.28b), the sandwich variance estimator $\hat\Sigma(\hat\beta)$ is model-robust: being consistent for estimating the rescaled asymptotic variance $\Sigma(\bar\beta)$ for $\hat\beta$, whether linear regression (1.3b) is correctly specified or not, in the random design. [The unscaled asymptotic variance of $\hat\beta$ is $\Sigma(\bar\beta)/n$.] Moreover, by the usual argument using Slutsky's theorem shows that with possible model misspecification, the Wald confidence interval (1.19) achieves asymptotic size $1-c$ in covering $\bar\beta_j$ (the jth element of $\bar\beta$). If linear regression (1.3b) is correctly specified, then $\bar\beta_j$ coincides with β_j^* and hence the confidence interval (1.19) automatically achieves the desired coverage about β_j^*. Therefore, the sandwich variance estimator $\hat\Sigma(\hat\beta)$ and the associated Wald confidence intervals enable model-robust uncertainty quantification of $\hat\beta$ against the target value $\bar\beta$.

Table 1.1 summarizes consistency (or lack of) for the LS estimator $\hat{\beta}$ and the variance estimators $\hat{\sigma}^2\hat{H}^{-1}$ and $\hat{\Sigma}(\hat{\beta})$ for $n\,\mathrm{var}(\hat{\beta})$. For simplicity, we do not distinguish consistency for $n\,\mathrm{var}(\hat{\beta})$ and the asymptotic variance of $\sqrt{n}(\hat{\beta}-\bar{\beta})$ (if exists). In the random design, there exists a clear difference in variance estimation: the variance estimator $\hat{\sigma}^2\hat{H}^{-1}$ is consistent only if linear regression is correctly specified with homoscedastic noises, whereas the variance estimator $\hat{\Sigma}(\hat{\beta})$ whether or not linear regression is correctly specified with possible heteroscedastic noises. Hence $\hat{\sigma}^2\hat{H}^{-1}$ is called a *model-based (and homoscedasticity-based) variance estimator* whereas $\hat{\Sigma}(\hat{\beta})$ and associated confidence interval (1.19) is called a *model-robust (and heteroscedasticity-robust) variance estimator*, in addition to being called a sandwich variance estimator. The associated Wald confidence interval (1.13) or (1.19) is said to be respectively model-based or model-robust.

An intriguing phenomenon from our discussion is that the impact of model misspecification on variance estimation (or uncertainty quantification) for the LS estimator $\hat{\beta}$ differs between the fixed and random designs. In the fixed design, $n\,\mathrm{var}(\hat{\beta})$ in general cannot be consistently estimated [by the sandwich variance estimator $\hat{\Sigma}(\hat{\beta})$] unless linear regression is correctly specified. In the random design, $n\,\mathrm{var}(\hat{\beta})$ can be consistently estimated [by $\hat{\Sigma}(\hat{\beta})$] whether linear regression is correctly specified or not. Conceptually, how can such a remarkable difference be explained? See Section 1.10.3 for further discussion.

Bibliographical note. Model-robust inference has been studied in various contexts. Early references on sandwich variance estimation include Huber (1967) and White (1982). Classical asymptotic theory with misspecified models is covered in Manski (1988). Recently, Buja, Berk, et al. (2019) and Buja, Brown, et al. (2019) provided systematic discussions on model-robust inference, where many intersections with our discussions in this section can be found.

1.5 Numerical demonstration: Linear regression

We demonstrate implementation of linear regression in R, mainly least squares estimation and associated variance estimation for regression coefficients. Although such tasks can be handled directly using the R software including contributed packages, one of our objectives is to demonstrate self-contained code to implement various statistical methods or algorithms, while calling only basic mathematical functions in R. We find and believe that the process of putting an algorithm in computer code (i.e., translating math into code) not only enables numerical application, but also enhances understanding of the algorithm. Any bug in computer code will directly result in an algorithm not working as desired.

TABLE 1.1
Consistency of LS estimator and variance estimators (after rescaled by n)

	Correctly specified linear regression		Misspecified linear regression	
	Fixed design	Random design	Fixed design	Random design
LS estimator $\hat{\beta}$	consistent for β^* in (1.2b)	consistent for β^* in (1.3b)	consistent for $\bar{\beta}$ in (1.26)	consistent for $\bar{\beta}$ in (1.27)
Variance estimator $\hat{\sigma}^2 \hat{H}^{-1}$	consistent under homoscedasticity	consistent under homoscedasticity	inconsistent	inconsistent
Variance estimator $\hat{\Sigma}(\hat{\beta})$ in (1.17a), (1.23b) and (1.28b)	consistent under heteroscedasticity	consistent under heteroscedasticity	inconsistent	consistent

Consider the data (`prostate.data`) taken from a study by Stamey et al. (1989) and used in Hastie et al. (2009), Section 3.2.1. The data include a response variable, the log level of prostate specific antigen (`lpsa`), and 8 covariates, log cancer volume (`lcavol`), log prostate weight (`lweight`), `age`, log of the amount of benign prostatic hyperplasia (`lbph`), seminal vesicle invasion (`svi`), log of capsular penetration (`lcp`), Gleason score (`gleason`), and percent of Gleason scores 4 or 5 (`pgg45`), in 97 men. For pre-processing, as in Hastie et al. (2009) and shown below, the covariates are standardized to have mean 0 and variance 1 (using the $1/(n-1)$ formula). The dataset is then split into a training set of size 67 and a test set of size 30.

```
> pcancer <- read.table("prostate.data", header=TRUE)
> pcancer[1:5,]
      lcavol  lweight age      lbph svi       lcp gleason pgg45
1 -0.5798185 2.769459  50 -1.386294   0 -1.386294       6     0
2 -0.9942523 3.319626  58 -1.386294   0 -1.386294       6     0
3 -0.5108256 2.691243  74 -1.386294   0 -1.386294       7    20
4 -1.2039728 3.282789  58 -1.386294   0 -1.386294       6     0
5  0.7514161 3.432373  62 -1.386294   0 -1.386294       6     0
        lpsa train
1 -0.4307829  TRUE
2 -0.1625189  TRUE
3 -0.1625189  TRUE
4 -0.1625189  TRUE
5  0.3715636  TRUE
>
> pcancer[,1:8] <- scale(pcancer[,1:8], center=TRUE, scale=TRUE)
>
> train <- pcancer[pcancer$train,1:9]  # size 67
> test <- pcancer[!pcancer$train,1:9]  # size 30
```

The following is the output from using the built-in R function `lm()` to fit linear regression of the response variable `lpsa` on the 8 covariates in addition to the intercept. [To save space, some lines are omitted in R output and indicated by "`......`".] From `summary(fit.lm)`, the column `Estimate` gives the individual components of the LS estimator by (1.9), whereas the column `Std.Error` gives the model-based standard errors by (1.14). The estimate $\hat{\sigma}$ is shown as `Residual standard error`, 0.7123.

```
> y <- train[,9]
> x <- as.matrix(train[,1:8])  #convert from data frame to data matrix
>
```

```
> fit.lm <- lm(y ~ x)
> summary(fit.lm)
......
Coefficients:
             Estimate Std. Error t value Pr(>|t|)
(Intercept)  2.46493    0.08931  27.598  < 2e-16 ***
xlcavol      0.67953    0.12663   5.366 1.47e-06 ***
xlweight     0.26305    0.09563   2.751  0.00792 **
xage        -0.14146    0.10134  -1.396  0.16806
xlbph        0.21015    0.10222   2.056  0.04431 *
xsvi         0.30520    0.12360   2.469  0.01651 *
xlcp        -0.28849    0.15453  -1.867  0.06697 .
xgleason    -0.02131    0.14525  -0.147  0.88389
xpgg45       0.26696    0.15361   1.738  0.08755 .
---
Signif. codes:  0 '***' 0.001 '**' 0.01 '*' 0.05 '.' 0.1 ' ' 1

Residual standard error: 0.7123 on 58 degrees of freedom
Multiple R-squared:  0.6944,    Adjusted R-squared:  0.6522
F-statistic: 16.47 on 8 and 58 DF,  p-value: 2.042e-12
```

To verify the above statement about `lm()` and accommodate model-robust inference, we write the following function, `my.lm()`, to implement linear regression, calling only basic mathematical functions in R. The function returns a matrix with 4 columns, giving respectively the LS estimates (`coef`), the model-based SEs (`se`) by formula (1.14), the unadjusted model-based SEs (`u.se`), same as `se` but with the adjustment of degrees of freedom for $\hat{\sigma}^2$ removed, and the model-robust SEs (`r.se`) by formula (1.20). Additional comments about coding can be found in the accompanying R code file.

```
my.lm <- function(y,x, intercept=TRUE) {
  if (intercept) {
    x <- cbind(1,x)
  } else {
    x <- cbind(x)
  }

  n <- dim(x)[1]
  p <- dim(x)[2]

  H <- t(x)%*%x/n
  bet <- solve(H, t(x)%*%y/n)
```

```
  res <- as.vector( y - x%*%bet )
  sig2 <- sum(res^2) /(n-p)

  # model-based variance
  evar <- sig2 * solve(H) /n

  # unadjusted for degrees of freedom
  u.evar <- (n-p)/n *evar

  # robust variance
  r.evar <- solve(H, t(x)%*%(res^2*x)/n) %*% solve(H) /n

  out <- cbind(bet, sqrt(diag(evar)),
               sqrt(diag(u.evar)), sqrt(diag(r.evar)))
  if (intercept)  rownames(out)[1] <- "(inter)"
  colnames(out) <- c("coef", "se", "u.se", "r.se")
  out
}
```

The following is the output from using the R function `my.lm()` to fit linear regression of `lpsa` on the 8 covariates in addition to the intercept.

```
> my.lm(y,x)
               coef         se       u.se       r.se
(inter)  2.46493292 0.08931498 0.08309998 0.08141220
lcavol   0.67952814 0.12662903 0.11781752 0.11277623
lweight  0.26305307 0.09562821 0.08897390 0.09970498
age     -0.14146483 0.10134245 0.09429051 0.07355824
lbph     0.21014656 0.10221904 0.09510610 0.10646284
svi      0.30520060 0.12360027 0.11499951 0.11034144
lcp     -0.28849277 0.15452934 0.14377638 0.11508690
gleason -0.02130504 0.14524723 0.13514017 0.12546113
pgg45    0.26695576 0.15361357 0.14292433 0.12093526
```

By comparison, the LS estimates (`coef`) match those (`Estimate`) from `lm()`, and the model-based SEs (`se`) match the reported SEs (`Std.Error`) from `lm()`. The model-robust SEs (`r.se`) from `my.lm()` exhibit various differences from the model-based SEs, ranging from being 27% smaller for `age` to 4% greater for `lweight`.

1.6 Residual on residual regression

In this section, we take an algorithmic perspective and study how the LE estimator $\hat{\beta}$ and the rescaled variance estimator, either $\hat{\sigma}^2\hat{H}^{-1}$ or $\hat{\Sigma}(\hat{\beta})$, for an individual coefficient can be algebraically determined in a scheme known as residual on residual (RonR) regression. In contrast with Sections 1.2–1.4, both the point and variance estimators are purely treated as data-based algorithms, without being concerned about statistical assumptions (fixed-design vs random-design, correctly specified vs misspecified models). See Tan (2025), Section 6.1, for a connection of RonR regression with debiased Lasso inference in high-dimensional settings (Zhang & Zhang, 2014; van de Geer et al., 2014; Javanmard & Montanari, 2014).

1.6.1 Algebraic study

From an algorithmic perspective, we write linear regression of Y_i on X_i, $i = 1, \ldots, n$, in an operational form (ignoring statistical assumptions) as

$$Y_i \sim X_i^{\mathrm{T}}\beta^*, \quad i = 1, \ldots, n.$$

In the matrix notation $(\mathbf{X}, \mathbf{Y})$, this can be expressed as $\mathbf{Y} \sim \mathbf{X}\beta^*$. To study an individual coefficient associated with a scalar covariate, we rewrite the specification as

$$Y_i \sim Z_i\theta^* + W_i^{\mathrm{T}}\alpha^*, \quad i = 1, \ldots, n,$$

where the covariate vector X_i and the cofficient vector β^* are partitioned as

$$X_i = \begin{pmatrix} Z_i \\ W_i \end{pmatrix}, \quad \beta^* = \begin{pmatrix} \theta^* \\ \alpha^* \end{pmatrix}.$$

Then the LS estimator $\hat{\beta}$ can be partitioned accordingly

$$\hat{\beta} = \begin{pmatrix} \hat{\theta} \\ \hat{\alpha} \end{pmatrix}.$$

It is convenient to work with the matrix notation:

$$\mathbf{X} = (\mathbf{Z}, \mathbf{W}), \quad \mathbf{Z} = \begin{pmatrix} Z_1 \\ \vdots \\ Z_n \end{pmatrix}, \quad \mathbf{W} = \begin{pmatrix} W_1^{\mathrm{T}} \\ \vdots \\ W_n^{\mathrm{T}} \end{pmatrix}.$$

For simplicity, as reflected in the preceding notation, assume that Z_i is a scalar and so is θ. Our subsequent discussion remains applicable even if Z_i is extended to a subvector of X_i.

The RonR regression is defined as follows:

- Run linear regression of $\mathbf{Y}$ on $\mathbf{W}$, i.e., $\mathbf{Y} \sim \mathbf{W}\eta^*$. Compute

$$\begin{aligned}\hat{\eta} &= (\mathbf{W}^{\mathrm{T}}\mathbf{W})^{-1}(\mathbf{W}^{\mathrm{T}}\mathbf{Y}),\\ R_Y &= \mathbf{Y} - \mathbf{W}\hat{\eta},\end{aligned}$$

  where R_Y is the residual vector, i.e., the elements of R_Y are called residuals, in the linear regression of $\mathbf{Y}$ on $\mathbf{W}$.

- Run linear regression of $\mathbf{Z}$ on $\mathbf{W}$, i.e., $\mathbf{Z} \sim \mathbf{W}\gamma^*$. Compute

$$\begin{aligned}\hat{\gamma} &= (\mathbf{W}^{\mathrm{T}}\mathbf{W})^{-1}(\mathbf{W}^{\mathrm{T}}\mathbf{Z}),\\ R_Z &= \mathbf{Z} - \mathbf{W}\hat{\gamma},\end{aligned}$$

  where R_Z is the residual vector in the linear regression of $\mathbf{Z}$ on $\mathbf{W}$.

- Run linear regression of R_Y on R_Z (without intercept). Compute

$$\tilde{\theta} = (R_Z^{\mathrm{T}} R_Z)^{-1}(R_Z^{\mathrm{T}} R_Y). \tag{1.30}$$

Algebraically, it can be shown that $\tilde{\theta}$ obtained from RonR regression coincides with $\hat{\theta}$, i.e., the first element of $\hat{\beta}$ in the original regression of $\mathbf{Y}$ on $\mathbf{X}$. This identity demonstrates that the coefficient $\hat{\theta}$ for a covariate Z_i in linear regression with multiple covariates is precisely determined by how the response Y_i is associated with the covariate Z_i, *after* both being adjusted for the remaining covariates in the regression.

Proposition 1.7 (Point estimation from RonR regression)**.** *In the preceding notation,* $\tilde{\theta} = \hat{\theta}$.

Proof. To express $\tilde{\theta}$ explicitly, we calculate

$$\begin{aligned}R_Z^{\mathrm{T}} R_Z &= (\mathbf{Z} - \mathbf{W}\hat{\gamma})^{\mathrm{T}}(\mathbf{Z} - \mathbf{W}\hat{\gamma}) = \mathbf{Z}^{\mathrm{T}}\mathbf{Z} - \hat{\gamma}^{\mathrm{T}}\mathbf{W}^{\mathrm{T}}\mathbf{W}\hat{\gamma},\\ R_Z^{\mathrm{T}} R_Y &= (\mathbf{Z} - \mathbf{W}\hat{\gamma})^{\mathrm{T}}(\mathbf{Y} - \mathbf{W}\hat{\eta}) = \mathbf{Z}^{\mathrm{T}}\mathbf{Y} - \hat{\gamma}^{\mathrm{T}}\mathbf{W}^{\mathrm{T}}\mathbf{W}\hat{\eta}.\end{aligned}$$

[Exercise: Provide details of these calculations.] Hence $\tilde{\theta}$ can be expressed as

$$\tilde{\theta} = (\mathbf{Z}^{\mathrm{T}}\mathbf{Z} - \hat{\gamma}^{\mathrm{T}}\mathbf{W}^{\mathrm{T}}\mathbf{W}\hat{\gamma})^{-1}(\mathbf{Z}^{\mathrm{T}}\mathbf{Y} - \hat{\gamma}^{\mathrm{T}}\mathbf{W}^{\mathrm{T}}\mathbf{W}\hat{\eta}). \tag{1.31}$$

It remains to express $\hat{\theta}$ explicitly. For pedagogical purposes, we give two proofs.

First proof. The estimator $\hat{\beta} = (\mathbf{X}^{\mathrm{T}}\mathbf{X})^{-1}(\mathbf{X}^{\mathrm{T}}\mathbf{Y})$ can be calculated as

$$\hat{\beta} = \begin{pmatrix}\hat{\theta}\\ \hat{\alpha}\end{pmatrix} = \begin{pmatrix}\mathbf{Z}^{\mathrm{T}}\mathbf{Z} & \mathbf{Z}^{\mathrm{T}}\mathbf{W}\\ \mathbf{W}^{\mathrm{T}}\mathbf{Z} & \mathbf{W}^{\mathrm{T}}\mathbf{W}\end{pmatrix}^{-1}\begin{pmatrix}\mathbf{Z}^{\mathrm{T}}\mathbf{Y}\\ \mathbf{W}^{\mathrm{T}}\mathbf{Y}\end{pmatrix}.$$

Then we use the following result for inverting a partitioned matrix (Exercise 1.3). For any matrices A, B, C, and D of appropriate sizes,

$$\begin{pmatrix}A & B\\ C & D\end{pmatrix}^{-1} = \begin{pmatrix}\tilde{A}^{-1} & -\tilde{A}^{-1}BD^{-1}\\ -D^{-1}C\tilde{A}^{-1} & D^{-1} + D^{-1}C\tilde{A}^{-1}BD^{-1}\end{pmatrix}, \tag{1.32}$$

provided that D and $\tilde{A} = A - BD^{-1}C$ are nonsingular. Taking $A = \mathbf{Z}^{\mathrm{T}}\mathbf{Z}$, $B = \mathbf{Z}^{\mathrm{T}}\mathbf{W}$, $C = B^{\mathrm{T}} = \mathbf{W}^{\mathrm{T}}\mathbf{Z}$, and $D = \mathbf{W}^{\mathrm{T}}\mathbf{W}$ yields

$$\begin{aligned}
\tilde{A} &= \mathbf{Z}^{\mathrm{T}}\mathbf{Z} - \mathbf{Z}^{\mathrm{T}}\mathbf{W}(\mathbf{W}^{\mathrm{T}}\mathbf{W})^{-1}\mathbf{W}^{\mathrm{T}}\mathbf{Z} \\
\hat{\theta} &= \tilde{A}^{-1}(\mathbf{Z}^{\mathrm{T}}\mathbf{Y} - \mathbf{Z}^{\mathrm{T}}\mathbf{W}(\mathbf{W}^{\mathrm{T}}\mathbf{W})^{-1}\mathbf{W}^{\mathrm{T}}\mathbf{Y}) \\
&= (\mathbf{Z}^{\mathrm{T}}\mathbf{Z} - \mathbf{Z}^{\mathrm{T}}\mathbf{W}(\mathbf{W}^{\mathrm{T}}\mathbf{W})^{-1}\mathbf{W}^{\mathrm{T}}\mathbf{Z})^{-1}(\mathbf{Z}^{\mathrm{T}}\mathbf{Y} - \mathbf{Z}^{\mathrm{T}}\mathbf{W}(\mathbf{W}^{\mathrm{T}}\mathbf{W})^{-1}\mathbf{W}^{\mathrm{T}}\mathbf{Y}).
\end{aligned} \tag{1.33}$$

Substituting the definitions of $\hat{\eta}$ and $\hat{\gamma}$ into the expression of $\tilde{\theta}$ in (1.31) and then comparing (1.31) with (1.33) concludes that $\tilde{\theta} = \hat{\theta}$.

Second proof. We give a direct proof without invoking the matrix inverse formula (1.32). The main problem is to invert the following matrix:

$$\mathbf{X}^{\mathrm{T}}\mathbf{X} = \begin{pmatrix} \mathbf{Z}^{\mathrm{T}} \\ \mathbf{W}^{\mathrm{T}} \end{pmatrix} (\mathbf{Z}, \mathbf{W}).$$

The idea, called orthogonalization, is to transform the two blocks, represented by $\mathbf{Z}$ and $\mathbf{W}$, in $\mathbf{X} = (\mathbf{Z}, \mathbf{W})$ such that the two blocks become orthogonal after the transformation. In fact, this is achieved by taking the residual vector after linear regression of $\mathbf{Z}$ on $\mathbf{W}$:

$$\mathbf{X}\begin{pmatrix} I & 0 \\ -\hat{\gamma} & I \end{pmatrix} = (\mathbf{Z} - \mathbf{W}\hat{\gamma}, \mathbf{W}) = (R_Z, \mathbf{W}).$$

Then [Exercise: Verify these calculations.]

$$\begin{aligned}
\begin{pmatrix} I & 0 \\ -\hat{\gamma} & I \end{pmatrix}^{\mathrm{T}} \mathbf{X}^{\mathrm{T}}\mathbf{X} \begin{pmatrix} I & 0 \\ -\hat{\gamma} & I \end{pmatrix} &= \begin{pmatrix} R_Z^{\mathrm{T}} \\ \mathbf{W}^{\mathrm{T}} \end{pmatrix} (R_Z, \mathbf{W}), \\
(\mathbf{X}^{\mathrm{T}}\mathbf{X})^{-1} &= \begin{pmatrix} I & 0 \\ -\hat{\gamma} & I \end{pmatrix} \begin{pmatrix} (R_Z^{\mathrm{T}}R_Z)^{-1} & 0 \\ 0 & (\mathbf{W}^{\mathrm{T}}\mathbf{W})^{-1} \end{pmatrix} \begin{pmatrix} I & 0 \\ -\hat{\gamma} & I \end{pmatrix}^{\mathrm{T}}.
\end{aligned} \tag{1.34}$$

Hence we obtain an explicit expression of $(\mathbf{X}^{\mathrm{T}}\mathbf{X})^{-1}$, through only the inversion of a block-diagonal matrix; the inverse of $\begin{pmatrix} I & 0 \\ -\hat{\gamma} & I \end{pmatrix}$ does not need to be explicitly calculated. Returning to the definition of $\hat{\beta}$ yields

$$\begin{aligned}
\hat{\beta} &= (\mathbf{X}^{\mathrm{T}}\mathbf{X})^{-1}(\mathbf{X}^{\mathrm{T}}\mathbf{Y}) \\
&= \begin{pmatrix} I & 0 \\ -\hat{\gamma} & I \end{pmatrix} \begin{pmatrix} (R_Z^{\mathrm{T}}R_Z)^{-1} & 0 \\ 0 & (\mathbf{W}^{\mathrm{T}}\mathbf{W})^{-1} \end{pmatrix} \begin{pmatrix} I & 0 \\ -\hat{\gamma} & I \end{pmatrix}^{\mathrm{T}} \begin{pmatrix} \mathbf{Z}^{\mathrm{T}}\mathbf{Y} \\ \mathbf{W}^{\mathrm{T}}\mathbf{Y} \end{pmatrix} \\
&= \begin{pmatrix} I & 0 \\ -\hat{\gamma} & I \end{pmatrix} \begin{pmatrix} (R_Z^{\mathrm{T}}R_Z)^{-1}(\mathbf{Z}^{\mathrm{T}}\mathbf{Y} - \hat{\gamma}^{\mathrm{T}}\mathbf{W}^{\mathrm{T}}\mathbf{Y}) \\ (\mathbf{W}^{\mathrm{T}}\mathbf{W})^{-1}(\mathbf{W}^{\mathrm{T}}\mathbf{Y}) \end{pmatrix} \\
&= \begin{pmatrix} (R_Z^{\mathrm{T}}R_Z)^{-1}(\mathbf{Z}^{\mathrm{T}}\mathbf{Y} - \hat{\gamma}^{\mathrm{T}}\mathbf{W}^{\mathrm{T}}\mathbf{Y}) \\ (\mathbf{W}^{\mathrm{T}}\mathbf{W})^{-1}(\mathbf{W}^{\mathrm{T}}\mathbf{Y}) - \hat{\gamma}(R_Z^{\mathrm{T}}R_Z)^{-1}(\mathbf{Z}^{\mathrm{T}}\mathbf{Y} - \hat{\gamma}^{\mathrm{T}}\mathbf{W}^{\mathrm{T}}\mathbf{Y}) \end{pmatrix}.
\end{aligned}$$

Reading the first row gives

$$\hat{\theta} = (R_Z^{\mathrm{T}}R_Z)^{-1}(\mathbf{Z}^{\mathrm{T}}\mathbf{Y} - \hat{\gamma}^{\mathrm{T}}\mathbf{W}^{\mathrm{T}}\mathbf{Y})$$

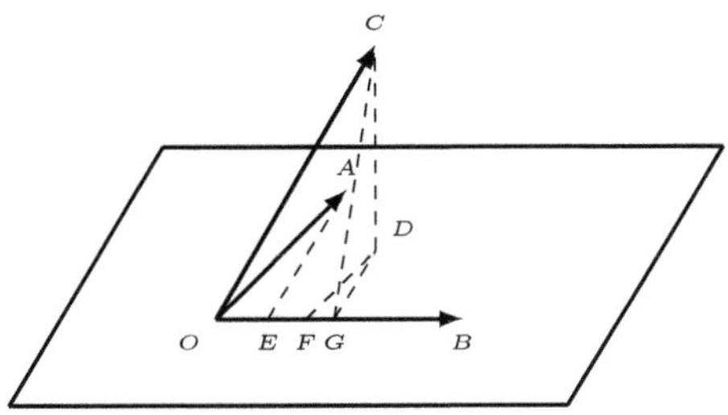

FIGURE 1.2
Geometric illustration of residual on residual regression.

$$= (\mathbf{Z}^{\mathrm{T}}\mathbf{Z} - \mathbf{Z}^{\mathrm{T}}\mathbf{W}(\mathbf{W}^{\mathrm{T}}\mathbf{W})^{-1}\mathbf{W}^{\mathrm{T}}\mathbf{Z})^{-1}(\mathbf{Z}^{\mathrm{T}}\mathbf{Y} - \mathbf{Z}^{\mathrm{T}}\mathbf{W}(\mathbf{W}^{\mathrm{T}}\mathbf{W})^{-1}\mathbf{W}^{\mathrm{T}}\mathbf{Y}),$$

the same expression of $\hat{\theta}$ in (1.33). Hence $\tilde{\theta} = \hat{\theta}$. Moreover, reading the second row gives

$$\hat{\alpha} = \hat{\eta} - \hat{\gamma}\hat{\theta}, \tag{1.35}$$

which will be used later in the proof of (1.36). □

Figure 1.2 provides a geometric illustration of residual on residual regression, where the vectors $\overrightarrow{OA}$, $\overrightarrow{OB}$, $\overrightarrow{OC}$ can be identified as $\mathbf{Z}$, $\mathbf{W}$, $\mathbf{Y}$. The vector $\overrightarrow{OD}$ is the projection of $\overrightarrow{OC}$ into the plane spanned by $\overrightarrow{OA}$ and $\overrightarrow{OB}$, and hence represents $\mathbf{X}\hat{\beta}\,(= \mathbf{Z}\hat{\theta} + \mathbf{W}\hat{\alpha})$. The vector $\overrightarrow{OD}$ can be decomposed as $\overrightarrow{FD} + \overrightarrow{OF}$, with $\overrightarrow{FD}$ representing $\mathbf{Z}\hat{\theta}$ and $\overrightarrow{OF}$ representing $\mathbf{W}\hat{\alpha}$. The vector $\overrightarrow{OG}$ is the projection of $\overrightarrow{OC}$ into the line determined by $\overrightarrow{OB}$, and hence $\overrightarrow{OG}$ represents $\mathbf{W}\hat{\eta}$ and $\overrightarrow{GC}$ represents R_Y. Similarly, the vector $\overrightarrow{OE}$ is the projection of $\overrightarrow{OA}$ into the line determined by $\overrightarrow{OB}$, and hence $\overrightarrow{OE}$ represents $\mathbf{W}\hat{\gamma}$ and $\overrightarrow{EA}$ represents R_Z. Then $\hat{\theta} = \tilde{\theta}$ is equivalent to $\frac{FD}{OA} = \frac{GD}{EA}$, where FD denotes the length of $\overrightarrow{FD}$, etc. [Exercise: Prove this statement and show that $\frac{FD}{OA} = \frac{GD}{EA}$ based on the geometry described.]

In addition to the matched coefficients, further relationship can be established between the RonR and original regression. The residual vector from RonR regression, $R_Y - R_Z\tilde{\theta}$, also coincides with that from the original regression, $\mathbf{Y} - \mathbf{X}\hat{\beta}$:

$$R_Y - R_Z\tilde{\theta} = \mathbf{Y} - \mathbf{X}\hat{\beta}. \tag{1.36}$$

This can be shown by direct calculation:

$$\begin{aligned} R_Y - R_Z\tilde{\theta} &= (\mathbf{Y} - \mathbf{W}\hat{\eta}) - (\mathbf{Z} - \mathbf{W}\hat{\gamma})\hat{\theta} \\ &= \mathbf{Y} - \mathbf{Z}\hat{\theta} - \mathbf{W}(\hat{\eta} - \hat{\gamma}\hat{\theta}) \\ &= \mathbf{Y} - \mathbf{Z}\hat{\theta} - \mathbf{W}\hat{\alpha} = \mathbf{Y} - \mathbf{X}\hat{\beta}, \end{aligned}$$

where the third equality follows from (1.35). In Figure 1.2, the vector $\overrightarrow{DC}$ corresponds to both the residual vector $\mathbf{Y} - \mathbf{X}\hat{\beta}$ and $R_Y - R_Z\tilde{\theta}$. From the matched residual vectors, the variance estimators from the RonR and original regression can also be matched. The rescaled model-based variance estimator for $\tilde{\theta}$ in RonR regression is

$$\tilde{\Sigma}_{\mathrm{b}}(\tilde{\theta}) = \tilde{\sigma}^2 (R_Z^{\mathrm{T}} R_Z)^{-1},$$

where $\tilde{\sigma}^2 = \|R_Y - R_Z\tilde{\theta}\|_2^2/(n-1)$ is the (homoscedastic) noise variance estimator without an adjustment of degrees of freedom, i.e., instantiation of $\hat{\sigma}^2$ in (1.10) to RonR regression. The rescaled model-robust variance estimator for $\tilde{\theta}$ in RonR regression is

$$\tilde{\Sigma}(\tilde{\theta}) = (R_Z^{\mathrm{T}} R_Z)^{-1} (R_Z^{\mathrm{T}} \tilde{D}(\tilde{\theta}) R_Z)(R_Z^{\mathrm{T}} R_Z)^{-1},$$

where $\tilde{D}(\tilde{\theta}) = \mathrm{diag}(\tilde{r}_1^2, \ldots, \tilde{r}_n^2)$ and $\tilde{r}_i$ is the ith element of the residual vector $R_Y - R_Z\tilde{\theta}$.

Proposition 1.8 (Variance estimation from RonR regression). *In the preceding notation,*

$$\tilde{\Sigma}_b(\tilde{\theta}) = \frac{n-p}{n-1}\{\hat{\Sigma}_b(\hat{\beta})\}_{11}, \quad \tilde{\Sigma}(\tilde{\theta}) = \{\hat{\Sigma}(\hat{\beta})\}_{11},$$

where $\hat{\Sigma}_b(\hat{\beta}) = \hat{\sigma}^2 \hat{H}^{-1}$ is the rescaled model-based variance estimator and $\hat{\Sigma}(\hat{\beta})$ is the rescaled model-robust variance estimator for $\hat{\beta}$ in the original regression, and A_{11} denotes the $(1,1)$th element of a matrix A.

Proof. The equality for model-based variance estimation holds because

$$\tilde{\sigma}^2 = \frac{n-p}{n-1}\hat{\sigma}^2,$$
$$\{(\mathbf{X}^{\mathrm{T}}\mathbf{X})^{-1}\}_{11} = (R_Z^{\mathrm{T}} R_Z)^{-1}.$$

The first equality follows from the definitions of $\tilde{\sigma}^2$ and $\hat{\sigma}^2$ and the residual matching (1.36). The second equality follows from (1.34). It is left as Exercise 1.7 to show the equality for model-robust variance estimation. □

From Propositions 1.7 and 1.8, both the point and variance estimators are algebraically matched between the RonR and original regression, except for an adjustment of degrees of freedom in model-based variance estimation. The latter difference exists only because such an adjustment is made in the noise variance estimators $\tilde{\sigma}^2$ and $\hat{\sigma}^2$.

Bibliographical note. Residual on residual regression has been discussed in various forms in linear regression. See Draper and Smith (1998), Section 4.5.4, and Hastie et al. (2009), Section 2.2.3. Our exposition addresses not only point estimation of coefficients, but also variance estimation. Moreover, the algebraic discussion here is also deeply connected to doubly robust estimation in partially linear regression, as shown in Section 1.7.

1.6.2 Numerical demonstration

We write the following function to implement residual on residual regression, calling the function `my.lm()` in Section 1.5. The three arguments `y`, `z`, `w`, correspond to **Y**, **Z**, **W** in the preceding discussion.

```
rr.lm <- function(y, z, w) {
  yOnw <- my.lm(y,w)
  alp <- yOnw [,"coef"]
  r.y <- as.vector(y - cbind(1,w)%*%alp)

  zOnw <- my.lm(z,w)
  gam <- zOnw [,"coef"]
  r.z <- as.vector(z - cbind(1,w)%*%gam)

  out <- my.lm(r.y, r.z, intercept=FALSE)
  out
}
```

For example, consider linear regression with the prostate cancer data as in Section 1.5. We apply the function `rr.lm()`, taking the response variable as `y`, any of the 8 covariates as `z`, and and the remaining covariates `w`. From the output below, the LS estimates, column `coef` in `rr.fit`, match those from `my.lm(y,x)` in Section 1.5. Moreover, the unadjusted model-based SEs, column `u.se`, and model-robust SEs, column `r.se`, also match the corresponding ones from `my.lm(y,x)`. Hence these results numerically confirm the algebraic identities in Propositions 1.7 and 1.8 about point and variance estimation respectively.
[Exercise: Show that `rr.lm()` can also be used to recover the intercept estimate and associated SEs from `my.lm(y,x)`.]

```
> rr.fit <- matrix(NA, dim(x)[2], 4)
> rownames(rr.fit) <- colnames(x)
> colnames(rr.fit) <- c("coef", "se", "u.se", "r.se")
> for (j in 1:dim(x)[2]) {
+   rr.fit[j,] <- rr.lm(y, z=x[,j], w=x[,-j])
+ }
> rr.fit
               coef         se       u.se       r.se
lcavol   0.67952814 0.11870672 0.11781752 0.11277623
lweight  0.26305307 0.08964541 0.08897390 0.09970498
age     -0.14146483 0.09500215 0.09429051 0.07355824
lbph     0.21014656 0.09582389 0.09510610 0.10646284
```

```
svi      0.30520060 0.11586745 0.11499951 0.11034144
lcp     -0.28849277 0.14486150 0.14377638 0.11508690
gleason -0.02130504 0.13616011 0.13514017 0.12546113
pgg45    0.26695576 0.14400302 0.14292433 0.12093526
```

1.7 Partially linear regression and doubly robust estimation

In Section 1.6, the residual-on-residual (RonR) regression is discussed from an algorithmic (or algebraic) perspective. In this section, we study statistical implications of RonR regression in the context of partially linear regression. See Tan (2025), Section 6.1, for a related discussion of debiased Lasso inference in high-dimensional settings (Zhang & Zhang, 2014; van de Geer et al., 2014; Javanmard & Montanari, 2014).

As in Section 1.6, suppose that the covariate vector is partitioned as $X_i = (Z_i, W_i^{\mathrm{T}})^{\mathrm{T}}$, where Z_i is a scalar covariate of the primary interest and W_i is the vector of remaining covariates in X_i. Consider a random design where $(X_1, Y_1), \ldots, (X_n, Y_n)$ are IID as (X, Y), with $X = (Z, W^{\mathrm{T}})^{\mathrm{T}}$. A partially linear regression model postulates that

$$E(Y|X) = Z\theta^* + g^*(W), \tag{1.37}$$

where θ^* is an unknown (scalar) parameter and $g^*(w)$ is an unknown function of w (considered a nuisance parameter). To estimate θ^*, there are at least two distinct statistical approaches.

- The first approach is to invoke certain smoothness assumptions on $g^*(w)$ and then apply smoothing (or nonparametric regression) techniques to estimate $g^*(w)$ together with θ^* (e.g., Speckman, 1988; Severini & Staniswalis, 1994).
- The second approach, called doubly robust estimation, is to introduce two parametric, *working models* on $g^*(w)$ and another nuisance function (to be specified later) and derive an estimator of θ^* which remains consistent if either working model is correctly specified and the other working model may be misspecified.

[We refer to a model as "a working model" to emphasize that the model is used to facilitate estimation of nuisance parameters such as g^* and allowed to be misspecified.] We mainly discuss the second approach, which is deeply connected to RonR regression.

As a working model for $g^*(W)$, consider a linear specification:

$$g^*(W) = W^{\mathrm{T}}\alpha^*,$$

or equivalently a full linear regression model of Y on (Z, W):

$$E(Y|Z,W) = X^{\mathrm{T}}\beta^* = Z\theta^* + W^{\mathrm{T}}\alpha^*, \tag{1.38}$$

where $\beta^* = (\theta^*, \alpha^{*\mathrm{T}})^{\mathrm{T}}$ and α^* is an unknown parameter vector, in addition to the parameter θ^*. Let $\hat\beta = (\hat\theta, \hat\alpha^{\mathrm{T}})^{\mathrm{T}}$ be the LS estimator of $\beta = (\theta, \alpha^{\mathrm{T}})^{\mathrm{T}}$ in full regression (1.38). As motivated by RonR regression, take the other nuisance function as $f^*(W) = E(Z|W)$. Consider a linear working model for $f^*(W)$:

$$E(Z|W) = W^{\mathrm{T}}\gamma^*, \tag{1.39}$$

where γ^* is an unknown parameter vector. Let $\hat\gamma$ be the LS estimator of γ in model (1.39). Model (1.38) or model (1.39) may be misspecified, in which case the true parameter value α^* or γ^* does not exist.

It turns out that $\hat\theta$ extracted from the LS estimator $\hat\beta$ in the full regression (1.38) is doubly robust for θ^*, as described in Proposition 1.9 later. To facilitate the subsequent discussion, we note that by Proposition 1.7, the estimator $\hat\theta$ can be expressed as

$$\begin{aligned}
\hat\theta &= (R_Z^{\mathrm{T}} R_Z)^{-1}(R_Z R_Y) \\
&= \frac{\tilde E\{(Z - W^{\mathrm{T}}\hat\gamma)(Y - W^{\mathrm{T}}\hat\eta)\}}{\tilde E\{(Z - W^{\mathrm{T}}\hat\gamma)(Z - W^{\mathrm{T}}\hat\gamma)\}} \\
&= \frac{\tilde E\{(Z - W^{\mathrm{T}}\hat\gamma)Y\}}{\tilde E\{(Z - W^{\mathrm{T}}\hat\gamma)Z\}} \\
&= \frac{\tilde E\{(Z - W^{\mathrm{T}}\hat\gamma)(Y - W^{\mathrm{T}}\hat\alpha)\}}{\tilde E\{(Z - W^{\mathrm{T}}\hat\gamma)Z\}},
\end{aligned} \tag{1.40}$$

where the $\tilde E(\cdot)$ notation is used, and $\hat\eta$ and $\hat\gamma$ are the LS estimators from RonR regression in Section 1.6. The last two equalities hold because $\tilde E\{(Z - W^{\mathrm{T}}\hat\gamma)W\} = 0$ by the definition of $\hat\gamma$ as a LS estimator.

1.7.1 Doubly robust point estimation

As preparation, we recall from Section 1.4 that LS estimators admit well-defined probability limits with possible model misspecification. As $n \to \infty$ and p is fixed, $(\hat\theta, \hat\alpha^{\mathrm{T}})^{\mathrm{T}} \to (\bar\theta, \bar\alpha^{\mathrm{T}})^{\mathrm{T}}$, where the target value $\bar\beta = (\bar\theta, \bar\alpha^{\mathrm{T}})^{\mathrm{T}}$ is defined in (1.27), i.e., $\bar\beta = \operatorname{argmin}_\beta E\{(Y - X^{\mathrm{T}}\beta)^2\}$. The target value $\bar\beta$ coincides with the true value β^* if model (1.38) is correctly specified but otherwise remains well-defined. Similarly, $\hat\gamma \to \bar\gamma$, where the target value $\bar\gamma$ is defined as $\bar\gamma = \operatorname{argmin}_\gamma E\{(Z - W^{\mathrm{T}}\gamma)^2\}$. The target value $\bar\gamma$ coincides with the true value γ^* if model (1.39) is correctly specified but otherwise remains well-defined.

Proposition 1.9 (Doubly robust point estimation)**.** *Suppose that partially linear regression (1.37) is correctly specified. Then the estimator $\hat\theta$ from the*

LS estimator $\hat{\beta} = (\hat{\theta}, \hat{\alpha}^{\mathrm{T}})^{\mathrm{T}}$ in the full regression (1.38) is doubly robust: if either model (1.38) or model (1.39) is correctly specified, then as $n \to \infty$ and p is fixed, $\hat{\theta} \xrightarrow[\mathcal{P}]{} \theta^*$, where θ^* is well-defined by (1.37) even if model (1.38) is misspecified.
[Equivalently, if either model (1.38) or model (1.39) is correctly specified, then $\bar{\theta} = \theta^*$, where $\bar{\theta}$ is from the target value $\bar{\beta} = (\bar{\theta}, \bar{\alpha}^{\mathrm{T}})^{\mathrm{T}}$.]

Proof. We notice that

$$\hat{\theta} \xrightarrow[\mathcal{P}]{} \frac{E\{(Z - W^{\mathrm{T}}\bar{\gamma})(Y - W^{\mathrm{T}}\bar{\alpha})\}}{E\{(Z - W^{\mathrm{T}}\bar{\gamma})Z\}} \; (= \bar{\theta}),$$

while allowing that model (1.38) or model (1.39) may be misspecified. The convergence of $\hat{\theta}$ in probability to the above ratio of expectations (without relating to $\bar{\theta}$) can be directly shown using the expression (1.40) for $\hat{\theta}$ and usual large-sample arguments. Alternatively, the result follows because $\hat{\theta} \xrightarrow[\mathcal{P}]{} \bar{\theta}$ as a subsequence from $\hat{\beta} \xrightarrow[\mathcal{P}]{} \bar{\beta}$, and $\bar{\theta}$ can be expressed in the stated form as a population version of (1.40). [Exercise: Prove these statements.]

To show the double robustness of $\hat{\theta}$, consider the following two cases. In the first case, suppose that model (1.38) is correctly specified. Then $\bar{\alpha} = \alpha^*$ and hence

$$\begin{aligned} \hat{\theta} &\xrightarrow[\mathcal{P}]{} \frac{E\{(Z - W^{\mathrm{T}}\bar{\gamma})(Y - W^{\mathrm{T}}\alpha^*)\}}{E\{(Z - W^{\mathrm{T}}\bar{\gamma})Z\}} \\ &= \theta^* + \frac{E\{(Z - W^{\mathrm{T}}\bar{\gamma})(Y - W^{\mathrm{T}}\alpha^* - Z\theta^*)\}}{E\{(Z - W^{\mathrm{T}}\bar{\gamma})Z\}} = \theta^*. \end{aligned}$$

The last equality holds because $E\{Z(Y - W^{\mathrm{T}}\alpha^* - Z\theta^*)\} = 0$ and $E\{W(Y - W^{\mathrm{T}}\alpha^* - Z\theta^*)\} = 0$ under correctly specified model (1.38).

In the second case, suppose that model (1.39) is correctly specified. Then $\bar{\gamma} = \gamma^*$ and hence

$$\begin{aligned} \hat{\theta} &\xrightarrow[\mathcal{P}]{} \frac{E\{(Z - W^{\mathrm{T}}\gamma^*)(Y - W^{\mathrm{T}}\bar{\alpha})\}}{E\{(Z - W^{\mathrm{T}}\gamma^*)Z\}} \\ &= \theta^* + \frac{E\{(Z - W^{\mathrm{T}}\gamma^*)(Y - W^{\mathrm{T}}\bar{\alpha} - Z\theta^*)\}}{E\{(Z - W^{\mathrm{T}}\gamma^*)Z\}} \\ &= \theta^* + \frac{E\{(Z - W^{\mathrm{T}}\gamma^*)(Y - g^*(W) - Z\theta^*)\}}{E\{(Z - W^{\mathrm{T}}\gamma^*)Z\}} = \theta^*. \end{aligned}$$

The second equality follows because $E\{(Z - W^{\mathrm{T}}\gamma^*)g(W)\} = 0$ for any function $g(W)$ under correctly specified model (1.39). The last equality follows because $E\{Z(Y - g^*(W) - Z\theta^*)\} = 0$ and $E\{W(Y - g^*(W) - Z\theta^*)\} = 0$ under partially linear regression (1.37). □

From the expression (1.40), $\hat{\theta}$ is a solution to the estimating equation in θ:

$$0 = \tilde{E}\{\varphi(Z, W, Y; \theta, \hat{\alpha}, \hat{\gamma})\}, \tag{1.41}$$

where $\hat{\alpha}$ and $\hat{\gamma}$ are the LS estimators and

$$\varphi(Z, W, Y; \theta, \alpha, \gamma) = (Z - W^{\mathrm{T}}\gamma)(Y - W^{\mathrm{T}}\alpha - Z\theta). \tag{1.42}$$

Then φ satisfies the following two properties of unbiasedness, which can be directly deduced from the proof of Proposition 1.9.

- If model (1.38) is correctly specified, then for any γ,

$$0 = E\{\varphi(Z, W, Y; \theta^*, \alpha^*, \gamma)\}. \tag{1.43}$$

- If model (1.39) is correctly specified, then for any α,

$$0 = E\{\varphi(Z, W, Y; \theta^*, \alpha, \gamma^*)\}. \tag{1.44}$$

In this sense, φ is said to be a doubly robust estimating function for θ^* with respect to working models (1.38) and (1.39) for nuisance functions $g^*(w)$ and $f^*(w)$. If exist, doubly robust estimating functions can often be derived from *semiparamtric influence functions* (Robins & Rotnitzky, 2001; Tsiatis, 2006). However, the existence and the derivation of doubly robust estimating functions are known only in specific problems, and a general theory remains open.

1.7.2 Doubly robust inference

In addition to the point double robustness of $\hat{\theta}$, the estimated variance for $\hat{\theta}$ extracted from the model-robust variance estimator $\hat{\Sigma}(\hat{\beta})$ in the full regression (1.38) is also doubly robust and hence leads to doubly robust inference about θ^*. Let

$$V(\theta^*, \bar{\alpha}, \bar{\gamma}) = \frac{E\{\varphi^2(Z, W, Y; \theta^*, \bar{\alpha}, \bar{\gamma})\}}{E^2\{Z(Z - W^{\mathrm{T}}\bar{\gamma})\}},$$

$$\hat{V}(\hat{\theta}, \hat{\alpha}, \hat{\gamma}) = \frac{\tilde{E}\{\varphi^2(Z, W, Y; \hat{\theta}, \hat{\alpha}, \hat{\gamma})\}}{\tilde{E}^2\{Z(Z - W^{\mathrm{T}}\hat{\gamma})\}}.$$

Proposition 1.10 (Doubly robust inference). *Suppose that partially linear regression (1.37) is correctly specified. If either model (1.38) or model (1.39) is correctly specified, then as $n \to \infty$ and p is fixed,*

$$\sqrt{n}(\hat{\theta} - \theta^*) \xrightarrow[\mathcal{D}]{} \mathrm{N}(0, V(\theta^*, \bar{\alpha}, \bar{\gamma})), \tag{1.45a}$$

$$\hat{V}(\hat{\theta}, \hat{\alpha}, \hat{\gamma}) \xrightarrow[\mathcal{P}]{} V(\theta^*, \bar{\alpha}, \bar{\gamma}), \tag{1.45b}$$

$$V(\theta^*, \bar{\alpha}, \bar{\gamma}) = \{\Sigma(\bar{\beta})\}_{11}, \quad \hat{V}(\hat{\theta}, \hat{\alpha}, \hat{\gamma}) = \{\hat{\Sigma}(\hat{\beta})\}_{11}, \tag{1.45c}$$

where θ^* is well-defined by (1.37) even if model (1.38) is misspecified, $\bar{\beta} = (\theta^*, \bar{\alpha}^{\mathrm{T}})^{\mathrm{T}}$, $\hat{\beta} = (\hat{\theta}, \hat{\alpha}^{\mathrm{T}})^{\mathrm{T}}$, and $\Sigma(\beta)$ and $\hat{\Sigma}(\beta)$ are defined as in (1.17a) and (1.22a).

Proof. We use the o_p and O_p notation. We show that, whether model (1.38) or (1.39) is correctly specified or not (including the case where both models are misspecified), $\hat\theta$ satisfies the asymptotic expansion:

$$\hat\theta - \bar\theta = \frac{\tilde E\{\varphi(Z, W, Y; \bar\theta, \bar\alpha, \bar\gamma)\}}{\tilde E\{Z(Z - W^{\mathrm{T}}\bar\gamma)\}} + o_p(n^{-1/2}), \tag{1.46}$$

where $\bar\theta$ is from the target value $\bar\beta = (\bar\theta, \bar\alpha^{\mathrm{T}})^{\mathrm{T}}$. This result is *stronger* than indicated by Proposition 1.10. If either model (1.38) or model (1.39) is correctly specified, then $\bar\theta = \theta^*$ by Proposition 1.9 and (1.46) reduces to

$$\hat\theta - \theta^* = \frac{\tilde E\{\varphi(Z, W, Y; \theta^*, \bar\alpha, \bar\gamma)\}}{\tilde E\{Z(Z - W^{\mathrm{T}}\bar\gamma)\}} + o_p(n^{-1/2}). \tag{1.47}$$

Moreover, by the central limit theorem with $E\{\varphi(Z, W, Y; \theta^*, \bar\alpha, \bar\gamma)\} = 0$ and the law of large numbers respectively,

$$\sqrt{n}\tilde E\{\varphi(Z, W, Y; \theta^*, \bar\alpha, \bar\gamma)\} \xrightarrow[\mathcal{D}]{} \mathrm{N}(0, E\{\varphi^2(Z, W, Y; \theta^*, \bar\alpha, \bar\gamma)\}),$$
$$\tilde E\{Z(Z - W^{\mathrm{T}}\bar\gamma)\} \xrightarrow[\mathcal{P}]{} E\{Z(Z - W^{\mathrm{T}}\bar\gamma)\}.$$

The first convergence depends on either model (1.38) or model (1.39) being correctly specified, whereas the second convergence holds even if both models (1.38) and (1.39) are misspecified. Hence the asymptotic normality (1.45a) follows by applying Slutsky's theorem in (1.47). [Exercise: Provide details of applying Slutsky's theorem.] The identity $\hat V(\hat\theta, \hat\alpha, \hat\gamma) = \{\hat\Sigma(\hat\beta)\}_{11}$ in (1.45c) follows directly from Proposition 1.8. It is left as an exercise to show the remaining conclusions in (1.45).

We give a direct proof of asymptotic expansion (1.46). From the expression (1.40),

$$\hat\theta - \bar\theta = \frac{\tilde E\{(Z - W^{\mathrm{T}}\hat\gamma)(Y - W^{\mathrm{T}}\hat\alpha - Z\bar\theta)\}}{\tilde E\{Z(Z - W^{\mathrm{T}}\hat\gamma)\}}.$$

Then it suffices to show that [Exercise: Justify this reasoning.]

$$\begin{aligned}&\tilde E\{(Z - W^{\mathrm{T}}\hat\gamma)(Y - W^{\mathrm{T}}\hat\alpha - Z\bar\theta)\}\\ &= \tilde E\{(Z - W^{\mathrm{T}}\bar\gamma)(Y - W^{\mathrm{T}}\bar\alpha - Z\bar\theta)\} + o_p(n^{-1/2}).\end{aligned} \tag{1.48}$$

This indicates that replacing $(\hat\alpha, \hat\gamma)$ by $(\bar\alpha, \bar\gamma)$ does not affect the asymptotic behavior of the left-hand side up to $o_p(n^{-1/2})$. See Section 1.10.4 for further discussion. In fact, by simple algebra, we obtain the decomposition

$$\begin{aligned}&\tilde E\{(Z - W^{\mathrm{T}}\hat\gamma)(Y - W^{\mathrm{T}}\hat\alpha - Z\bar\theta)\}\\ &= \tilde E\{(Z - W^{\mathrm{T}}\bar\gamma)(Y - W^{\mathrm{T}}\bar\alpha - Z\bar\theta)\}\\ &\quad - \tilde E\{(\hat\gamma - \bar\gamma)^{\mathrm{T}} W (Y - W^{\mathrm{T}}\bar\alpha - Z\bar\theta)\} - \tilde E\{(Z - W^{\mathrm{T}}\bar\gamma) W^{\mathrm{T}}(\hat\alpha - \bar\alpha)\}\\ &\quad + \tilde E\{(\hat\gamma - \bar\gamma)^{\mathrm{T}} W W^{\mathrm{T}}(\hat\alpha - \bar\alpha)\}.\end{aligned}$$

The second, third, and fourth terms on the right-hand side are $o_p(n^{-1/2})$, because by Proposition 1.6, $\hat{\alpha} - \bar{\alpha} = O_p(n^{-1/2})$ and $\hat{\gamma} - \bar{\gamma} = O_p(n^{-1/2})$, and by the central limit theorem with the mean-zero vectors $(Y - W^{\mathrm{T}}\bar{\alpha} - Z\bar{\theta})W$ and $(Z - W^{\mathrm{T}}\bar{\gamma})W$,

$$\tilde{E}\{(Y - W^{\mathrm{T}}\bar{\alpha} - Z\bar{\theta})W\} = O_p(n^{-1/2}),$$
$$\tilde{E}\{(Z - W^{\mathrm{T}}\bar{\gamma})W\} = O_p(n^{-1/2}).$$

Hence the asymptotic expansion (1.48) holds.
[Exercise: Show that (1.46) can also be deduced by calculating the first component in the *joint* asymptotic expansion (1.29) of $\hat{\beta}$ about $\bar{\beta}$ in the proof of Proposition 1.6. Therefore, the asymptotic expansion (1.46) results from combining Proposition 1.6, which gives (1.46), and Proposition 1.9, which gives $\bar{\theta} = \theta^*$ if model (1.38) or model (1.39) is correctly specified. See Tan (2025), Section 6.1.1, for a related discussion.] □

From Proposition 1.10, the Wald confidence interval (1.19) with $j = 1$, i.e.,

$$\hat{\theta} \pm z_{c/2}\sqrt{\frac{\{\hat{\Sigma}(\hat{\beta})\}_{11}}{n}}, \tag{1.49}$$

achieves asymptotic size $1 - c$ in covering θ^* in partially linear regression (1.37) if either model (1.38) or model (1.39) is correctly specified. Hence the model-robust Wald confidence interval for an individual coefficient in linear regression is valid in three distinct manners:

- [Heteroscedasticity robust] being valid in covering the true value in correctly specified linear regression with heteroscedastic noises (Proposition 1.4);
- [Model robust] being valid in covering the target value in possibly misspecified linear regression (Proposition 1.6);
- [Doubly robust] being valid in covering the true value in the associated partially linear regression, if either the linear regression is correctly specified or the linear regression of the associated covariate on the remaining covariates is correctly specified (Proposition 1.10).

We provide additional comments about the double robustness.

First, the double robustness is distinct from the "model robustness" discussed in Section 1.4, which says that the model-robust Wald confidence interval (1.49) remains valid about the target value $\bar{\theta}$ (defined as the first element of $\bar{\beta}$), while linear regression (1.38) may be misspecified. The parameter θ^* in partially linear regression (1.37) is more descriptive than the target value $\bar{\theta}$, but at the same time the double robustness is more restricted in depending on at least one of the two models, (1.38) and (1.39), being correctly specified. Nevertheless, the two properties are also related. The double robustness of the confidence interval (1.49) in covering θ^* can be deduced by combining the model robustness of (1.49) in covering $\bar{\theta}$ and the fact that if model (1.38) or (1.39) is correctly specified, then $\bar{\theta} = \theta^*$ by Proposition 1.9. Hence the confidence interval (1.49) can be interpreted in term of θ^* in partially linear regression (1.37) if model (1.38) or (1.39) is correctly specified, but otherwise remains valid for the target value $\bar{\theta}$. This relationship can also be seen from asymptotic expansions (1.46) and (1.47) in the proof of Propsotion 1.10.

Second, underlying the double robustness above, there is a more basic layer of model-dependency: the parameter θ^* is defined only if partially linear regression (1.37) is correctly specified. See Section 1.10.5 for further discussion on how the double robustness can be reformulated while allowing partially linear regression (1.37) to be misspecified.

Third, the double robustness above is achieved for the linear coefficient in partially linear regression (1.37) at no additional cost ("for free") by the corresponding element in the LS estimator $\hat{\beta}$ together with the model-robust variance estimator $\hat{\Sigma}(\hat{\beta})$ from the full regression (1.38). This feature is special, due to the fact that linear regression (1.39) is used as a working model for the nuisance function $f^*(W) = E(Z|W)$. Alternatively, for a binary covariate Z, a more appropriate working model for $f^*(W)$ is logistic regression:

$$P(Z = 1|W) = \text{expit}(W^{\mathrm{T}}\gamma^*), \tag{1.50}$$

where $\text{expit}(u) = (1 + \mathrm{e}^{-u})^{-1}$. The estimator $\hat{\theta}$ from $\hat{\beta}$ is no longer doubly robust for θ^* with respect to working models (1.38) and (1.50). In fact, instead of φ in (1.42), a doubly robust estimating function for θ^* with respect to working models (1.38) and (1.50) is

$$\varphi(Z, W, Y; \theta, \alpha, \gamma) = (Z - \text{expit}(W^{\mathrm{T}}\gamma))(Y - W^{\mathrm{T}}\alpha - Z\theta), \tag{1.51}$$

satisfying the two properties (1.43) and (1.44) with model (1.39) replaced by (1.50). A doubly robust estimator for θ^* can be defined as a solution to (1.41) with the estimating function φ in (1.51) and some estimators $(\hat{\alpha}, \hat{\gamma})$ based on models (1.38) and (1.50) respectively. See Section 1.10.4 for further discussion on how the estimators $(\hat{\alpha}, \hat{\gamma})$ can be constructed to achieve similar results as in Propositions 1.9 and 1.10.

The estimating function φ in (1.42) or (1.51) is stated directly in terms of the parameters α and γ in working models for g^* and f^*. The two cases can be unified as

$$\varphi(Z, W, Y; \theta, \alpha, \gamma) = (Z - f(W; \gamma))(Y - Z\theta - g(W; \alpha)),$$

where $g(W; \alpha) = W^{\mathrm{T}}\alpha$ from working model (1.38) for g^* and $f(W; \gamma) = W^{\mathrm{T}}\gamma$ or $\text{expit}(W^{\mathrm{T}}\gamma)$, from working model (1.39) or (1.50) for f^*. Equivalently, the estimating function φ can be stated as $\varphi(Z, W, Y; \theta, \alpha, \gamma) = \varphi_0(Z, W, Y; \theta, g(\cdot; \alpha), f(\cdot; \gamma))$ with

$$\varphi_0(Z, W, Y; \theta, g, f) = (Z - f(W))(Y - Z\theta - g(W)),$$

where $g(W)$ and $f(W)$ are functions of W. The estimating function φ_0 is doubly robust with respect to the choices of g and f, in the sense of satisfying the following two properties similar to (1.43) and (1.44):

$$0 = E\{\varphi_0(Z, W, Y; \theta^*, g^*, f)\} \quad \text{for any } f, \tag{1.52}$$

$$0 = E\{\varphi_0(Z, W, Y; \theta^*, g, f^*)\} \quad \text{for any } g, \tag{1.53}$$

provided the expectations involved are finite. As a result, doubly robust estimating functions can be derived from φ_0 using other types of working models for $f^*(W) = E(Z|W)$, for example, log-linear regression with nonnegative Z. For $g^*(W)$, other types of working models can also be used instead of linear regression (1.38), but a motivation seems lacking.

Bibliographical note. The double robustness of the least squares estimator (Proposition 1.9) for point estimation of θ^* in partially linear regression (1.37) is identified in Robins et al. (1992), Section 3. The double robustness of the sandwich variance estimator (Proposition 1.10) is discussed in Ghosh and Tan (2022), Example 5. Doubly robust estimation has been developed in various semiparametric problems, including partially linear and log-linear models (Robins & Rotnitzky, 2001), logistic partially linear models (Chen, 2007; Tchetgen Tchetgen et al., 2010; Tan, 2019), and missing-data problems and causal inference (Robins et al., 1994; Scharfstein et al., 1999), among others. For estimation of population means with missing data or that of average treatment effects, comparisons between conventional and doubly robust methods can be found in Kang and Schafer (2007) and Tan (2007, 2010). To obtain doubly robust variance estimation and inference, careful methods for estimating nuisance parameters have been discussed in various settings by Kim and Haziza (2014), Vermeulen and Vansteelandt (2015), Tan (2020a), and Ghosh and Tan (2022). See Section 1.10.4 for a discussion of calibrated estimation.

1.8 Penalized linear regression: Algorithms

The previous Sections 1.2–1.7 study least squares estimation for linear regression from various angles. In this section, we study penalized least squares estimation for linear regression. Our discussion is mainly about basic properties and computational algorithms, without dealing with associated asymptotic theory, for penalized linear regression.

Consider linear regression in an operational form as in Section 1.6:

$$Y_i \sim X_i^{\mathrm{T}} \beta^*, \quad i = 1, \ldots, n,$$

where $X_i = (X_{i1}, \ldots, X_{ip})^{\mathrm{T}}$ and $\beta^* = (\beta_1^*, \ldots, \beta_p^*)^{\mathrm{T}}$. The least squares (LS) estimator of β is defined as in (1.8):

$$\hat{\beta} = \text{argmin}_\beta \frac{1}{n}\sum_{i=1}^{n}(Y_i - X_i^{\mathrm{T}}\beta)^2.$$

A penalized least squares estimator is defined in the form

$$\hat{\beta}_\lambda = \text{argmin}_\beta \left\{ \frac{1}{2n}\sum_{i=1}^{n}(Y_i - X_i^{\mathrm{T}}\beta)^2 + \lambda\,\text{pen}(\beta) \right\}, \tag{1.54}$$

where $\lambda \geq 0$ is a tuning parameter and pen(β) is a penalty function in β. For algebraic convenience later, an additional factor of $1/2$ is introduced to the average of squares term. This affects the scale of λ for $\lambda > 0$. Two popular choices of the penalty is

$$\text{pen}(\beta) = \frac{1}{2}\|\beta\|_2^2 = \frac{1}{2}\sum_{j=1}^{p}\beta_j^2,$$

$$\text{pen}(\beta) = \|\beta\|_1 = \sum_{j=1}^{p}|\beta_j|,$$

which are called L_2 or ridge penalty and, respectively, L_1 or Lasso penalty. For algebraic convenience, an additional factor of $1/2$ is also introduced in the ridge penality. The corresponding estimator is denoted as $\hat{\beta}_{\lambda,L_2}$ or $\hat{\beta}_{\lambda,L_1}$, and called the ridge-penalized or Lasso-penalized LS estimator of β or, for short, the ridge or Lasso estimator of β. [Question: What can be said for the choice pen$(\beta) = \|\beta\|_2$?]

1.8.1 Penalized estimation of a normal mean

To understand the logic of penalized estimation, we consider the simple problem of estimating a normal mean (with a known variance, for example, being 1). Suppose that a single observation is obtained, $Z \sim \mathrm{N}(\mu^*, 1)$, where μ^* is an unknown parameter. The discussion can be readily extended when independent observations, $Z_1, \ldots, Z_n$, are obtained from $\mathrm{N}(\mu^*, 1)$. [Exercise: Provide details of the extension.]

To estimate μ^*, the LS estimator (or equivalently the ML estimator), defined as

$$\hat{\mu} = \text{argmin}_\mu \frac{1}{2}(Z - \mu)^2,$$

gives $\hat{\mu} = Z$. The ridge estimator of μ^* is defined as

$$\hat{\mu}_{\lambda,L_2} = \text{argmin}_\mu \left\{ \frac{1}{2}(Z - \mu)^2 + \lambda\frac{\mu^2}{2} \right\},$$

and the Lasso estimator of μ^* is defined as

$$\hat{\mu}_{\lambda,L_1} = \text{argmin}_{\mu} \left\{ \frac{1}{2}(Z-\mu)^2 + \lambda|\mu| \right\}. \tag{1.55}$$

Solving the two minimization problems yields closed-form expressions for $\hat{\mu}_{\lambda,L_2}$ and $\hat{\mu}_{\lambda,L_1}$.

Proposition 1.11 (Penalized estimation of normal mean). *In the preceding notation, $\hat{\mu}_{\lambda,L_2}$ and $\hat{\mu}_{\lambda,L_1}$ can be determined as follows:*

$$\hat{\mu}_{\lambda,L_2}(Z) = \frac{Z}{1+\lambda},$$

and

$$\hat{\mu}_{\lambda,L_1}(Z) = \begin{cases} Z-\lambda, & \text{if } Z > \lambda, \\ 0, & \text{if } |Z| \le \lambda, \\ Z+\lambda, & \text{if } Z < -\lambda, \end{cases}$$

where the dependency on Z is made explicit in the notation.

Proof. Setting the derivative of $\frac{1}{2}(Z-\mu)^2 + \lambda\frac{\mu^2}{2}$ to 0,

$$0 = \frac{\mathrm{d}}{\mathrm{d}\mu}\left\{ \frac{1}{2}(Z-\mu)^2 + \lambda\frac{\mu^2}{2} \right\} = -(Z-\mu) + \lambda\mu,$$

and solving for μ directly yields the expression for $\hat{\mu}_{\lambda,L_2}(Z)$.

For $\hat{\mu}_{\lambda,L_1}(Z)$, the penalized objective function $f(\mu) = \frac{1}{2}(Z-\mu)^2 + \lambda|\mu|$ is convex but not everywhere differentiable in μ. We use several properties of convex functions, which are described in Section 1.10.6. The sub-differential of f (i.e., the set of sub-derivatives) can be calculated as

$$\partial_\mu \left\{ \frac{1}{2}(Z-\mu)^2 + \lambda|\mu| \right\} = -(Z-\mu) + \lambda\partial_\mu|\mu|, \tag{1.56}$$

where $\partial_\mu|\mu|$ is the sub-differential of $|\mu|$ (see Figure 1.3):

$$\partial_\mu|\mu| \begin{cases} = 1, & \text{if } \mu > 0, \\ = [-1,1], & \text{if } \mu = 0, \\ = -1, & \text{if } \mu < 0. \end{cases}$$

In a slightly different way, it is convenient to write

$$\partial_\mu|\mu| \begin{cases} = 1, & \text{if } \mu > 0, \\ \in [-1,1], & \text{if } \mu = 0, \\ = -1, & \text{if } \mu < 0. \end{cases}$$

We interpret $\partial_\mu|\mu|$ as a possible sub-derivative (instead of the set of sub-derivatives), i.e., $\partial_\mu|\mu| = 1$ if $\mu > 0$, $\partial_\mu|\mu| = -1$ if $\mu < 0$, or $\partial_\mu|\mu| = \nu$ for

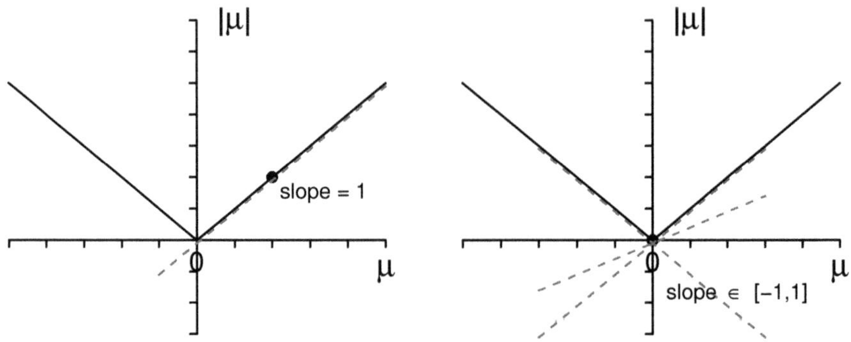

FIGURE 1.3
Sub-derivatives of $|\mu|$ (as slopes) at $\mu > 0$ (left) and $\mu = 0$ (right).

some value $\nu \in [-1, 1]$ which can be arbitrarily specified if $\mu = 0$. Similarly, (1.56) gives a possible sub-derivative of $f(\mu)$, depending on a possible sub-derivative of $|\mu|$.

To determine $\hat{\mu}_{\lambda,L_1}(Z)$, it suffices to solve the stationary condition, i.e., find μ and, if $\mu = 0$, $\nu \in [-1, 1]$ satisfying the equation,

$$
\begin{aligned}
0 &= \partial_\mu \left\{ \frac{1}{2}(Z-\mu)^2 + \lambda|\mu| \right\} \\
&= \begin{cases} -(Z-\mu)+\lambda, & \text{if } \mu > 0, \quad \text{(Case I)} \\ -(Z-\mu)+\lambda\nu, & \text{if } \mu = 0, \quad \text{(Case II)} \\ -(Z-\mu)-\lambda, & \text{if } \mu < 0. \quad \text{(Case III)} \end{cases}
\end{aligned}
$$

From Case I, if $Z > \mu$, then $\mu = Z - \lambda$. From Case III, if $Z < -\mu$, then $\mu = Z + \lambda$. From case II, $0 = \mu = Z - \lambda\nu$ and hence $Z = \lambda\nu \in [-\lambda, \lambda]$. In other words, if $Z \in [-\lambda, \lambda]$, then the preceding equation is satisfied by $\mu = 0$ and $\nu = Z/\lambda$. By combining the three cases, the preceding equation is satisfied by $\hat{\mu}_{\lambda,L_1}(Z)$ as stated in Proposition 1.11. □

Figure 1.4 shows the two estimators $\hat{\mu}_{\lambda,L_2}$ and $\hat{\mu}_{\lambda,L_1}$ as functions of Z for fixed λ. Both estimators shrink the observation Z toward 0, but in different manners. The ridge estimator $\hat{\mu}_{\lambda,L_2}(Z)$ shrinks Z linearly toward 0, whereas the Lasso estimator $\hat{\mu}_{\lambda,L_1}(Z)$ shrinks Z in a nonlinear, "thresholding" manner such that the estimator reduces to 0 for small enough Z (more precisely, $|Z| \le \lambda$). The function of Z represented by $\hat{\mu}_{\lambda,L_1}(Z)$ is commonly called the soft-thresholding function at level λ.

The penalized estimators $\hat{\mu}_{\lambda,L_2}$ and $\hat{\mu}_{\lambda,L_1}$ can also be interpreted from a Bayesian perspective. Given μ, it is assumed that $Z \sim \mathrm{N}(\mu, 1)$. For the normal prior $\mu \sim \mathrm{N}(0, \lambda^{-1})$, the posterior log density is $-(Z-\mu)^2/2 - \lambda\mu^2/2$ up to an

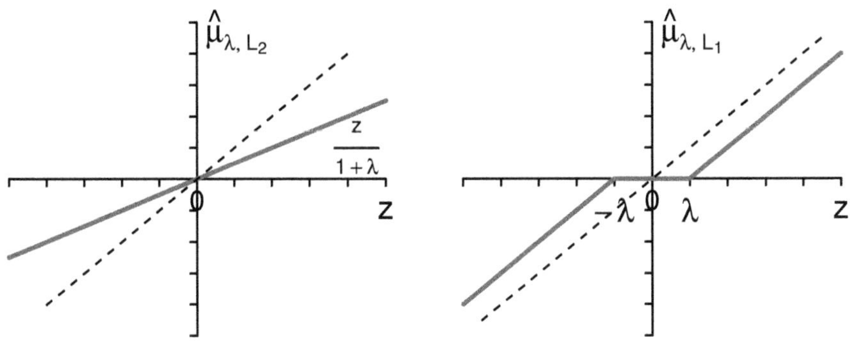

FIGURE 1.4
The ridge estimator $\hat{\mu}_{\lambda,L_2}$ (linear shrinkage) and Lasso estimator $\hat{\mu}_{\lambda,L_1}$ (soft-thresholding). The dashed line is the ML estimator $\hat{\mu}(z) = z$.

additive constant, and the ridge estimator $\hat{\mu}_{\lambda,L_2}(Z)$ is the posterior mean and mode of μ. For the Laplace prior $\mu \sim \text{Laplace}(0, \lambda^{-1})$, the posterior log density is $-(Z-\mu)^2/2 - \lambda|\mu|$ up to an additive constant, and the Lasso estimator $\hat{\mu}_{\lambda,L_1}(Z)$ is the posterior mode (but not posterior mean) of μ.

1.8.2 Ridge linear regression

We return to penalized least squares estimation for linear regression. The ridge estimator is defined as $\hat{\beta}_{\lambda,L_2} = \text{argmin}_\beta K_{2,n}(\beta)$, with

$$K_{2,n}(\beta) = \frac{1}{2n}\sum_{i=1}^{n}(Y_i - X_i^{\mathrm{T}}\beta)^2 + \frac{\lambda}{2}\|\beta\|_2^2,$$

which is quadratic in β. Setting the gradient of the penalized objective $K_{2,n}(\beta)$ to 0,

$$0 = \frac{\partial}{\partial\beta}K_{2,n}(\beta) = -\frac{1}{n}\sum_{i=1}^{n}(Y_i - X_i^{\mathrm{T}}\beta)X_i + \lambda\beta,$$

and solving for β gives a closed-form expression provided $\lambda > 0$:

$$\begin{aligned}\hat{\beta}_{\lambda,L_2} &= \left(\frac{1}{n}\sum_{i=1}^{n}X_iX_i^{\mathrm{T}} + \lambda I\right)^{-1}\left(\frac{1}{n}\sum_{i=1}^{n}X_iY_i\right)\\ &= (\mathbf{X}^{\mathrm{T}}\mathbf{X} + n\lambda I)^{-1}\mathbf{X}^{\mathrm{T}}\mathbf{Y}. \qquad (1.57)\end{aligned}$$

Note that $\mathbf{X}^{\mathrm{T}}\mathbf{X} + n\lambda I$ is nonsingular with $\lambda > 0$ even when $\mathbf{X}^{\mathrm{T}}\mathbf{X}$ is singular, with linear dependency among columns of $\mathbf{X}$. The vector of fitted values is

$$\mathbf{X}\hat{\beta}_{\lambda,L_2} = \underbrace{\mathbf{X}(\mathbf{X}^{\mathrm{T}}\mathbf{X} + n\lambda I)^{-1}\mathbf{X}^{\mathrm{T}}}_{\mathbf{S}_\lambda}\mathbf{Y},$$

that is, the fitted vector $\mathbf{X}\hat{\beta}_{\lambda,L_2}$ is $\mathbf{Y}$ left-multiplied by $\mathbf{S}_\lambda$ which depends on $\mathbf{X}$ but not $\mathbf{Y}$. By analogy with smoothing splines, $\mathbf{S}_\lambda$ is called the smoother matrix, and $\mathrm{tr}(\mathbf{S}_\lambda)$, the trace of $\mathbf{S}_\lambda$, is called the degrees of freedom. See Section 3.3 and Section 3.9.1 for further discussion about these concepts. For $\lambda > 0$, it can be shown that $\mathbf{S}_\lambda$ satisfies the following properties [Exercise]:

(i) $\mathbf{S}_\lambda$ is symmetric and nonnegative definite;

(ii) $\mathbf{S}_\lambda^2 \le \mathbf{S}_\lambda$, i.e., $\mathbf{S}_\lambda - \mathbf{S}_\lambda^2$ is nonnegative definite;

(iii) $\mathrm{tr}(\mathbf{S}_\lambda) \le \min(p, n)$, and $\mathrm{tr}(\mathbf{S}_\lambda)$ is decreasing in λ.

Taking $\lambda = 0$ above recovers the LS estimator $(\mathbf{X}^{\mathrm{T}}\mathbf{X})^{-1}\mathbf{X}^{\mathrm{T}}\mathbf{Y}$, provided that $n \ge p$ and $\mathbf{X}^{\mathrm{T}}\mathbf{X}$ is nonsingular. In this case, the smoother matrix becomes $\mathbf{S}_0 = \mathbf{X}(\mathbf{X}^{\mathrm{T}}\mathbf{X})^{-1}\mathbf{X}^{\mathrm{T}}$ and satisfies $\mathbf{S}_0^2 = \mathbf{S}_0$, and the degrees of freedom reduces to p: $\mathrm{tr}(\mathbf{S}_0) = p$.

For $\lambda > 0$, there is an alternative but equivalent expression for $\hat{\beta}_{\lambda,L_2}$:

$$\hat{\beta}_{\lambda,L_2} = \mathbf{X}^{\mathrm{T}}(\mathbf{X}\mathbf{X}^{\mathrm{T}} + n\lambda I)^{-1}\mathbf{Y}, \tag{1.58}$$

and accordingly the smoother matrix can be expressed as $\mathbf{S}_\lambda = \mathbf{X}\mathbf{X}^{\mathrm{T}}(\mathbf{X}\mathbf{X}^{\mathrm{T}} + n\lambda I)^{-1}$. The equivalence can be verified using the identity, for $\lambda > 0$,

$$(\mathbf{X}^{\mathrm{T}}\mathbf{X} + n\lambda I)^{-1}\mathbf{X}^{\mathrm{T}} = \mathbf{X}^{\mathrm{T}}(\mathbf{X}\mathbf{X}^{\mathrm{T}} + n\lambda I)^{-1}.$$

This identity can be proved by noting $\mathbf{X}^{\mathrm{T}}(\mathbf{X}\mathbf{X}^{\mathrm{T}} + n\lambda I) = (\mathbf{X}^{\mathrm{T}}\mathbf{X} + n\lambda I)\mathbf{X}^{\mathrm{T}}$ and using the fact that if $\lambda > 0$, then both $\mathbf{X}^{\mathrm{T}}\mathbf{X} + n\lambda I$ and $\mathbf{X}\mathbf{X}^{\mathrm{T}} + n\lambda I$ are nonsingular even when $\mathbf{X}^{\mathrm{T}}\mathbf{X}$ or $\mathbf{X}\mathbf{X}^{\mathrm{T}}$ is singular. See Example 3.10 for the derivation of (1.58) through reproducing-kernel-Hilbert-space regression, and Exercise 1.5 for the meaning of (1.58) with $\lambda = 0$.

For further understanding and numerical implementation, consider a singular value decomposition of the covariate matrix $\mathbf{X}$ in two distinct cases, $n \ge p$ or $n < p$. In the case of $n \ge p$, a (compact) singular value decomposition of $\mathbf{X}$ states

$$\mathbf{X} = UDV^{\mathrm{T}},$$

where U and V are $n \times p$ and $p \times p$ orthogonal matrices ($U^{\mathrm{T}}U = V^{\mathrm{T}}V = I$), and D is a $p \times p$ diagonal matrix with nonnegative diagonal elements, $d_1, \ldots, d_p$, which are called the singular values of $\mathbf{X}$. Note that $VV^{\mathrm{T}} = I$ but $UU^{\mathrm{T}} \ne I$ with $n > p$. Then $\mathbf{X}^{\mathrm{T}}\mathbf{X} = VD^2V^{\mathrm{T}}$ gives an eigenvalue decomposition of $\mathbf{X}^{\mathrm{T}}\mathbf{X}$, with eigenvalues $d_1^2, \ldots, d_p^2$. By expression (1.57), the ridge estimator can be calculated as

$$\hat{\beta}_{\lambda,L_2} = V\text{diag}(\frac{d_1}{d_1^2 + n\lambda}, \cdots, \frac{d_p}{d_p^2 + n\lambda})U^{\mathrm{T}}\mathbf{Y},$$

and the vector of fitted values can be obtained as

$$\mathbf{X}\hat{\beta}_{\lambda,L_2} = U\text{diag}(\frac{d_1^2}{d_1^2 + n\lambda}, \cdots, \frac{d_p^2}{d_p^2 + n\lambda})U^{\mathrm{T}}\mathbf{Y}.$$

When fitting ridge regression with multiple λ values, these expressions are convenient in avoiding repeated matrix inversion. See Hastie et al. (2009), Section 3.4.1, for further discussion about the interpretation of these expressions (in case of $n \geq p$).

In the case of $n < p$, a (compact) singular value decomposition of $\mathbf{X}$ states

$$\mathbf{X} = UDV^{\mathrm{T}},$$

where U and V are $n \times n$ and $p \times n$ orthogonal matrices ($U^{\mathrm{T}}U = V^{\mathrm{T}}V = I$), and D is a $n \times n$ diagonal matrix with nonnegative diagonal elements, $d_1, \ldots, d_n$. Note that $UU^{\mathrm{T}} = I$ but $VV^{\mathrm{T}} \neq I$ with $n < p$. Then $\mathbf{X}\mathbf{X}^{\mathrm{T}} = UD^2U^{\mathrm{T}}$ gives an eigenvalue decomposition of $\mathbf{X}\mathbf{X}^{\mathrm{T}}$, with eigenvalues $d_1^2, \ldots, d_n^2$. By expression (1.58), the ridge estimator can be calculated as

$$\hat{\beta}_{\lambda,L_2} = V\text{diag}(\frac{d_1}{d_1^2 + n\lambda}, \cdots, \frac{d_n}{d_n^2 + n\lambda})U^{\mathrm{T}}\mathbf{Y},$$

and the vector of fitted values can be obtained as

$$\mathbf{X}\hat{\beta}_{\lambda,L_2} = U\text{diag}(\frac{d_1^2}{d_1^2 + n\lambda}, \cdots, \frac{d_n^2}{d_n^2 + n\lambda})U^{\mathrm{T}}\mathbf{Y}.$$

The expressions of $\hat{\beta}_{\lambda,L_2}$ and $\mathbf{X}\hat{\beta}_{\lambda,L_2}$ in case of $n < p$ differ from those in case of $n \geq p$ in a subtle manner. Nevertheless, these expressions can also be used after a one-time singular value decomposition when fitting ridge regression with multiple λ values.

1.8.3 Lasso linear regression: Basics

For linear regression, the Lasso estimator is defined as $\hat{\beta}_{\lambda,L_1} = \text{argmin}_\beta K_{1,n}(\beta)$ with

$$K_{1,n}(\beta) = \frac{1}{2n}\sum_{i=1}^{n}(Y_i - X_i^{\mathrm{T}}\beta)^2 + \lambda \|\beta\|_1,$$

which is convex in β. Setting the sub-gradient of $K_{1,n}(\beta)$ to 0 gives the stationary condition that for $j = 1, \ldots, p$,

$$0 = \partial_{\beta_j} K_{1,n}(\beta) = -\frac{1}{n}\sum_{i=1}^{n}(Y_i - X_i^{\mathrm{T}}\beta)X_{ij} + \lambda\partial_{\beta_j}|\beta_j|$$

or equivalently

$$\frac{1}{n}\sum_{i=1}^{n}(Y_i - X_i^{\mathrm{T}}\beta)X_{ij} = \lambda\partial_{\beta_j}|\beta_j| \begin{cases} = \lambda, & \text{if } \beta_j > 0, \\ \in [-\lambda, \lambda], & \text{if } \beta_j = 0, \\ = -\lambda, & \text{if } \beta_j < 0. \end{cases} \tag{1.59}$$

For a helpful interpretation, the Lasso estimator $\hat{\beta}_{\lambda,L_1}$ satisfies the "box constraints"

$$\left|\frac{1}{n}\sum_{i=1}^{n}(Y_i - X_i^{\mathrm{T}}\hat{\beta}_{\lambda,L_1})X_{ij}\right| \le \lambda, \quad j = 1, \ldots, p, \tag{1.60}$$

i.e., the inner product of the residual vector $\mathbf{Y} - \mathbf{X}\hat{\beta}_{\lambda,L_1}$ and jth covariate column $\mathbf{X}_j$ is bounded by $n\lambda$ for $j = 1, \ldots, p$. Equivalently, the box constraints (1.60) can be expressed as

$$\left\|\frac{1}{n}\mathbf{X}^{\mathrm{T}}(\mathbf{Y} - \mathbf{X}\hat{\beta}_{\lambda,L_1})\right\|_{\infty} \le \lambda.$$

[Similarly, a "ball constraint" can be derived for the root ridge estimator with $\text{pen}(\beta) = \|\beta\|_2$ in (1.54), although not for the ridge estimator with $\text{pen}(\beta) = \|\beta\|_2^2/2$. See Exercise 1.16.] Taking $\lambda = 0$ recovers the LS estimator for β^*. The stationary condition (1.59) is, however, more complex than the box constraints (1.60), in further associating the sign of $(\hat{\beta}_{\lambda,L_1})_j$ with whether the corresponding box constraint is attained for each j.

Similarly as in the normal mean problem, Lasso estimation has a sparsity property that depending on the tuning parameter λ, some components of $\hat{\beta}_{\lambda,L_1}$ may be exactly zero, so that the corresponding covariates are effectively removed from the linear regression. In this sense, Lasso penalized estimation is said to enable automatic variable selection.

The stationary condition (1.59) does not in general admit a closed-form solution. Nevertheless, the sparsity property of Lasso estimation is readily seen from special cases. For example, when the columns of $\mathbf{X}$ are orthogonal or equivalently $\mathbf{X}^{\mathrm{T}}\mathbf{X}$ is a diagonal matrix, the Lasso estimator $\hat{\beta}_{\lambda,L_1}$ can be determined componentwise via soft thresholding. [Exercise: Verify this statement.] As another special case, the following result can be deduced from (1.59), which gives a simple condition to decide $\hat{\beta}_{\lambda,L_1} = 0$. [Exercise: Prove this result.]

Corollary 1.1 (Zero solution for Lasso). *If and only if* $\lambda \ge \max_{j=1,\ldots,p} |n^{-1}\sum_{i=1}^{n} Y_i X_{ij}|$, *then* $\hat{\beta}_{\lambda,L_1} = 0$, *i.e.,* 0 *is a minimizer of* $K_{1,n}(\beta)$.

To compute the Lasso estimator $\hat{\beta}_{\lambda,L_1}$, iterative algorithms are in general required (when $\lambda < \max_{j=1,\ldots,p} |n^{-1}\sum_{i=1}^{n} Y_i X_{ij}|$). There are at least three algorithms of different types: coordinate descent, proximal gradient, and active-set descent. We mainly discuss the first two algorithms.

Algorithm 1 Coordinate descent for Lasso linear regression

Input: Assume $\|\mathbf{X}_j\|_n = 1$ for $j = 1, \ldots, p$. Set termination level δ and initial value β^1.
repeat
  Set $\beta^0 = \beta^1$.
  for $j = 1, 2, \ldots, p$ **do**
    Update $\beta^1_j = \hat{\beta}_j$ while fixing $\beta^1_{-j} = (\beta^1_k : k \neq j)^{\mathrm{T}}$ as β_{-j} in (1.61).
  end for
until $K_{1,n}(\beta^0) - K_{1,n}(\beta^1) \leq \delta$.
Return: Lasso estimator β^1

Bibliographical note. Coordinate descent for Lasso regression has been proposed and studied by Fu (1998) and Friedman et al. (2007) among others. Numerical implementation is provided by the R package `glmnet` (Friedman et al., 2021). General theory for coordinate descent can be found in Tseng (2001). Proximal gradient for Lasso regression has been proposed by Beck and Teboulle (2009). A useful survey on proximal algorithms is Parikh and Boyd (2014). Active-set descent for Lasso regression has been proposed by Osborne et al. (2000). Numerical implementation is included in the R package `RCAL` (Tan & Sun, 2020).

1.8.4 Coordinate descent for Lasso regression

The main idea of coordinate descent is to minimize the penalized objective $K_{1,n}(\beta)$ over one coordinate of β at a time, while keeping the remaining coordinates fixed. The algorithm is convenient in the current problem, because there is a closed-form solution in the coordinate-wise minimization as we show next. Justification of coordinate descent for Lasso regression, however, requires some care, because the objective function $K_{1,n}(\beta)$ is not differentiable at some β due to the Lasso penalty. See Exercise 1.15 for related information.

Without loss of generality, we fix $\beta_{-1} = (\beta_2, \ldots, \beta_p)^{\mathrm{T}}$ and rewrite the objective $K_{1,n}(\beta)$ as a function of β_1 only:

$$
\begin{aligned}
K_{1,n}(\beta) &= \frac{1}{2n} \sum_{i=1}^{n} (r_i^{(1)}(\beta_{-1}) - X_{i1}\beta_1)^2 + \lambda|\beta_1| + \text{constant} \\
&= \frac{1}{2} \left\{ \|\mathbf{X}_1\|_n^2 \beta_1^2 - \frac{2}{n} \sum_{i=1}^{n} r_i^{(1)}(\beta_{-1}) X_{i1}\beta_1 \right\} + \lambda|\beta_1| + \text{constant} \\
&= \frac{1}{2} \|\mathbf{X}_1\|_n^2 \left\{ \beta_1 - \frac{1}{\|\mathbf{X}_1\|_n^2} \frac{1}{n} \sum_{i=1}^{n} r_i^{(1)}(\beta_{-1}) X_{i1} \right\}^2 + \lambda|\beta_1| + \text{constant},
\end{aligned}
$$

where $r_i^{(1)}(\beta_{-1}) = Y_i - \sum_{k=2}^{p} X_{ik}\beta_k$, $\mathbf{X}_1 = (X_{11}, \ldots, X_{n1})^{\mathrm{T}}$, $\|\mathbf{X}_1\|_n^2 = n^{-1}\sum_{i=1}^{n} X_{i1}^2$, and $\|\mathbf{X}_1\|_n$ is called the empirical L_2 norm of $\mathbf{X}_1$. To minimize over β_1 is similar to solving the Lasso problem (1.55) in the normal-mean example.

For simplicity, we assume $\|\mathbf{X}_j\|_n = 1$, where $\mathbf{X}_j$ is the jth column of $\mathbf{X}$ for $j = 1, \ldots, p$. Otherwise, this can be obtained by rescaling the columns of $\mathbf{X}$, the coefficient vector β, and the tuning parameter λ. [See Exercise 1.14 for details.] From the preceding calculation, $K_{1,n}(\beta)$ as a function of β_j for fixed $\beta_{-j} = (\beta_k : k \neq j)^{\mathrm{T}}$ can be rewritten as

$$K_{1,n}(\beta) = \frac{1}{2}\left\{\beta_j - \frac{1}{n}\sum_{i=1}^{n} r_i^{(j)}(\beta_{-j})X_{ij}\right\}^2 + \lambda|\beta_j| + \text{constant},$$

where $r_i^{(j)}(\beta_{-j}) = Y_i - \sum_{k\neq j} X_{ik}\beta_k$. Then by the Lasso solution in the normal mean problem (Proposition 1.11), $K_{1,n}(\beta)$ is minimized over β_j at

$$\hat{\beta}_j = S_\lambda\left(\frac{1}{n}\sum_{i=1}^{n} r_i^{(j)}(\beta_{-j})X_{ij}\right), \tag{1.61}$$

where $S_\lambda(\cdot) = \hat{\mu}_{\lambda,L_1}(\cdot)$ is the soft-thresholding function as in Proposition 1.11, i.e.,

$$S_\lambda(z) = \begin{cases} z - \lambda, & \text{if } z > \lambda, \\ 0, & \text{if } |z| \leq \lambda, \\ z + \lambda, & \text{if } z < -\lambda. \end{cases} \tag{1.62}$$

Running through the coordinates $\beta_1, \ldots, \beta_p$ one by one and iterating gives the coordinate descent algorithm for computing the Lasso estimator (Algorithm 1).

1.8.5 Majorization-Minimization algorithm

As a preparation and a digression, we discuss the Majorization-Minimization (MM) algorithm, which provides a simple and general template (or framework) to derive and interpret various optimization algorithms including gradient descent and proximal gradient descent algorithms. A useful survey on MM algorithms is Wu and Lange (2010).

For a scalar function $h(\beta)$, consider the minimization problem

$$\min_{\beta}\ h(\beta).$$

Suppose that for any fixed $\tilde{\beta}$, a function $g(\cdot;\tilde{\beta})$ can be constructed such that

$$g(\tilde{\beta};\tilde{\beta}) = h(\tilde{\beta}), \tag{1.63a}$$

$$g(\beta;\tilde{\beta}) \geq h(\beta) \quad \text{for all } \beta, \tag{1.63b}$$

i.e., $g(\beta;\tilde{\beta})$ coincides with $h(\beta)$ at $\beta = \tilde{\beta}$ and is no smaller than $h(\beta)$ for all β. Such a function $g(\cdot;\tilde{\beta})$ satisfying (1.63a)–(1.63b) is called a majorizing function for $h(\cdot)$ at $\tilde{\beta}$. Then the MM algorithm proceeds as follows (see Figure 1.5).

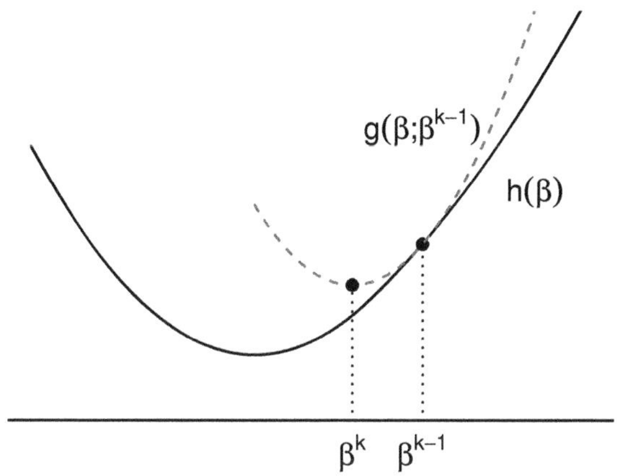

FIGURE 1.5
An iteration of the MM algorithm.

- Pick an initial value β^0.
- Compute $\beta^k = \text{argmin}_\beta\, g(\beta; \beta^{k-1})$ for $k = 1, 2, \ldots$.

The MM algorithm is known to satisfy a monotonicity property.

Proposition 1.12 (Monotonicity of MM algorithm). *For the MM algorithm,*

$$h(\beta^k) \leq h(\beta^{k-1}), \quad k = 1, 2, \ldots.$$

The property remains valid provided that β^k satisfies the descent condition $g(\beta^k; \beta^{k-1}) \leq g(\beta^{k-1}; \beta^{k-1})$ instead of minimizing $g(\cdot; \beta^{k-1})$ for $k \geq 1$.

Proof. The result follows from the chain of inequalities:

$$\begin{aligned} h(\beta^k) &\leq g(\beta^k; \beta^{k-1}) && \text{[by (1.63b)]} \\ &\leq g(\beta^{k-1}; \beta^{k-1}) && \text{[by the definition } \beta^k = \text{argmin}_\beta\, g(\beta; \beta^{k-1})] \\ &= h(\beta^{k-1}). && \text{[by (1.63a)]} \end{aligned}$$

□

The main challenge in using the MM algorithm lies in constructing a suitable majorizing function such that evaluation of β^k is easy and the convergence of β^k to a minimizer is fast. Majorizing functions are often constructed through manipulation of inequalities.

As an example, we show how gradient descent can be obtained as an MM algorithm using a quadratic majorization technique, known as the quadratic lower bound principle (Bohning & Lindsay, 1988). Suppose that $h(\beta)$ is twice-differentiable and

$$\frac{\partial^2 h}{\partial\beta\partial\beta^{\mathrm{T}}}(\beta) \le M \quad \text{for all } \beta,$$

where M is a constant, positive-definite matrix. For two matrices A_1 and A_2, $A_1 \le A_2$ indicates that $A_2 - A_1$ is nonnegative-definite. Then a quadratic majorizing function can be obtained from the Taylor expansion:

$$\begin{aligned} h(\beta) &= h(\tilde{\beta}) + \frac{\partial h}{\partial\beta^{\mathrm{T}}}(\tilde{\beta})(\beta-\tilde{\beta}) + \frac{1}{2}(\beta-\tilde{\beta})^{\mathrm{T}}\frac{\partial^2 h}{\partial\beta\partial\beta^{\mathrm{T}}}(\dot{\beta})(\beta-\tilde{\beta}) \\ &\le \underbrace{h(\tilde{\beta}) + \frac{\partial h}{\partial\beta^{\mathrm{T}}}(\tilde{\beta})(\beta-\tilde{\beta}) + \frac{1}{2}(\beta-\tilde{\beta})^{\mathrm{T}}M(\beta-\tilde{\beta})}_{g(\beta;\tilde{\beta})}, \end{aligned}$$

where $\dot{\beta}$ lies on the line segment between β and $\tilde{\beta}$. [Here $\frac{\partial h}{\partial\beta^{\mathrm{T}}}$ is the transpose of $\frac{\partial h}{\partial\beta}$, hence a row vector.] Apparently, $g(\tilde{\beta};\tilde{\beta}) = h(\tilde{\beta})$. Hence $g(\beta;\tilde{\beta})$ is a valid majorizing function for $h(\beta)$. To minimize $g(\beta;\beta^{k-1})$, we solve

$$0 = \frac{\partial g}{\partial\beta}(\beta;\beta^{k-1}) = \frac{\partial h}{\partial\beta}(\beta^{k-1}) + M(\beta-\beta^{k-1}).$$

and obtain the MM update

$$\beta^k = \beta^{k-1} - M^{-1}\frac{\partial h}{\partial\beta}(\beta^{k-1}).$$

In the special case of $M = \tau^{-1}I$ for a constant scalar $\tau > 0$, i.e.,

$$\frac{\partial^2 h}{\partial\beta\partial\beta^{\mathrm{T}}}(\beta) \le \tau^{-1}I \quad \text{for all } \beta, \tag{1.64}$$

the MM update reduces to the gradient descent (GD):

$$\beta^k = \beta^{k-1} - \tau\frac{\partial h}{\partial\beta}(\beta^{k-1}), \tag{1.65}$$

where τ is called a step size. By Proposition 1.12, the preceding discussion indicates that if (1.64) holds, then $h(\beta^k)$ from GD (1.65) is non-increasing in $k \ge 0$.

As a further example, we also show how proximal gradient can be obtained as an MM algorithm using the quadratic lower bound principle. Suppose that $h(\beta)$ is *not* differentiable at all β but can be decomposed into two functions:

$$h(\beta) = h_0(\beta) + h_1(\beta),$$

where h_1 may not be differentiable, but h_0 is twice-differentiable satisfying

$$\frac{\partial^2 h_0}{\partial\beta\partial\beta^{\mathrm{T}}}(\beta) \le \tau^{-1} I \quad \text{for all } \beta. \tag{1.66}$$

for a constant scalar $\tau > 0$. Then a quadratic majorizing function can be obtained from the Taylor expansion of $h_0(\beta)$ while keeping $h_1(\beta)$ unchanged:

$$h(\beta) = h_0(\beta) + h_1(\beta) \le \underbrace{h_0(\tilde{\beta}) + \frac{\partial h_0}{\partial\beta^{\mathrm{T}}}(\tilde{\beta})(\beta - \tilde{\beta}) + \frac{1}{2\tau}\|\beta - \tilde{\beta}\|_2^2 + h_1(\beta)}_{g(\beta;\tilde{\beta})},$$

Apparently, $g(\tilde{\beta}; \tilde{\beta}) = h(\tilde{\beta})$. Hence $g(\beta; \tilde{\beta})$ above is a valid majorizing function for $h(\beta)$. The MM update can be reexpressed as

$$\begin{aligned} \beta^k &= \mathrm{argmin}_\beta\, g(\beta; \beta^{k-1}) \\ &= \mathrm{argmin}_\beta\, \frac{1}{2\tau}\|\beta - \beta^{k-1} + \tau\frac{\partial h_0}{\partial\beta}(\beta^{k-1})\|_2^2 + h_1(\beta) \\ &= \mathrm{prox}_{\tau h_1}\left(\beta^{k-1} - \tau\frac{\partial h_0}{\partial\beta}(\beta^{k-1})\right), \end{aligned} \tag{1.67}$$

which is known as the proximal gradient algorithm. For a scalar function f, $\mathrm{prox}_f(\cdot)$ is the proximal mapping of f, defined as

$$\mathrm{prox}_f(u) = \mathrm{argmin}_\beta\, f(\beta) + \frac{1}{2}\|\beta - u\|_2^2.$$

By Proposition 1.12, if (1.66) holds, then $h(\beta^k)$ from (1.67) is non-increasing in $k \ge 0$. The proximal gradient algorithm is useful when the proximal mapping, $\mathrm{prox}_{\tau h_1}(\cdot)$, can be easily evaluated in a closed form, for example, in an application to Lasso regression.

1.8.6 Proximal gradient for Lasso regression

We apply the proximal gradient algorithm to minimize the Lasso objective

$$\underbrace{K_{1,n}(\beta)}_{h(\beta)} = \underbrace{\frac{1}{2n}\sum_{i=1}^{n}(Y_i - X_i^{\mathrm{T}}\beta)^2}_{h_0(\beta)} + \underbrace{\lambda\,\|\beta\|_1}_{h_1(\beta)}.$$

The proximal mapping $\mathrm{prox}_{\tau h_1}(\cdot)$ can be evaluated using Proposition 1.11:

$$\begin{aligned} \mathrm{prox}_{\tau h_1}(u) &= \mathrm{argmin}_\beta\, \frac{1}{2}\|\beta - u\|_2^2 + \tau\lambda\|\beta\|_1 \\ &= S_{\tau\lambda}(u), \end{aligned}$$

where for $u \in \mathbb{R}^p$, $S_{\tau\lambda}(u)$ is defined by applying the soft-thresholding function (1.62) at level $\tau\lambda$ elementwise to u, i.e., $S_{\tau\lambda}(u) = (S_{\tau\lambda}(u_j) : j = 1, \ldots, p)^{\mathrm{T}}$. [Exercise: Explain why the elementwise evaluation is valid.] The gradient of h_0 is $\frac{\partial h_0}{\partial \beta}(\beta) = -n^{-1}\sum_{i=1}^n (Y_i - X_i^{\mathrm{T}}\beta)X_i$. Hence the proximal gradient update (1.67) becomes

$$\beta^k = S_{\tau\lambda}\left(\beta^{k-1} + \tau\frac{1}{n}\sum_{i=1}^n (Y_i - X_i^{\mathrm{T}}\beta^{k-1})X_i\right), \tag{1.68}$$

which is also called iterative soft-thresolding algorithm (ISTA) (Beck & Teboulle 2008). If $\lambda = 0$, then the update (1.68) reduces to GD for computing the LS estimator.

Compared with coordinate descent, the step size τ in the proximal gradient algorithm needs to be carefully chosen (or tuned) to ensure reasonable convergence of the algorithm. A guideline can be obtained from condition (1.66), which reduces to

$$\frac{1}{n}\sum_{i=1} X_i X_i^{\mathrm{T}} = \frac{1}{n}\mathbf{X}^{\mathrm{T}}\mathbf{X} \le \tau^{-1}I,$$

i.e., τ is no greater than the reciprocal of the largest eigenvalue of $n^{-1}\mathbf{X}^{\mathrm{T}}\mathbf{X}$.

1.9 Numerical demonstration: Penalized linear regression

We demonstrate implementation of penalized linear regression in R, mainly ridge and Lasso estimation for regression coefficients. A popular R package for performing the computation is `glmnet` (Friedman et al., 2021). Similarly as in Section 1.5, we provide self-contained code to implement ridge and Lasso regression and demonstrate replication of results from using `glmnet`. The code may not be computationally optimized, but written in a direct manner to illustrate numerical implementation of the algorithms.

For example, we use the prostate cancer data and consider linear regression of the response `lpsa` on the 8 covariates (in addition to the intercept), as in Section 1.5. The original dataset is split into a training set of size 67 and a test set of size 30. Given a regression model, the training set is used to fit the model (i.e., decide the coefficient estimates), whereas the test set is used to evaluate the prediction performance of fitted models.

```
> pcancer[,1:8] <- scale(pcancer[,1:8], center=TRUE, scale=TRUE)
> train <- pcancer[pcancer$train,1:9]  # size 67
> test <- pcancer[!pcancer$train,1:9]  # size 30
```

```
>
> y <- train[,9]
> x <- as.matrix(train[,1:8])
>
> y.test <- test[,9]
> x.test <- as.matrix(test[,1:8])
```

For penalized regression, particularly when the covariates have considerably different scales (standard deviations), a common practice is to standardize the covariates before fitting (with the unweighted ridge or Lasso penalty) and then return the coefficient estimates in the original scale. This amounts to using a weighted penalty on the original coefficients, with the weights depending on the scales of the covariates, as described in Exercise 1.13. The covariate standardization is executed in `glmnet` with the option `standardize=T`. For the prostate cancer data, the covariates are standardized in the full dataset (training and test combined) after pre-processing. Hence in the following discussion, no standardization is applied to the covariates, and `glmnet` is used with `standardize=F`.

1.9.1 Ridge linear regression

As a modification from Section 1.8, ridge regression with an intercept term is typically fitted with the intercept non-penalized. In other words, the ridge objective is modified as

$$K_{2,n}(\beta) = \frac{1}{2n}\sum_{i=1}^{n}(Y_i - \beta_0 - X_{i,1:p}^{\mathrm{T}}\beta_{1:p})^2 + \frac{\lambda}{2}\|\beta_{1:p}\|_2^2, \tag{1.69}$$

where $\beta = (\beta_0, \beta_{1:p}^{\mathrm{T}})^{\mathrm{T}}$, $X_{i,1:p} = (X_{i1}, \ldots, X_{ip})^{\mathrm{T}}$, and $\lambda \geq 0$ is a tuning parameter. The ridge estimator of β, defined as a minimizer of $K_{2,n}(\beta)$, can be computed through fitting ridge regression by formula (1.57) after centering the response and covariates by sample means, as described in Exercise 1.12. For numerical comparison, the following output using `glmnet` (together with some manual scaling and unscaling) gives the ridge estimates corresponding to the minimizer of the objective function (1.69) with the choices of λ in an arithmetic sequence from 0 to 2 with increment 0.4.

For ridge linear regression, `glmnet` chooses to standardize the response `y` to have unit variance (i.e., divides by sample standard deviation of `y`) before fitting with the λ-sequence and then unstandardizes the resulting coefficients (i.e., multiplies by sample standard deviation of `y`). Therefore, the manual scaling and unrescaling is needed to obtain the minimizer of (1.69). Additional comments can be found in the accompanying R code file.

```
> # this does not give the ridge estimates as specified
> #fit.ridge <- glmnet(x, y, family="gaussian", alpha=0, standardize=F,
```

```
> #                         lambda=seq(2,0,by=-.4))
> #rbind(fit.ridge $a0, fit.ridge $bet) [,6:1]
>
> sd.y <- sqrt(mean((y-mean(y))^2))   # using 1/n rather than 1/(n-1)
> y2 <- y / sd.y
>
> fit.ridge2 <- glmnet(x, y2, family="gaussian", alpha=0, standardize=F,
+                         lambda=seq(2,0,by=-.4))
> # use "6:1", for increasing lambda
> rbind(fit.ridge2 $a0, fit.ridge2 $bet) [,6:1] * sd.y
......
                 s5          s4           s3          s2
         2.46494109  2.46359869  2.459375234 2.456865783
lcavol   0.67957627  0.40698286  0.318147099 0.268054245
lweight  0.26319187  0.23530606  0.205064382 0.181100277
age     -0.14151421 -0.04229073 -0.008231857 0.007503174
lbph     0.21005038  0.15820535  0.128333505 0.108730653
svi      0.30524194  0.22226380  0.187982537 0.165523777
lcp     -0.28866573  0.01032626  0.060142338 0.074846537
gleason -0.02115815  0.04258764  0.050053984 0.052124271
pgg45    0.26691750  0.12869130  0.109573803 0.100241512
                s1         s0
        2.45530968 2.45428930
lcavol  0.23419155 0.20918667
lweight 0.16210497 0.14676237
age     0.01564419 0.02009700
lbph    0.09466628 0.08398396
svi     0.14878812 0.13558333
lcp     0.07905710 0.07922818
gleason 0.05219889 0.05130335
pgg45   0.09342666 0.08772110
```

We write the following function, `my.ridge.path()`, to implement the ridge estimator as a minimizer of $K_{2,n}(\beta)$ in (1.69) for a sequence of λ choices, allowing the intercept to be either present and non-penalized or absent. This function returns a list including the coefficient estimates (`bet`) as a matrix, with each column giving the coefficients for a λ choice, the degrees of freedom (`df`) excluding the intercept (see Exercise 1.12), and the training errors (`err.train`). The training error is the mean squared prediction error in the training data (as defined in Section 1.9.3). For simplicity, we use the matrix inverse function `solve()` repeatedly for different λ choices. In principle, it would be more efficient to compute the ridge estimator with different λ values, after performing a singular value decomposition of the covariate matrix, which amounts to matrix inversion for only one time. See Section 1.8.2.

```
my.ridge.path <- function(y,x, lam.seq, intercept=T) {
  x <- cbind(x)
  n <- dim(x)[1]
  p <- dim(x)[2]

  if (intercept) {
    xbar <- apply(x,2,mean)
    x <- x - rep(xbar, each=n)
    ybar <- mean(y)
    y <- y-ybar
  }

  H <- t(x)%*%x/n

  # loop over the tuning parameter choices
  m <- length(lam.seq)
  bet <- matrix(NA, nrow=p, ncol=m)
  rownames(bet) <- colnames(x)
  df <- rep(NA, m)
  err.train <- rep(NA,m)

  for (j in 1:m) {
    bet[,j] <- solve(H +diag(lam.seq[j],p), t(x)%*%y/n)
    df[j] <- sum(diag( solve(H +diag(lam.seq[j],p), t(x)%*%x/n) ))
    err.train[j] <- mean((y-x%*%bet[,j])^2)
  }

  if (intercept)
    bet <- rbind(ybar - t(xbar)%*%bet, bet)

  list(bet=bet, df=df, err.train=err.train)
}
```

The following output using `my.ridge.path()` gives the ridge estimates with the sequence of λ choices from 0 to 2 with increment 0.4. The coefficient estimates match those from **glmnet** except in last few digits, presumably up to machine precision.

```
> out.ridge <- my.ridge.path(y,x, seq(0,2,.4))
> out.ridge
$bet
               [,1]        [,2]         [,3]         [,4]
         2.46493292  2.46360405  2.459375214 2.456865420
lcavol   0.67952814  0.40693704  0.318147141 0.268054412
lweight  0.26305307  0.23529366  0.205052283 0.181097724
age     -0.14146483 -0.04232374 -0.008227323 0.007508687
lbph     0.21014656  0.15822075  0.128339738 0.108731794
svi      0.30520060  0.22221076  0.187982997 0.165529767
lcp     -0.28849277  0.01036551  0.060156549 0.074852514
gleason -0.02130504  0.04262358  0.050049873 0.052123999
pgg45    0.26695576  0.12869630  0.109570126 0.100237867
              [,5]       [,6]
        2.45530871 2.45428972
lcavol  0.23421258 0.20917531
lweight 0.16210740 0.14675760
age     0.01565794 0.02009184
lbph    0.09466363 0.08398670
svi     0.14880334 0.13557580
lcp     0.07906081 0.07923040
gleason 0.05219023 0.05130709
pgg45   0.09341964 0.08772348

$df
[1] 8.000000 4.821408 3.668664 3.013654 2.577582 2.261457

$err.train
[1] 0.4391998 0.5033129 0.5617563 0.6118247 0.6563576 0.6965253
```

Figure 1.6 shows the ridge estimates and the training and test errors versus the degrees of freedom, computed from a more dense sequence of λ choices. The test error is the mean squared prediction error in the test data (as defined in Section 1.9.3). See the accompanying R code file for details. The coefficient estimates shrink toward 0 (or expand) as the tuning parameter λ increases (or decreases) and the degree of freedom decreases (or increases). In addition, as λ decreases, the training error keeps decreasing, whereas the test error first decreases and then increases, hence exhibiting a U-shaped pattern, which is typical of test errors versus tuning parameters.

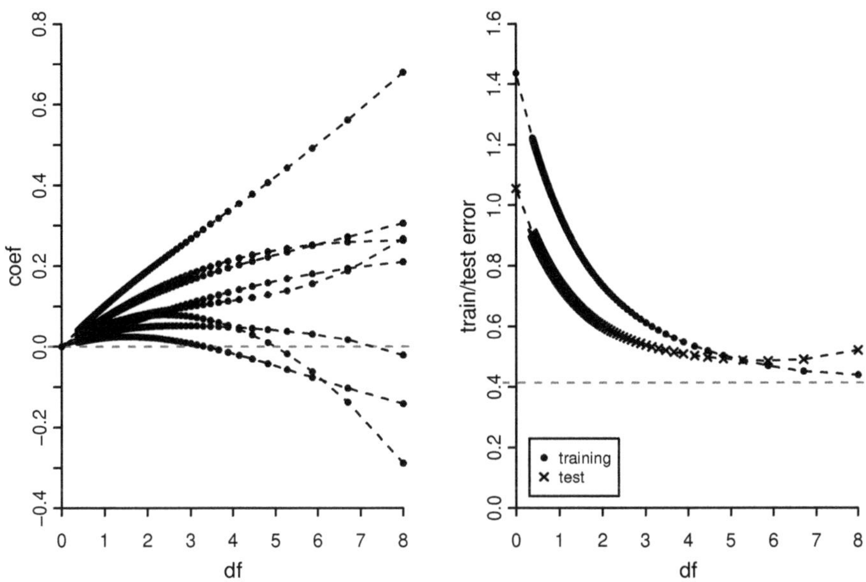

FIGURE 1.6
Plots of the ridge estimates versus the degrees of freedom (left) and the training and test errors versus the degrees of freedom (right), as the λ choice is varied, in the prostate cancer example. A horizontal line is also placed at the test error for the (infeasible) LS estimator trained on the training and test data combined.

1.9.2 Lasso linear regression

Similarly as ridge regression, Lasso regression with an intercept term is also usually fitted with the intercept non-penalized. In other words, the Lasso objective is modified as

$$K_{1,n}(\beta) = \frac{1}{2n} \sum_{i=1}^{n} (Y_i - \beta_0 - X_{i,1:p}^{\mathrm{T}} \beta_{1:p})^2 + \lambda \|\beta_{1:p}\|_1, \tag{1.70}$$

where $\beta = (\beta_0, \beta_{1:p}^{\mathrm{T}})^{\mathrm{T}}$, $X_{i,1:p} = (X_{i1}, \ldots, X_{ip})^{\mathrm{T}}$, and $\lambda \geq 0$ is a tuning parameter. The Lasso estimator of β, defined as a minimizer of $K_{1,n}(\beta)$, can be computed through fitting Lasso regression after centering the response and covariates by sample means, as described in Exercise 1.12. For numerical comparison, the following output using `glmnet` gives the Lasso estimates corresponding to the minimizer of the objective function (1.70) with the choices of λ in an arithmetic sequence from 0 to 1 with increment 0.2. As shown in the accompanying R code file, the same result can also be obtained using R package `RCAL`, which implements the active-set descent algorithm for Lasso regression (Tan & Sun, 2020),

For Lasso linear regression, unlike for ridge regression, the coefficients obtained from `glmnet` are the minimizer of (1.70), *without* being affected by the choice of `glmnet` to standardize the response y before fitting and then unstandardizes the resulting coefficients. [Exercise: Explain why.] Additional comments can be found in the accompanying R code file.

```
> fit.lasso <- glmnet(x, y, family="gaussian", alpha=1, standardize=F,
+                     lambda=seq(1,0,by=-.2))
> rbind(fit.lasso $a0, fit.lasso $bet) [,6:1]  #increasing lambda
......
                 s5         s4         s3        s2        s1       s0
         2.46487803 2.46803639 2.46682611 2.4613902 2.4557306 2.452345
lcavol    0.67953377 0.53975471 0.45243367 0.2919335 0.1092682 .
lweight   0.26301892 0.19366013 0.06996797 .         .         .
age      -0.14143958 .          .          .         .         .
lbph      0.21018348 .          .          .         .         .
svi       0.30529048 0.08937739 .          .         .         .
lcp      -0.28880634 .          .          .         .         .
gleason  -0.02173182 .          .          .         .         .
pgg45     0.26741340 0.01074909 .          .         .         .
```

We write the following functions, `soft()` and `my.lasso()`, to implement the coordinate descent algorithm (Algorithm 1) for computing the Lasso estimator as a minimizer of $K_{1,n}(\beta)$ in (1.70) for a single choice of λ, allowing the intercept to be either present and non-penalized or absent. [Implementation of the proximal gradient algorithm for Lasso regression is left as an exercise.] The function `soft()` performs the soft-thresholding $\hat{\mu}_{\lambda,L_1}()$ as stated in Proposition 1.11. Because the covariate columns are required to have unit empirical L_2 norms, $\|\cdot\|_n$, in Algorithm 1, rescaling and unscaling are performed as described in Exercise 1.14, in addition to the centering of the response and covariates described in Exercise 1.12. The function `my.lasso()` returns a list including the coefficient estimates (`bet`), convergence status (`conv`, taking value 1 if convergence is obtained or 0 if exceeding the maximum number of iterations), number of iterations executed (`iter`), the penalized objective values along the iterations (`obj`), and the training error based on the final iteration (`err.train`).

```
soft <- function(x, lam) {
  if (x>lam) x-lam
  else if (x< -lam) x+lam
  else 0
}
```

```
my.lasso <- function(y, x, lam, init=NULL, intercept=T,
                     niter=100, eps=1e-6) {
  x <- cbind(x)
  n <- dim(x)[1]
  p <- dim(x)[2]

  if (is.null(init))
    init <- rep(0,p)

  if (intercept) {
    xbar <- apply(x,2,mean)
    x <- x - rep(xbar, each=n)
    ybar <- mean(y)
    y <- y - ybar
  }
  sd.x <- sqrt(apply(x^2, 2, mean))
  x <- x /rep(sd.x, each=n)
  lam2 <- lam /sd.x

  # coordinate descent iterations
  bet <- init *sd.x
  obj <- rep(NA,niter)
  res <- as.vector(y-x%*%bet)
  err.train <- mean(res^2)
  obj[1] <- err.train/2 + sum(lam2*abs(bet))

  conv <- 0
  i <- 1
  while (i< niter & conv<= 0) {
    for (j in 1:p) {
      incr <- soft(bet[j]+mean(res*x[,j]), lam2[j]) -bet[j]
      bet[j] <- bet[j] +incr
      res <- res -incr*x[,j]
    }
    err.train <- mean(res^2)
    obj[i+1] <- err.train/2 + sum(lam2*abs(bet))

    if  (obj[i]-obj[i+1] < eps)
      conv <- 1
    i <- i+1
  }
```

```
  bet <- bet /sd.x
  if (intercept)
    bet <- c(ybar - sum(xbar*bet), bet)

  list(bet=bet, conv=conv, iter=i, obj=obj[1:i], err.train=err.train)
}
```

The following output using `my.lasso()` gives the Lasso estimates with the choice $\lambda = 0.4$. The stationary condition for minimizing the Lasso objective (1.70), similar to (1.59), can be verified to hold (up to machine precision) by computing the empirical inner products of the residual vector with the covariate columns (i.e., inner products divided by n). The coefficient estimates associated with `lcavol` and `lweight` are positive, and the empirical inner products are $\lambda = 0.4$, whereas the coefficient estimates associated with the remaining 6 covariates are 0, and the empirical inner products fall in the interval $[-\lambda, \lambda] = [-0.4, 0.4]$.

```
> out.lasso <- my.lasso(y,x, lam=.4)
> out.lasso
$bet
                lcavol     lweight         age        lbph         svi
2.46682654 0.45244844 0.06996376 0.00000000 0.00000000 0.00000000
       lcp    gleason       pgg45
0.00000000 0.00000000 0.00000000

$conv
[1] 1

$iter
[1] 5

$obj
[1] 0.7185182 0.5927381 0.5924953 0.5924933 0.5924933

$err.train
[1] 0.7670569

>
> # inner products of residual with covariates
> apply(x*as.vector(y-cbind(1,x)%*%out.lasso$bet), 2, mean)
   lcavol    lweight         age        lbph         svi         lcp
```

```
0.3999838 0.4000000 0.1128815 0.2518598 0.3754343 0.2452776
  gleason     pgg45
0.2014723 0.3120550
```

Similarly to `my.ridge.path()` for ridge regression, the following function `my.lasso.path()` computes the Lasso estimator using `my.lasso()`, for a sequence of λ choices (`lam.seq`) specified in an increasing order. Compared with `my.ridge.path()`, a difference is that the Lasso estimator is computed over the λ choices in an decreasing order, using `my.lasso()` with initial value (`init`) for the current λ set to the final value from the previous larger λ, where the initial value for the largest λ is set to a zero vector. This process may speed up the convergence of coordinate descent iterations. The function `my.lasso.path()` returns a list including the coefficient estimates (`bet`) as a matrix, with each column giving the coefficients for a λ choice, the convergence indicators (`conv`), the numbers of iterations executed (`iter`), and the training errors (`err.train`), each as a vector over the λ choices in an increasing order.

```
my.lasso.path <- function(y, x, lam.seq, init=NULL, intercept=T,
                          niter=100, eps=1e-6) {
  x <- cbind(x)
  n <- dim(x)[1]
  p <- dim(x)[2]

  if (is.null(init))
    init <- rep(0,p)

  if (intercept) {
    xbar <- apply(x,2,mean)
    x <- x - rep(xbar, each=n)
    ybar <- mean(y)
    y <- y - ybar
  }

  # loop over the tuning parameter choices
  m <- length(lam.seq)
  bet <- matrix(NA, nrow=dim(x)[2], ncol=m)
  rownames(bet) <- colnames(x)
  conv <- rep(NA,m)
  iter <- rep(NA,m)
  err.train <- rep(NA,m)

  for (j in m:1) {     # use m:1, for decreasing lambda
```

```
    if (j<m) init <- bet[,j+1]
    out <- my.lasso(y, x, lam=lam.seq[j], init, intercept=F, niter, eps)
    bet[,j] <- out $bet
    conv[j] <- out $conv
    iter[j] <- out $iter
    err.train[j] <- out $err.train
  }

  if (intercept)
    bet <- rbind(ybar - t(xbar)%*%bet, bet)

  list(bet=bet, conv=conv, iter=iter, err.train=err.train)
}
```

The following output using `my.lasso.path()` gives the Lasso estimates with the sequence of λ choices from 0 to 1 with increment 0.2. The coefficient estimates match those from `glmnet` except in last few digits, due to machine precision and the termination choice (`eps`).

```
> out.lasso <- my.lasso.path(y,x, lam.seq=seq(0,1,.2))
> out.lasso
$bet
              [,1]       [,2]       [,3]      [,4]      [,5]     [,6]
        2.46501172 2.46804269 2.46682654 2.4613902 2.4557306 2.452345
lcavol  0.67933564 0.53983546 0.45244844 0.2919335 0.1092682 0.000000
lweight 0.26345865 0.19363221 0.06996376 0.0000000 0.0000000 0.000000
age    -0.14174667 0.00000000 0.00000000 0.0000000 0.0000000 0.000000
lbph    0.20993088 0.00000000 0.00000000 0.0000000 0.0000000 0.000000
svi     0.30499623 0.08948647 0.00000000 0.0000000 0.0000000 0.000000
lcp    -0.28845167 0.00000000 0.00000000 0.0000000 0.0000000 0.000000
gleason -0.02050477 0.00000000 0.00000000 0.0000000 0.0000000 0.000000
pgg45   0.26658432 0.01066043 0.00000000 0.0000000 0.0000000 0.000000

$conv
[1] 1 1 1 1 1 1

$iter
[1] 14  8  5  3  3  2

$err.train
[1] 0.4392001 0.5834858 0.7670569 0.9934033 1.2491347 1.4370365
```

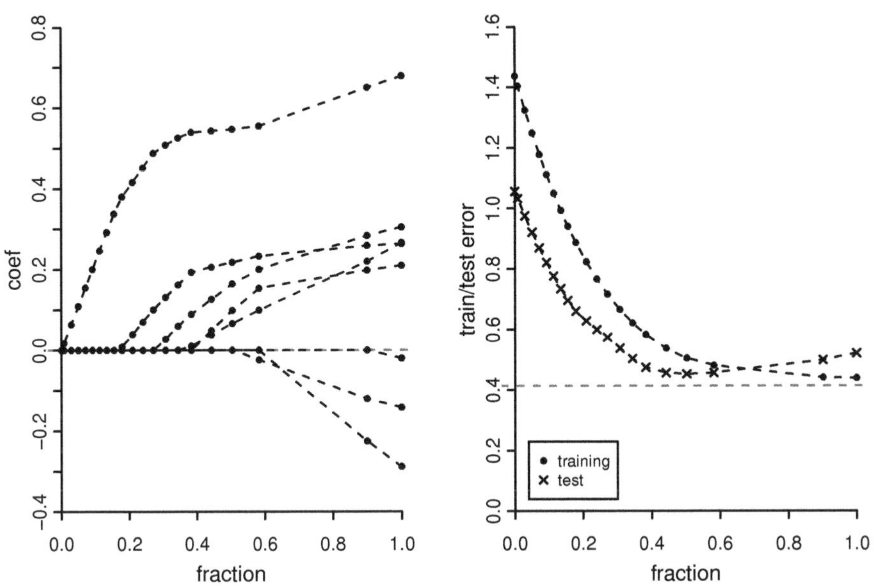

FIGURE 1.7
Plots of the Lasso estimates versus the fraction of L_1 norms (left) and the training and test errors versus the shrinkage fraction (right), as the λ choice is varied, in the prostate cancer example. The shrinkage fraction is $\frac{\|\hat{\beta}_{1:p}(\lambda)\|_1}{\|\hat{\beta}_{1:p}(0)\|_1}$, where $\hat{\beta}_{1:p}(\lambda)$ is the Lasso estimator given λ, and $\hat{\beta}_{1:p}(0)$ is the LS estimator. A horizontal line is also placed at the test error for the (infeasible) LS estimator trained on the training and test data combined.

Figure 1.7 shows the Lasso estimates and the training and test errors versus the shrinkage fraction of L_1 norms, computed from a more dense sequence of λ choices. See the accompanying R code file for details. As λ decreases (and hence the fraction of L_1 norms increases), the number of nonzero coefficient estimates increases from 0 to 8, reflecting the sparsity of Lasso estimation. In addition, as λ decreases, the training error keeps decreasing, whereas the test error exhibits a U-shaped pattern, similarly as in Figure 1.6.

1.9.3 Cross validation

For penalized regression, an important question (which we have not addressed) is how to select the tuning parameter λ. As illustrated by the U-shaped pattern of test errors in Figures 1.6 and 1.7, the tuning parameter λ needs to be selected in an appropriate range, not too small or too large, in order to optimize the prediction performance. To select the tuning parameter, a practically useful method is validation, either one-time or multi-hold. The latter is also called cross validation (CV).

To formalize the ideas, suppose that the available dataset of size N is randomly split into a training set of size n and a test set of size $N-n$, denoted respectively as $\{(X_i, Y_i) : i = 1, \ldots, n\}$ and $\{(X_i, Y_i) : i = n+1, \ldots, N\}$. A penalized LS estimator is defined as in (1.54):

$$\hat{\beta}_\lambda = \text{argmin}_\beta \left\{ \frac{1}{2n} \sum_{i=1}^{n} (Y_i - X_i^{\mathrm{T}} \beta)^2 + \lambda \, \text{pen}(\beta) \right\},$$

where $\lambda \geq 0$ is a tuning parameter and $\text{pen}(\beta)$ is a penalty function in β. Some components of β (notably the intercept) may be non-penalized as in (1.69) and (1.70). Given λ and $\hat{\beta}_\lambda$, the training error is defined as the mean squared prediction error in the training set, $\text{err}_\lambda = n^{-1} \sum_{i=1}^{n} (Y_i - X_i^{\mathrm{T}} \hat{\beta}_\lambda)^2$, and the test error is defined as that in the test set, $\widehat{\text{Err}}_\lambda = (N-n)^{-1} \sum_{i=n+1}^{N} (Y_i - X_i^{\mathrm{T}} \hat{\beta}_\lambda)^2$. See Section 3.9.1 for further information about assessment of prediction errors. The training and test errors may also be rescaled by $\frac{1}{2}$, to match the non-penalized objective function used for $\hat{\beta}_\lambda$.

The idea of (one-time) validation proceeds as follows, for selecting the tuning parameter.

- Randomly split the training set into a validation subset $\{(X_i, Y_i) : i \in \mathcal{I}\}$ and a training subset $\{(X_i, Y_i) : i \in \mathcal{I}^c\}$, where $\mathcal{I} \subset \{1, 2, \ldots, n\}$ and $\mathcal{I}^c$ is the complement of $\mathcal{I}$.
- For each λ, compute

$$\tilde{\beta}_\lambda = \text{argmin}_\beta \left\{ \frac{1}{2|\mathcal{I}^c|} \sum_{i \in \mathcal{I}^c} (Y_i - X_i^{\mathrm{T}} \beta)^2 + \lambda \, \text{pen}(\beta) \right\},$$

  and then $\widetilde{\text{Err}}_\lambda = \frac{1}{|\mathcal{I}|} \sum_{i \in \mathcal{I}} (Y_i - X_i^{\mathrm{T}} \tilde{\beta}_\lambda)^2$, called the validation error, where $|\cdot|$ denotes the size (or cardinality) of a set.
- Select λ by minimizing $\widetilde{\text{Err}}_\lambda$. In practice, this is done by searching over a pre-specified grid of values, i.e., selecting λ with the smallest $\widetilde{\text{Err}}_\lambda$ over a grid of values of λ.
- For the selected λ, compute $\hat{\beta}_\lambda$ and report err_λ and $\widehat{\text{Err}}_\lambda$.

Cross validation extends the idea of validation by alternating the assignment of training and validation subsets and aggregating the validation errors. Formally, K-hold cross validation proceeds as follows, where a typical choice of K is 5 or 10 (see Figure 1.8).

- Randomly split the training set into K approximately equal-sized parts, i.e., split $\{1, 2, \ldots, n\}$ into K approximately equal-sized subsets $\mathcal{I}_1, \ldots, \mathcal{I}_K$.

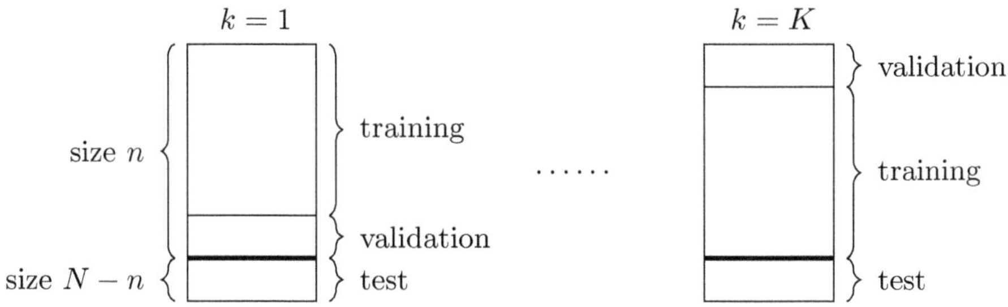

FIGURE 1.8
Illustration of K-fold cross validation.

- For each λ and $k = 1, \ldots, K$, compute

$$\tilde{\beta}_\lambda^{(k)} = \text{argmin}_\beta \left\{ \frac{1}{2|\mathcal{I}_k^c|} \sum_{i \in \mathcal{I}_k^c} (Y_i - X_i^{\mathrm{T}} \beta)^2 + \lambda \, \text{pen}(\beta) \right\},$$

and then $\widetilde{\text{Err}}_\lambda^{(k)} = \frac{1}{|\mathcal{I}_k|} \sum_{i \in \mathcal{I}_k} (Y_i - X_i^{\mathrm{T}} \tilde{\beta}_\lambda^{(k)})^2$. Compute the CV error

$$\text{CV}(\lambda) = \frac{1}{K} \sum_{k=1}^{K} \widetilde{\text{Err}}_\lambda^{(k)}.$$

- Select λ by minimizing $\text{CV}(\lambda)$ or, in practice, select λ with the smallest $\text{CV}(\lambda)$ over a grid of values $\lambda_1, \ldots, \lambda_m$, i.e., $\hat{\lambda} = \text{argmin}_{\lambda_1, \ldots, \lambda_m} \text{CV}(\lambda)$.
- For the selected λ, compute $\hat{\beta}_\lambda$ and report err_λ and $\widehat{\text{Err}}_\lambda$.

An alternative approach, called "one standard error rule" (Breiman et al., 1984; Hastie et al., 2009), is to select λ giving the most parsimonious model within one standard error from the minimum cross validation error, i.e., select the largest $\lambda \in \{\lambda_1, \ldots, \lambda_m\}$ such that

$$\text{CV}(\lambda) \leq \text{CV}(\hat{\lambda}) + \frac{\text{SD}(\hat{\lambda})}{\sqrt{K}},$$

where $\text{SD}(\lambda) = [\frac{1}{K-1} \sum_{k=1}^{K} \{\widetilde{\text{Err}}_\lambda^{(k)} - \text{CV}(\lambda)\}^2]^{1/2}$, the sample standard deviation of the validation errors $\widetilde{\text{Err}}_\lambda^{(1)}, \ldots, \widetilde{\text{Err}}_\lambda^{(K)}$. The term $\frac{\text{SD}(\lambda)}{\sqrt{K}}$ is referred to as the SE of $\text{CV}(\lambda)$.

In the remainder of this subsection, we return to the prostate cancer example and demonstrate application of cross validation. For ridge regression, we first write the following function, `my.ridge.path.b()`, by extending the previous function `my.ridge.path()` to incorporate calculation of test errors (if given a test set) for a sequence of λ choices. This function can be directly used to produce Figure 1.6, as shown in the accompanying R code file.

```
my.ridge.path.b <- function(y,x, lam.seq, intercept=T, test=NULL) {
  if (is.null(test)) {
    out.path <- my.ridge.path(y,x, lam.seq, intercept)
    list(bet=out.path$bet, df=out.path$df,
         err.train=out.path$err.train, err.test=NULL)
  } else {
    x <- as.matrix(x)
    y.test <- y[test]
    x.test <- x[test, , drop=F]
    y <- y[-test]
    x <- x[-test, , drop=F]

    out.path <- my.ridge.path(y,x, lam.seq, intercept)
    if (intercept)
      x.test <- cbind(1,x.test)
    err.test <- apply((y.test-x.test%*%out.path$bet)^2, 2, mean)

    list(bet=out.path$bet, df=out.path$df,
         err.train=out.path$err.train, err.test=err.test)
  }
}
```

We write the following functions, `split()` and `my.ridge.cv()`, to implement cross validation (with K coded as `fold`) for ridge regression with a sequence of λ choices. A random split of the training data `(y,x)` is achieved by combining a fixed split executed by `split()` and a random permutation executed by `sample()`. For ease of reproducing results, a permutation (`permut`) is allowed to be specified as an argument. The function `my.ridge.cv()` returns a list including the CV errors (`err.ave`) and their associated standard errors (`err.se`), each as a vector of the length of `lam.seq`, two λ values (`lam`) selected by minimization of the CV error or the one standard error rule, and results of ridge regression (`out`) with the training data `(y,x)` for the two selected λ values.

```
split <- function(n, fold) {
  n.test <- rep(floor(n/fold), fold)
  if (n%%fold >0)
    n.test[1:(n%%fold)] <- n.test[1] +1
  rep(1:fold, times=n.test)
}

my.ridge.cv <- function(fold, y,x, lam.seq, intercept=T,
```

```
                              permut=NULL) {
  x <- cbind(x)
  n <- dim(x)[1]
  if (is.null(permut))
    permut <- sample(1:n)

  fold.id <- split(n, fold)
  fold.id <- fold.id[permut]

  m <- length(lam.seq)
  err <- matrix(NA, nrow=m, ncol=fold)
  for (j in 1:fold) {
    test <- which(fold.id==j)
    out <- my.ridge.path.b(y,x, lam.seq, intercept, test)
    err[,j] <- out $err.test
  }
  err.ave <- apply(err, 1, mean)  # cv error
  err.se <- apply(err, 1, sd)/sqrt(fold)

  sel <- which.min(err.ave)
  sel <- c(sel, max(which(err.ave <= err.ave[sel] + err.se[sel])))
  out.all <- my.ridge.path.b(y,x, lam.seq[sel], intercept, test=NULL)

  list(out=out.all, sel=sel, lam=lam.seq[sel],
       err.ave=err.ave, err.se=err.se)
}
```

The following output using `my.ridge.cv()` gives the results of 10-fold cross validation for ridge regression in the prostate cancer example. To save space, some lines are omitted in the output below. Figure 1.9, extending Figure 1.6, shows the CV errors and their SEs and the two selected λ values, $\lambda = 0$ (no ridge penalty, selected by minimum CV error) or $\lambda = 1$ (selected by the one standard error rule).

```
> lam.seq <- c(seq(0,20,.1), 1e+4)
> # for ease of replication, a fixed permutation is used
> out.ridge.cv <- my.ridge.cv(fold=10, y,x, lam.seq, intercept=T,
+                              permut=1:n)
> out.ridge.cv
$out
$out$bet
```

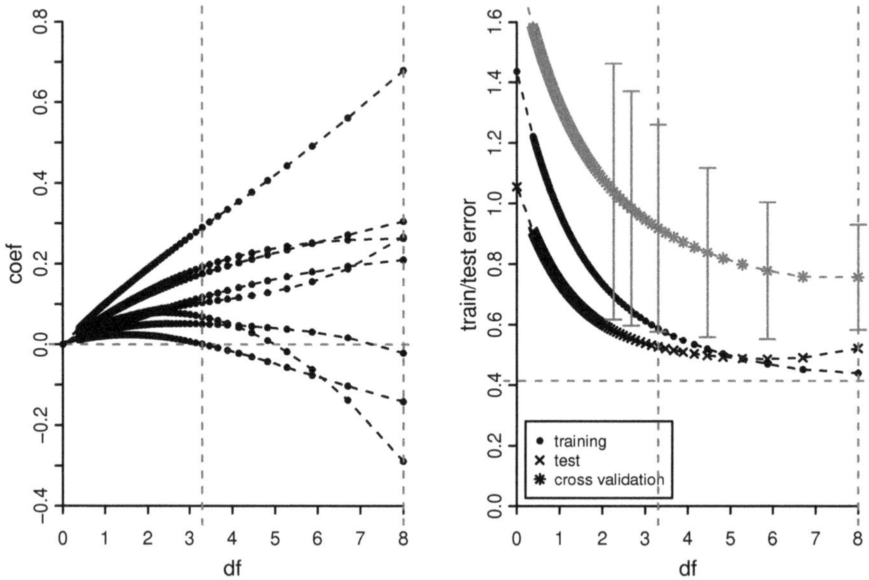

FIGURE 1.9
Extension of Figure 1.6, showing the 10-fold CV errors and their SEs versus the degrees of freedom (right). Two vertical lines are also placed at the degrees of freedom corresponding to the two selected λ values.

```
                [,1]         [,2]
          2.46493292 2.4579625995
lcavol    0.67952814 0.2902810273
lweight   0.26305307 0.1923575607
age      -0.14146483 0.0009934769
lbph      0.21014656 0.1176402248
svi       0.30520060 0.1758173973
lcp      -0.28849277 0.0695732668
gleason  -0.02130504 0.0514481255
pgg45     0.26695576 0.1044066302

$out$df
[1] 8.000000 3.303552
......
$lam
[1] 0 1
......
```

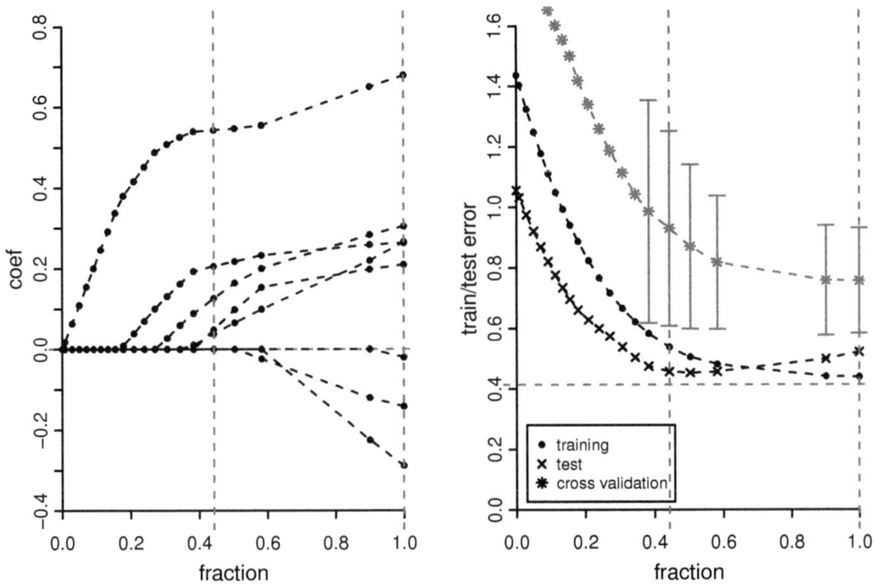

FIGURE 1.10
Extension of Figure 1.7, showing the 10-fold CV errors and their SEs versus the shrinkage fraction (right). Two vertical lines are also placed at the shrinkage fraction corresponding to the two selected λ values.

For Lasso regression, similarly as for ridge regression, we write the function `my.lasso.path.b()` by extending `my.lasso.path()` to incorporate calculation of test errors and then function `my.lasso.cv()` to implement cross validation with a sequence of λ choices. See the accompanying R code file for details.

The following output using `my.lasso.cv()` gives the results of 10-fold cross validation for Lasso regression in the prostate cancer example. Figure 1.10, extending Figure 1.7, shows the CV errors and their SEs and the two selected λ values, $\lambda = 0$ (no Lasso penalty, selected by minimum CV error) or $\lambda = .15$ (selected by the one standard error rule).

```
> lam.seq <- c(0, .01, seq(.05,1,.05))
> out.lasso.cv <- my.lasso.cv(fold=10, y,x, lam.seq, intercept=T,
+                                  permut=1:n)
> out.lasso.cv
$out
$out$bet
                   [,1]        [,2]
             2.46499639 2.46668124
```

```
lcavol   0.67939362 0.54372269
lweight  0.26340596 0.20606697
age     -0.14169682 0.00000000
lbph     0.20995195 0.04897418
svi      0.30504726 0.12691934
lcp     -0.28850504 0.00000000
gleason -0.02064529 0.00000000
pgg45    0.26665800 0.03864397
......
$lam
[1] 0.00 0.15
......
```

1.10 Appendices

1.10.1 Wald confidence intervals and tests

Wald confidence intervals and hypothesis tests can be understood as follows. Let $\bar{\theta}$ be a scalar (unknown) population value, for example, the jth true coefficient β_j^* if linear regression is correctly specified, or the jth target coefficient $\bar{\beta}_j$ if linear regression may be misspecified. Suppose that $\hat{\theta}$ is a consistent and asymptotically normal estimator of $\bar{\theta}$ as $n \to \infty$:

$$\sqrt{n}(\hat{\theta} - \bar{\theta}) \underset{\mathcal{D}}{\to} \mathrm{N}(0, V), \tag{1.71}$$

where n is a sample size, and $V\ (> 0)$ is the asymptotic variance. Note that (1.71) implies $\hat{\theta} \underset{\mathcal{P}}{\to} \bar{\theta}$ as $n \to \infty$. Moreover, suppose that $\hat{V}$ is a consistent estimator of V:

$$\hat{V} \underset{\mathcal{P}}{\to} V. \tag{1.72}$$

The estimated standard deviation, also called the standard error, for $\hat{\theta}$ is $\mathrm{SE}(\hat{\theta}) = \sqrt{\hat{V}/n}$. By Slutsky's theorem, the two properties (1.71) and (1.72) imply that

$$\frac{\sqrt{n}(\hat{\theta} - \bar{\theta})}{\sqrt{\hat{V}}} \underset{\mathcal{D}}{\to} \mathrm{N}(0, 1).$$

Then the confidence interval $\hat{\theta} \pm z_{c/2}\mathrm{SE}(\hat{\theta}) = \hat{\theta} \pm z_{c/2}\sqrt{\hat{V}/n}$, called a Wald confidence interval, achieves an asymptotic size $1 - c$ in covering $\bar{\theta}$:

$$P\left(\bar{\theta} \in [\hat{\theta} - z_{c/2}\sqrt{\hat{V}/n},\, \hat{\theta} + z_{c/2}\sqrt{\hat{V}/n}]\right) \to 1 - c.$$

By the usual relationship between confidence intervals and hypothesis tests (Casella & Berger, 2001), Wald hypothesis tests can be similarly derived.

1.10.2 Asymptotic normality in fixed design

We comment on how the asymptotic normality (1.12a) and confidence interval (1.13) can be shown under homoscedastic linear regression in the fixed design. A basic identity underlying (1.12a) is, as also shown in (1.24),

$$\hat{\beta} - \beta^* = \hat{H}^{-1}\frac{1}{n}\sum_{i=1}^{n} X_i^{\mathrm{T}}(Y_i - X_i^{\mathrm{T}}\beta^*) = \hat{H}^{-1}\frac{1}{n}\sum_{i=1}^{n} X_i^{\mathrm{T}}\varepsilon_i, \tag{1.73}$$

and hence

$$\sqrt{n}\hat{H}^{1/2}(\hat{\beta} - \beta^*) = \hat{H}^{-1/2}\frac{1}{\sqrt{n}}\sum_{i=1}^{n} X_i^{\mathrm{T}}\varepsilon_i. \tag{1.74}$$

Under suitable conditions on $\{(X_i, \varepsilon_i) : i = 1, \ldots, n\}$, triangular-array central limit theorems can be used to deduce (1.12a) from (1.74).

If $\hat{H}$ is assumed to converge to a limit H^* as $n \to \infty$ and p is fixed, then $\sqrt{n}(\hat{\beta} - \beta^*) \xrightarrow{\mathcal{D}} \mathrm{N}(0, \sigma^{*2}H^{*-1})$ and, as a component, $\sqrt{n}(\hat{\beta}_j - \beta^*_j) \xrightarrow{\mathcal{D}} \mathrm{N}(0, \sigma^{*2}(H^{*-1})_{jj})$. With this and $\hat{\sigma}^2 \xrightarrow{\mathcal{P}} \sigma^{*2}$ in (1.12b), the usual argument using Slutsky's theorem shows that the confidence interval (1.13) is of asymptotic size $1 - c$. If $\hat{H}$ is not required to converge to any limit, then the validity of (1.13) can be shown by a more careful argument. For any fixed $1 \le j \le p$, let $h_j = \hat{H}^{-1/2}e_j$, where $e_j = (0, \cdots, 0, 1, 0, \ldots, 0)^{\mathrm{T}}$ is the vector of all zeros except for the jth element being 1. Then $h_j^{\mathrm{T}}h_j = e_j^{\mathrm{T}}\hat{H}^{-1}e_j = (\hat{H}^{-1})_{jj}$ and hence from (1.74)

$$\begin{aligned}\sqrt{n}\frac{\hat{\beta}_j - \beta^*_j}{\sqrt{(\hat{H}^{-1})_{jj}}} &= \frac{h_j^{\mathrm{T}}}{\|h_j\|_2}\sqrt{n}\hat{H}^{1/2}(\hat{\beta} - \beta^*)\\ &= \frac{h_j^{\mathrm{T}}}{\|h_j\|_2}\hat{H}^{-1/2}\frac{1}{\sqrt{n}}\sum_{i=1}^{n} X_i^{\mathrm{T}}\varepsilon_i,\end{aligned}$$

which can be shown to converge to $\mathrm{N}(0,1)$ using the first equality and the joint asymptotic normality (1.12a) or, at a more basic level, using the second equality and triangular-array central limit theorems.

Similarly, the basic identity (1.73) can be exploited while using triangular-array central limit theorems to establish the asymptotic normality (1.18a) and confidence interval (1.19) under heteroscedastic linear regression, with additional technical complexity (White, 1980).

1.10.3 Fixed vs random design with model misspecification

We discuss the difference between the fixed and random designs in how the variance estimation for the LS estimator $\hat{\beta}$ is affected by model misspecification. As mentioned in Section 1.1, the fixed design can be considered to be derived from a random design, with $(X_1, \ldots, X_n)$ being conditioned on. By the law of total variance, $\mathrm{var}(\hat\beta)$ in the random design can be decomposed as

$$\begin{aligned}\mathrm{var}(\hat\beta) &= \mathrm{var}\{E(\hat\beta|X_1, \ldots, X_n)\} + E\{\mathrm{var}(\hat\beta|X_1, \ldots, X_n)\} \\ &= \mathrm{var}\{\bar\beta(X_1, \ldots, X_n)\} + E\{\mathrm{var}(\hat\beta|X_1, \ldots, X_n)\}, \end{aligned} \tag{1.75}$$

where $\bar\beta(X_1, \ldots, X_n)$ is defined as $\bar\beta$ in (1.26), and $\mathrm{var}(\hat\beta|X_1, \ldots, X_n)$ is the variance of $\hat\beta$ in the fixed design, i.e., conditional on $(X_1, \ldots, X_n)$. [The expression of $\mathrm{var}(\hat\beta|X_1, \ldots, X_n)$ in Proposition 1.3 remains valid with model misspecification.] By the law of large numbers, as $n \to \infty$ and p is fixed, $\bar\beta(X_1, \ldots, X_n)$ converges almost surely to $\bar\beta$ in (1.27), and $n\,\mathrm{var}(\hat\beta|X_1, \ldots, X_n)$ converges almost surely to a constant matrix

$$H^{-1} E\left\{\sigma^{*2}(X) X X^{\mathrm{T}}\right\} H^{-1}.$$

It is informative to compare the two sides in (1.75). On one hand, $\mathrm{var}(\hat\beta)$ on the left-hand side of (1.75), after rescaled by n, can be consistently estimated [by $\hat\Sigma(\hat\beta)$]. On the other hand, $\mathrm{var}(\hat\beta|X_1, \ldots, X_n)$ in general cannot be consistently estimated, suggesting that $E\{\mathrm{var}(\hat\beta|X_1, \ldots, X_n)\}$ cannot be consistently estimated, each after rescaled by n. Hence neither of the two terms on the right-hand side of (1.75), after rescaled by n, can be consistently estimated. The lack of a consistent estimator for $\mathrm{var}(\hat\beta|X_1, \ldots, X_n)$ is also the lack of a consistent estimator for $\mathrm{var}\{\bar\beta(X_1, \ldots, X_n)\}$. However, if linear regression is correctly specified, then $\bar\beta(X_1, \ldots, X_n)$ reduces to β^* independently of $(X_1, \ldots, X_n)$, so that $\mathrm{var}\{\bar\beta(X_1, \ldots, X_n)\} = 0$. This, together with the convergence of $n\,\mathrm{var}(\hat\beta|X_1, \ldots, X_n)$ to a constant matrix, allows that both $n\,\mathrm{var}(\hat\beta)$ and $n\,\mathrm{var}(\hat\beta|X_1, \ldots, X_n)$ can be consistently estimated [by the same estimator $\hat\Sigma(\hat\beta)$]. In general, if linear regression is misspecified, then $\mathrm{var}\{\bar\beta(X_1, \ldots, X_n)\}$ is nonzero and of order $O(n^{-1})$, but cannot be consistently estimated after rescaled by n.

1.10.4 Calibrated estimation with doubly robust estimating functions

Given a doubly robust estimating function, we describe a general method, called calibrated estimation (Tan, 2020a; Ghosh & Tan, 2022), for constructing estimators of nuisance parameters such that doubly robust inference can be obtained for the parameter of interest, similarly as shown for partially linear regression in Section 1.7.2.

Let $\{U_i : i = 1, \ldots, n\}$ be IID observations as U, which is assumed to include a covariate vector W. Consider semi-parametric estimation using an estimating function $\varphi(U; \theta, \alpha, \gamma)$, where θ is the parameter of interest (for simplicity, assumed to be scalar), and (α, γ) are two nuisance parameters defined through working models $g(W; \alpha)$ and $f(W; \gamma)$ for unknown functions $g^*(W)$ and $f^*(W)$. Suppose that $\varphi(U; \theta, \alpha, \gamma)$ is doubly robust in satisfying the following two properties: if model $g(\cdot; \alpha)$ is correctly specified, then

$$0 = E\{\varphi(U; \theta^*, \alpha^*, \gamma)\} \text{ for any } \gamma, \tag{1.76}$$

and if model $g(\cdot; \alpha)$ is correctly specified, then

$$0 = E\{\varphi(U; \theta^*, \alpha, \gamma^*)\} \text{ for any } \alpha, \tag{1.77}$$

where θ^* is the true value of θ, and α^* or γ^* is the true value of α or γ defined such that $g(W; \alpha^*) = g^*(W)$ or $f(W; \gamma^*) = f^*(W)$ if model $g(\cdot; \alpha)$ or $f(\cdot; \gamma)$ is correctly specified. Typically, doubly robust estimation using φ proceeds in two stages. First, some estimators $(\hat{\alpha}, \hat{\gamma})$ are defined, for example, by maximum likelihood or variations associated with models $g(\cdot; \alpha)$ or $f(\cdot; \gamma)$. Then an estimator for θ^*, denoted as $\hat{\theta}(\hat{\alpha}, \hat{\gamma})$, is defined as a solution to the estimating equation

$$0 = \tilde{E}\{\varphi(U; \theta, \hat{\alpha}, \hat{\gamma})\} = \frac{1}{n} \sum_{i=1}^{n} \varphi(U_i; \theta, \hat{\alpha}, \hat{\gamma}). \tag{1.78}$$

The asymptotic behavior of $\hat{\theta}(\hat{\alpha}, \hat{\gamma})$ depends on the choices of $(\hat{\alpha}, \hat{\gamma})$ in the first stage. With possible model misspecification, the estimators $(\hat{\alpha}, \hat{\gamma})$ are expected to satisfy the following properties as $n \to \infty$ and the dimensions of (α, γ) are fixed under suitable regularity conditions. See asymptotic theory of M-estimation and Z-estimation in Section 2.3.1.

(i) The estimator $\hat{\alpha}$ converges at rate $O_p(n^{-1/2})$ to a target value $\bar{\alpha}$, which coincides with the true value α^* (i.e., $\hat{\alpha}$ is consistent) if model $g(\cdot; \alpha)$ is correctly specified but remains well-defined even though α^* is undefined if model $g(\cdot; \alpha)$ is misspecified.

(ii) The estimator $\hat{\gamma}$ converges at rate $O_p(n^{-1/2})$ to a target value $\bar{\gamma}$, which coincides with the true value γ^* (i.e., $\hat{\gamma}$ is consistent) if model $g(\cdot; \gamma)$ is correctly specified but remains well-defined even though γ^* is undefined if model $g(\cdot; \gamma)$ is misspecified.

By the consistency properties of $(\hat{\alpha}, \hat{\gamma})$ above, the unbiasedness properties (1.76)–(1.77) can be used to show that $\hat{\theta}(\hat{\alpha}, \hat{\gamma})$ is pointwise doubly robust, i.e., $\hat{\theta}(\hat{\alpha}, \hat{\gamma})$ is consistent for θ^* if either model $g(\cdot; \alpha)$ or model $f(\cdot; \gamma)$ is correctly specified. Moreover, it can be shown that if model $g(\cdot; \alpha)$ is correctly specified with $\bar{\alpha} = \alpha^*$ or model $f(\cdot; \gamma)$ is correctly specified with $\bar{\gamma} = \gamma^*$, then $\hat{\theta}(\hat{\alpha}, \hat{\gamma})$ admits the asymptotic expansion,

$$\hat{\theta}(\hat{\alpha},\hat{\gamma}) - \theta^* =$$
$$-E^{-1}\left(\frac{\partial\varphi}{\partial\theta}\right)\left\{\tilde{E}(\varphi) + E^{\mathrm{T}}\left(\frac{\partial\varphi}{\partial\alpha}\right)(\hat{\alpha}-\bar{\alpha}) + E^{\mathrm{T}}\left(\frac{\partial\varphi}{\partial\gamma}\right)(\hat{\gamma}-\bar{\gamma})\right\} + o_p(n^{-1/2}), \tag{1.79}$$

where $\varphi = \varphi(U;\theta,\alpha,\gamma)$ and its partial derivatives, $\partial\varphi/\partial\theta, \partial\varphi/\partial\alpha, \partial\varphi/\partial\gamma$, are evaluated at $(\theta^*,\bar{\alpha},\bar{\gamma})$. The expansion (1.79) indicates how the asymptotic behavior of $\hat{\theta}(\hat{\alpha},\hat{\gamma})$ is affected by the estimators $(\hat{\alpha},\hat{\gamma})$ through the second and third terms in the curly brackets.

The calibrated estimators, $(\hat{\alpha}_{\mathrm{CAL}},\hat{\gamma}_{\mathrm{CAL}})$, are defined as jointly a solution to the estimating equations

$$0 = \tilde{E}\left\{\frac{\partial\varphi}{\partial\gamma}(U;\tilde{\theta},\alpha,\gamma)\right\}, \tag{1.80a}$$
$$0 = \tilde{E}\left\{\frac{\partial\varphi}{\partial\alpha}(U;\tilde{\theta},\alpha,\gamma)\right\}, \tag{1.80b}$$

where $\tilde{\theta}$ is a preliminary doubly robust estimator of θ^*, for example, $\tilde{\theta} = \hat{\theta}(\hat{\alpha},\hat{\gamma})$ with $(\hat{\alpha},\hat{\gamma})$ defined by maximum likelihood. Equations (1.80a) and (1.80b) are called sample calibration equations. Under suitable regularity conditions, if model $g(\cdot;\alpha)$ or model $f(\cdot;\gamma)$ is correctly specified, then $(\hat{\alpha}_{\mathrm{CAL}},\hat{\gamma}_{\mathrm{CAL}})$ are expected to converge in probability to target values $(\bar{\alpha}_{\mathrm{CAL}},\bar{\gamma}_{\mathrm{CAL}})$ satisfying the simultaneous equations

$$0 = E\left\{\frac{\partial\varphi}{\partial\gamma}(U;\theta^*,\alpha,\gamma)\right\}, \tag{1.81a}$$
$$0 = E\left\{\frac{\partial\varphi}{\partial\alpha}(U;\theta^*,\alpha,\gamma)\right\}, \tag{1.81b}$$

that is, the coefficients of $\hat{\gamma}-\bar{\gamma}$ and $\hat{\alpha}-\bar{\alpha}$ are set to 0 in the expansion (1.79). Equations (1.81a) and (1.81b) are called population calibration equations. By the unbiasedness property (1.76), if model $g(\cdot;\alpha)$ is correctly specified, then (1.81a) is an unbiased population estimating equation for α^* with any fixed γ:

$$0 = E\left\{\frac{\partial\varphi}{\partial\gamma}(U;\theta^*,\alpha^*,\gamma)\right\}.$$

By the unbiasedness property (1.77), if model $f(\cdot;\gamma)$ is correctly specified, then (1.81b) is an unbiased population estimating equation for γ^* with any fixed α:

$$0 = E\left\{\frac{\partial\varphi}{\partial\alpha}(U;\theta^*,\alpha,\gamma^*)\right\}.$$

(An interesting asymmetry is that differentiation of τ with respect to γ leads to an estimating equation in α, whereas that of τ with respect to α leads to an estimating equation in γ.) Therefore $(\hat{\alpha}_{\text{CAL}}, \hat{\gamma}_{\text{CAL}})$ are expected to satisfy the consistency properties (i) and (ii). By the earlier discussion, the resulting estimator $\hat{\theta}_{\text{CAL}} = \hat{\theta}(\hat{\alpha}_{\text{CAL}}, \hat{\gamma}_{\text{CAL}})$ is pointwise doubly robust for θ^*. More importantly, if model $g(\cdot;\alpha)$ or model $f(\cdot;\gamma)$ is correctly specified, then $\hat{\theta}_{\text{CAL}}$ admits the simple asymptotic expansion

$$\hat{\theta}_{\text{CAL}} - \theta^* = -E^{-1}\left(\frac{\partial\varphi}{\partial\theta}\right)\tilde{E}(\varphi) + o_p(n^{-1/2}), \tag{1.82}$$

where φ and $\partial\varphi/\partial\theta$ are evaluated at $(\theta^*, \bar{\alpha}_{\text{CAL}}, \bar{\gamma}_{\text{CAL}})$. Due to (1.81a) and (1.81b), the two terms associated with $(\hat{\alpha} - \bar{\alpha}, \hat{\gamma} - \bar{\gamma})$ can be removed from (1.79). In other words, the asymptotic behavior of $\hat{\theta}_{\text{CAL}}$ is, up to $o_p(n^{-1/2})$, equivalent to the estimator that would be obtained as a solution to (1.78) with $(\hat{\alpha}, \hat{\gamma})$ fixed at $(\bar{\alpha}_{\text{CAL}}, \bar{\gamma}_{\text{CAL}})$. Let

$$V = E^{-2}\left\{\frac{\partial\varphi}{\partial\theta}(U;\theta^*, \bar{\alpha}_{\text{CAL}}, \bar{\gamma}_{\text{CAL}})\right\} E\{\varphi^2(U;\theta^*, \bar{\alpha}_{\text{CAL}}, \bar{\gamma}_{\text{CAL}})\},$$

$$\hat{V} = \tilde{E}^{-2}\left\{\frac{\partial\varphi}{\partial\theta}(U;\hat{\theta}_{\text{CAL}}, \hat{\alpha}_{\text{CAL}}, \hat{\gamma}_{\text{CAL}})\right\} \tilde{E}\{\varphi^2(U;\hat{\theta}_{\text{CAL}}, \hat{\alpha}_{\text{CAL}}, \hat{\gamma}_{\text{CAL}})\}.$$

From (1.82), if model $g(\cdot;\alpha)$ or model $f(\cdot;\gamma)$ is correctly specified, then $\hat{\theta}_{\text{CAL}}$ satisfies the asymptotic normality: $\sqrt{n}(\hat{\theta}_{\text{CAL}} - \theta^*) \xrightarrow{\mathcal{D}} \text{N}(0, V)$. Moreover, $\hat{V}$ can be shown to be a consistent estimator of V: $\hat{V} \xrightarrow{\mathcal{P}} V$. As a result, the Wald confidence interval

$$\hat{\theta}_{\text{CAL}} \pm z_{c/2}\sqrt{\hat{V}/n} \tag{1.83}$$

is doubly robust in achieving asymptotic size $1 - c$ for covering θ^* if either model $g(\cdot;\alpha)$ or model $f(\cdot;\gamma)$ is correctly specified.

For any $(\hat{\alpha}, \hat{\gamma})$ satisfying the consistency properties, the estimator $\hat{\theta}(\hat{\alpha}, \hat{\gamma})$ is pointwise doubly robust for θ^*. In low-dimensional settings, Wald confidence intervals based on $\hat{\theta}(\hat{\alpha}, \hat{\gamma})$ can be derived from asymptotic expansion (1.79) and shown to be doubly robust for θ^*. But the asymptotic expansion (1.79) may in general fail in high-dimensional settings, where $(\hat{\alpha}, \hat{\gamma})$ may converge to target values at rates slower than $O_p(n^{-1/2})$. In contrast, the simple expansion (1.82) for calibrated estimation satisfying (1.81a)–(1.81b) can be shown to remain valid under suitable conditions in high-dimensional settings (Tan, 2020a; Ghosh & Tan, 2022).

Example 1.1. For partially linear regression (1.37), a doubly robust estimating function for θ^* with respect to working models $g^*(W) = W^{\text{T}}\alpha^*$ and $f^*(W) = E(Z|W) = W^{\text{T}}\gamma^*$ is, as stated in (1.42),

$$\varphi(U;\theta,\alpha,\gamma) = (Z - W^{\text{T}}\gamma)(Y - W^{\text{T}}\alpha - Z\theta),$$

where $U = (Z, W, Y)$. Given *any* estimators $(\hat{\alpha}, \hat{\gamma})$, the second-stage estimator for θ^* is

$$\hat{\theta}(\hat{\alpha}, \hat{\gamma}) = \frac{\tilde{E}\{(Z - W^{\text{T}}\hat{\gamma})(Y - W^{\text{T}}\hat{\alpha})\}}{\tilde{E}\{(Z - W^{\text{T}}\hat{\gamma})Z\}}, \tag{1.84}$$

which is of the same form as the expression (1.40) for the LS estimator $\hat{\theta}$ in model (1.38), an observation to be used below. The calibration equation (1.80a)–(1.80b) are

$$0 = \tilde{E}\left\{(Y - W^{\mathrm{T}}\alpha - Z\tilde{\theta})W\right\}, \tag{1.85a}$$

$$0 = \tilde{E}\left\{(Z - W^{\mathrm{T}}\gamma)W\right\}, \tag{1.85b}$$

where $\tilde{\theta}$ is a preliminary doubly robust estimator of θ^*. For example, we take $\tilde{\theta} = \hat{\theta}(\hat{\alpha}, \hat{\gamma})$, where $\hat{\alpha}$ is from the LS estimator $\hat{\beta} = (\hat{\theta}, \hat{\alpha}^{\mathrm{T}})^{\mathrm{T}}$ in model (1.38), and $\hat{\gamma}$ is the LS estimator in model (1.39). Then $\tilde{\theta} = \hat{\theta}$ by the expression (1.84) for $\hat{\theta}(\hat{\alpha}, \hat{\gamma})$ and (1.40) for $\hat{\theta}$. Moreover, it can be shown that as a solution to (1.85a)–(1.85b), the calibrated estimators $(\hat{\alpha}_{\mathrm{CAL}}, \hat{\gamma}_{\mathrm{CAL}})$ are the same as $(\hat{\alpha}, \hat{\gamma})$, and the resulting estimator $\hat{\theta}_{\mathrm{CAL}}$ is also the same as $\hat{\theta}$. [Exercise: Prove this statement.] Therefore, Propositions 1.9 and 1.10 are recovered from the general results in this section.

Example 1.2. For partially linear regression (1.37), a doubly robust estimating function for θ^* with respect to working models $g^*(W) = W^{\mathrm{T}}\alpha^*$ and $f^*(W) = E(Z|W) = \mathrm{expit}(W^{\mathrm{T}}\gamma^*)$ is, as stated in (1.51),

$$\varphi(U; \theta, \alpha, \gamma) = (Z - \mathrm{expit}(W^{\mathrm{T}}\gamma))(Y - W^{\mathrm{T}}\alpha - Z\theta),$$

where $U = (Z, W, Y)$. Given *any* estimators $(\hat{\alpha}, \hat{\gamma})$, the second-stage estimator for θ^* is

$$\hat{\theta}(\hat{\alpha}, \hat{\gamma}) = \frac{\tilde{E}\{(Z - \mathrm{expit}(W^{\mathrm{T}}\hat{\gamma}))(Y - W^{\mathrm{T}}\hat{\alpha})\}}{\tilde{E}\{(Z - \mathrm{expit}(W^{\mathrm{T}}\hat{\gamma}))Z\}},$$

which is similar to (1.84) with $W^{\mathrm{T}}\hat{\gamma}$ replaced by $\mathrm{expit}(W^{\mathrm{T}}\hat{\gamma})$. The calibration equations (1.80a)–(1.80b) are

$$0 = \tilde{E}\left\{(Y - W^{\mathrm{T}}\alpha - Z\tilde{\theta})\mathrm{expit}'(W^{\mathrm{T}}\gamma)W\right\}, \tag{1.86a}$$

$$0 = \tilde{E}\left\{(Z - \mathrm{expit}(W^{\mathrm{T}}\gamma))W\right\}, \tag{1.86b}$$

where $\mathrm{expit}'(u) = \mathrm{expit}(u)(1 - \mathrm{expit}(u))$ is the derivative of $\mathrm{expit}(u)$, and $\tilde{\theta}$ is a preliminary doubly robust estimator of θ^*. For example, we take $\tilde{\theta} = \hat{\theta}(\hat{\alpha}, \hat{\gamma})$, where $\hat{\alpha}$ is from the LS estimator $\hat{\beta} = (\hat{\theta}, \hat{\alpha}^{\mathrm{T}})^{\mathrm{T}}$ in linear model (1.38), and $\hat{\gamma}$ is the ML estimator in logistic model (1.50) and hence a solution to estimating equation (1.86b) (see Example 2.5 in Section 2.3.3). The calibrated estimator $\hat{\gamma}_{\mathrm{CAL}}$, as a solution to (1.86b), is the same as $\hat{\gamma}$, but the calibrated estimator $\hat{\alpha}_{\mathrm{CAL}}$ in general differs from $\hat{\alpha}$. The resulting estimator $\hat{\theta}_{\mathrm{CAL}}$ also differs from $\hat{\theta}$ and $\tilde{\theta}$. Nevertheless, $\hat{\theta}_{\mathrm{CAL}}$ is pointwise doubly robust and the associated Wald confidence interval (1.83) is doubly robust for θ^* with respect to linear model (1.38) and logistic model (1.50).

1.10.5 Double robustness for nonparametric estimands

The doubly robust estimation in the context of partially linear regression (1.37) in Section 1.7 can be reformulated in terms of a nonparametric estimand, without requiring partially linear regression to be correctly specified (Robins et al., 2009). In this case, both θ^* and $g^*(\cdot)$ need to be redefined.

As in Section 1.7, consider a random design where $(X_1, Y_1), \ldots, (X_n, Y_n)$ are IID as (X, Y), with $X = (Z, W^{\mathrm{T}})^{\mathrm{T}}$. The parameter of interest θ^* is defined as

$$\theta^* = \frac{E\{\mathrm{cov}(Y, Z|W)\}}{E\{\mathrm{var}(Z|W)\}}. \tag{1.87}$$

This parameter reduces to θ^* in partially linear regression (1.37) if correctly specified, but otherwise remains meaningful in capturing how Y is (linearly) related to Z after adjusting for W. [Question: How can this definition of θ^* be discovered in the first place?] The quantity θ^* can be equivalently determined in two distinct manners.

Lemma 1.1 (Nonparametric estimand).

(i) The quantity θ^* in (1.87) satisfies

$$E\{(Y - \theta^* Z - g^*(W))Z\} = 0, \tag{1.88}$$

where $g^*(W) = E(Y - \theta^* Z|W)$, which in general may differ from g^* in partially linear regression (1.37). As a result, $\theta^* Z + g^*(W)$ is the projection of Y on the set of all functions in the form $\theta Z + g(W)$, i.e.,

$$(\theta^*, g^*) = \mathrm{argmin}_{(\theta, g)}\ E\{(Y - \theta Z - g(W))^2\}. \tag{1.89}$$

(ii) θ^* also satisfies

$$E\{(Y - \theta^* Z)(Z - f^*(W))\} = 0, \tag{1.90}$$

where $f^*(W) = E(Z|W)$, the same as in Section 1.7.

Proof. We show that solving (1.88) or (1.90) leads to (1.87). The proof of (1.89) is left as an exercise. First, substituting the definition of g^* into (1.88) yields

$$\begin{aligned} 0 &= E\{(Y - \theta^* Z - E(Y - \theta^* Z|W))Z\} \\ &= E\{(Y - E(Y|W))Z - \theta^*(Z - E(Z|W))Z\}, \end{aligned}$$

and

$$\theta^* = \frac{E\{(Y - E(Y|W))Z\}}{E\{(Z - E(Z|W))Z\}},$$

which gives θ^* in (1.87) by the law of iterated expectations. Second, solving (1.90) yields

$$\begin{aligned}\theta^* &= \frac{E\{Y(Z - f^*(W)\}}{E\{Z(Z - f^*(W)\}} \\ &= \frac{E\{Y(Z - E(Z|W)\}}{E\{Z(Z - E(Z|W)\}}\end{aligned}$$

which also gives θ^* in (1.87) by the law of iterated expectations. □

From Lemma 1.1, the estimating function $\varphi_0(Z, W, Y; \theta, g, f) = (Y - Z\theta - g(W))(Z - f(W))$ is doubly robust in satisfying the following two properties as in (1.52) and (1.53):

$$\begin{aligned}0 &= E\{(Y - \theta^* Z - g^*(W))(Z - f(W))\} \quad \text{for any } f, \\ 0 &= E\{(Y - \theta^* Z - g(W))(Z - f^*(W))\} \quad \text{for any } g,\end{aligned}$$

provided the expectations are finite, where g^* and f^* are defined in Lemma 1.1, allowing partially linear regression (1.37) to be misspecified. [Exercise: Verify the two properties.] In other words, if either $g = g^*$ or $f = f^*$, then

$$0 = E\{(Y - \theta^* Z - g(W))(Z - f(W))\}.$$

The double robustness can also be stated in terms of parametric working models for g^* and f^*. For example, consider two linear working models: $g^*(W) = W^{\mathrm{T}}\alpha^*$ and $f^*(W) = W^{\mathrm{T}}\gamma^*$. Then the estimating function φ in (1.42), $\varphi(Z, W, Y; \theta, \alpha, \gamma) = (Z - W^{\mathrm{T}}\gamma)(Y - W^{\mathrm{T}}\alpha - Z\theta)$, satisfies the following two properties as in (1.43) and (1.44).

- If model $g^*(W) = W^{\mathrm{T}}\alpha^*$ is correctly specified, then for any γ,

$$0 = E\{\varphi(Z, W, Y; \theta^*, \alpha^*, \gamma)\}.$$

- If model $f^*(W) = W^{\mathrm{T}}\gamma^*$ is correctly specified, then for any α,

$$0 = E\{\varphi(Z, W, Y; \theta^*, \alpha, \gamma^*)\}.$$

Hence φ is a doubly robust estimating function for θ^* in (1.87), with respect to working models $g^*(W) = W^{\mathrm{T}}\alpha^*$ and $f^*(W) = W^{\mathrm{T}}\gamma^*$.

From this point, Propositions 1.9 and 1.10 can be shown to remain valid with the following modification: allowing partially linear regression (1.37) to be misspecified, replacing model (1.38) with $g^*(W) = E(Y - \theta^* Z|W) = W^{\mathrm{T}}\alpha^*$, while retaining model (1.39), $f^*(W) = E(Z|W) = W^{\mathrm{T}}\gamma^*$. [Note: By the characterization of (θ^*, g^*) in (1.89), the working model $g^*(W) = W^{\mathrm{T}}\alpha^*$ says that the projection of Y on all functions in the form $\theta Z + g(W)$ is $\theta^* Z + W^{\mathrm{T}}\alpha^*$, which is weaker than the linear regression model (1.38), which says that the projection of Y on all functions in the form $h(Z, W)$ is $\theta^* Z + W^{\mathrm{T}}\alpha^*$.]

1.10.6 Convex functions and sub-gradients

We describe some properties of convex functions, related to sub-gradients. More information can be found in Hiriart-Urruty and Lemaréchal (2001). Let $\mathcal{X} \subset \mathbb{R}^p$ be a convex, open set. A function $f : \mathcal{X} \to \mathbb{R}$ is said to be convex if for any $x, \tilde{x} \in \mathcal{X}$ and $0 \le \alpha \le 1$,

$$f(\alpha x + (1-\alpha)\tilde{x}) \le \alpha f(x) + (1-\alpha) f(\tilde{x}).$$

For a (possibly non-convex) function $f : \mathcal{X} \to \mathbb{R}$, a vector $g \in \mathbb{R}^p$ is said to be a sub-gradient of f at $\tilde{x} \in \mathcal{X}$ if for any $x \in \mathcal{X}$,

$$f(x) \ge f(\tilde{x}) + g^{\mathrm{T}}(x - \tilde{x}).$$

The affine function $h(x; \tilde{x}) = f(x) + g^{\mathrm{T}}(x - \tilde{x})$ coincides with $f(x)$ at $x = \tilde{x}$ and is no greater than $f(x)$ for all x:

$$\begin{aligned} h(\tilde{x}; \tilde{x}) &= f(\tilde{x}), \\ h(x; \tilde{x}) &\le h(x) \text{ for all x}, \end{aligned}$$

Hence $h(x; \tilde{x})$ is a minorizing function for f at $\tilde{x}$, in a similar sense as a majorizing function used in an MM algorithm discussed in Section 1.8.5. The set of sub-gradients of f at $\tilde{x}$ is called the sub-differential of f at $\tilde{x}$ and is denoted as $\partial f(\tilde{x})$.

For a (finite-valued) convex function $f : \mathcal{X} \to \mathbb{R}$, the sub-differential of f at any x is non-empty (i.e., there always exists a sub-gradient), closed and convex, and bounded. A convex function f is differentiable at $\tilde{x}$ if and only if the sub-differential is a singleton set, which then consists of the gradient of f at $\tilde{x}$.

In the scalar case ($p = 1$), a sub-gradient is also called a sub-derivative. For a scalar convex function f on an open interval $\mathcal{X}$, the sub-differential at $\tilde{x} \in \mathcal{X}$ can be shown to be a closed interval $[a, b]$, where

$$a = \lim_{x \to \tilde{x}-} \frac{f(x) - f(\tilde{x})}{x - \tilde{x}}, \quad b = \lim_{x \to \tilde{x}+} \frac{f(x) - f(\tilde{x})}{x - \tilde{x}},$$

i.e., a and b are respectively the left- and right-derivative of f at $\tilde{x}$. If and only if f is differentiable at $\tilde{x}$, then $a = b$, equal to the derivative $f'(\tilde{x})$. See Figure 1.11.

For optimization, a point x^* is a (global) minimizer of f if and only if $0 \in \partial f(x^*)$, i.e., the zero vector is a sub-gradient of f at x^*. [Exercise: Prove this statement.] While this property holds for non-convex functions, its application is limited in that case because the sub-differential (possibly empty) may be difficult to determine at any $\tilde{x} \in \mathcal{X}$ for a non-convex function.

There are calculus rules with sub-differentials for convex functions, which play similar roles as in ordinary differential calculus. For example, for a convex function f and $c \ge 0$,

$$\partial(cf)(x) = c\,\partial f(x).$$

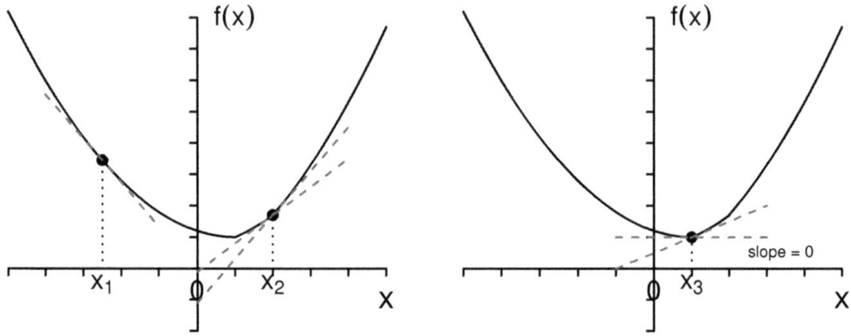

FIGURE 1.11
Sub-derivatives of a convex function $f(x)$, which is differentiable at x_1 but not at x_2 and x_3 (the minimizer of f).

For two convex functions f_1 and f_2,

$$\partial(f_1 + f_2)(x) = \partial f_1(x) + \partial f_2(x),$$

where $\partial f_1(x)+\partial f_2(x)$ is defined as the set $\{g_1+g_2 : g_1 \in \partial f_1(x), g_2 \in \partial g_2(x)\}$. See Section D.4 in Hiriart-Urruty and Lemaréchal (2001) for further results.

Exercises

Ex 1.1. Let $x = (x_1, \ldots, x_p)^{\mathrm{T}}$ be a $p \times 1$ vector. For a scalar function $f(x)$, $\frac{\partial f(x)}{\partial x}$ is defined as $(\frac{\partial f(x)}{\partial x_1}, \ldots, \frac{\partial f(x)}{\partial x_p})^{\mathrm{T}}$, where $\frac{\partial f(x)}{\partial x_j}$ is the partial derivative of $f(x)$ with respect to x_j. If $f(\cdot)$ is differentiable at x, then $\frac{\partial f(x)}{\partial x}$ is called the gradient of $f(\cdot)$ at x. Find

(i) $\frac{\partial}{\partial x}(a^{\mathrm{T}}x)$, where a is a constant $p \times 1$ vector,

(ii) $\frac{\partial}{\partial x}(x^{\mathrm{T}}Ax)$, where A is a constant $p \times p$ matrix.

Ex 1.2. Let $x = (x_1, \ldots, x_p)^{\mathrm{T}}$ be a $p \times 1$ vector. For a $q \times 1$ function $f(x) = (f_1(x), \ldots, f_p(x))^T$, $\frac{\partial f(x)}{\partial x^{\mathrm{T}}}$ is defined as a $q \times p$ matrix with (j,k)th element $\frac{\partial f_j(x)}{\partial x_k}$, and $\frac{\partial f^{\mathrm{T}}(x)}{\partial x}$ is defined as a $p \times q$ matrix with (j,k)th element $\frac{\partial f_k(x)}{\partial x_j}$, where f^{T} is a $1 \times q$ function.

(i) Show that for a scalar, twice-differentiable function $h(x)$,

$$\frac{\partial}{\partial x^{\mathrm{T}}}\left(\frac{\partial h(x)}{\partial x}\right) = \frac{\partial}{\partial x}\left(\frac{\partial h(x)}{\partial x^{\mathrm{T}}}\right),$$

which is then called the Hessian of h and denoted as $\frac{\partial^2}{\partial x \partial x^{\mathrm{T}}} h(x)$.

(ii) Find $\frac{\partial^2}{\partial x \partial x^{\mathrm{T}}}(x^{\mathrm{T}} A x)$, where A is a constant $p \times p$ matrix (possibly not symmetric).

Ex 1.3. Consider any matrices A, B, C, and D of appropriate sizes.

(i) Show that if D and $\tilde{A} = A - BD^{-1}C$ are nonsingular, then $\begin{pmatrix} A & B \\ C & D \end{pmatrix}$ is nonsingular and (1.32) holds, i.e.,

$$\begin{pmatrix} A & B \\ C & D \end{pmatrix}^{-1} = \begin{pmatrix} \tilde{A}^{-1} & -\tilde{A}^{-1}BD^{-1} \\ -D^{-1}C\tilde{A}^{-1} & D^{-1} + D^{-1}C\tilde{A}^{-1}BD^{-1} \end{pmatrix}.$$

Give a direct proof using the orthogonalization idea or variation.

(ii) Show that if D and $\begin{pmatrix} A & B \\ C & D \end{pmatrix}$ are nonsingular, then $\tilde{A} = A - BD^{-1}C$ is nonsingular and (1.32) holds.

(iii) Show that if A, D, and $\tilde{A} = A - BD^{-1}C$ are nonsingular, then $\tilde{D} = D - CA^{-1}B$ is nonsingular and

$$\tilde{D}^{-1} = D^{-1} + D^{-1}C\tilde{A}^{-1}BD^{-1}. \tag{1.91}$$

As an application, if $u^{\mathrm{T}}u \neq 1$ for any column vector u, then $I - uu^{\mathrm{T}}$ is nonsingular and

$$(I - uu^{\mathrm{T}})^{-1} = I + (1 - u^{\mathrm{T}}u)^{-1}uu^{\mathrm{T}}.$$

[Note: By exchanging the roles of A and D in (i), it also holds that if A and $\tilde{D} = D - CA^{-1}B$ are nonsingular, then $\begin{pmatrix} A & B \\ C & D \end{pmatrix}$ is nonsingular and

$$\begin{pmatrix} A & B \\ C & D \end{pmatrix}^{-1} = \begin{pmatrix} A^{-1} + A^{-1}B\tilde{D}^{-1}CA^{-1} & -A^{-1}B\tilde{D}^{-1} \\ -\tilde{D}^{-1}CA^{-1} & \tilde{D}^{-1} \end{pmatrix}. \tag{1.92}$$

Relating the $(2,2)$ sub-matrices on the right-hand side of (1.32) and (1.92) gives (1.91). A complete proof of (iii) needs to take care of the conditions involved.]

Ex 1.4. Consider any matrices A, B, C, and D of appropriate size. Show that if D is nonsingular, then

$$\det \begin{pmatrix} A & B \\ C & D \end{pmatrix} = \det(\tilde{A}) \det(D),$$

where $\tilde{A} = A - BD^{-1}C$, and $\det(\cdot)$ denotes the determinant.

Ex 1.5. Let $\mathbf{X}$ be $n \times p$ covariate matrix, and $\mathbf{Y}$ be $n \times 1$ response vector.

(i) Show that if $n \geq p$ and $\mathbf{X}$ is of rank p, then $\mathbf{X}^{\mathrm{T}}\mathbf{X}$ is nonsingular, and $\hat{\beta} = (\mathbf{X}^{\mathrm{T}}\mathbf{X})^{-1}\mathbf{X}^{\mathrm{T}}\mathbf{Y}$ is the unique solution to $\min_{\beta \in \mathbb{R}^p} \|\mathbf{Y} - \mathbf{X}\beta\|_2^2$.

(ii) Show that if $n \leq p$ and $\mathbf{X}$ is of rank n, then $\mathbf{X}\mathbf{X}^{\mathrm{T}}$ is nonsingular, and $\tilde{\beta} = \mathbf{X}^{\mathrm{T}}(\mathbf{X}\mathbf{X}^{\mathrm{T}})^{-1}\mathbf{Y}$ is the unique solution to $\min_{\beta \in \mathbb{R}^p} \|\beta\|_2$ subject to $\mathbf{Y} = \mathbf{X}\beta$. In this case, $\tilde{\beta}$ is called the minimum-norm interpolator.

(iii) Show that regardless of the rank of $\mathbf{X}$,

$$(\mathbf{X}^{\mathrm{T}}\mathbf{X})^{+}\mathbf{X}^{\mathrm{T}}\mathbf{Y} = \mathbf{X}^{\mathrm{T}}(\mathbf{X}\mathbf{X}^{\mathrm{T}})^{+}\mathbf{Y}, \tag{1.93}$$

where for a $k \times k$ matrix A, A^{+} denotes the Moore–Penrose inverse of A. As a result, the two estimators $\hat{\beta}$ and $\tilde{\beta}$ in (i) and (ii) can be unified as $\hat{\beta} = \tilde{\beta}$, both defined by (1.93).

[Note: For a $k \times k$ matrix A, let $A = Q_1 B Q_2^{\mathrm{T}}$ be a singular value decomposition, where Q_1 and Q_2 are $k \times k$ orthogonal matrices and B is a $k \times k$ diagonal matrix $\mathrm{diag}(b_1, \ldots, b_k)$. The Moore–Penrose inverse of A is defined as $A^{+} = Q_2 B^{+} Q_1^{\mathrm{T}}$, where $B^{+} = \mathrm{diag}(b_1^{+}, \ldots, b_k^{+})$, with $b_j^{+} = b_j^{-1}$ if $b_j \neq 0$ or 0 if $b_j = 0$ for $j = 1, \ldots, k$.]

Ex 1.6. Consider a linear regression model:

$$Y_i \sim X_i^{\mathrm{T}}\beta, \quad i = 1, \ldots, n,$$

where $X_i = (1, X_{i,1:p}^{\mathrm{T}})^{\mathrm{T}}$ and $X_{i,1:p} = (X_{i1}, \ldots, X_{ip})^{\mathrm{T}}$. [For simplicity, the true value is denoted as β instead of β^*.] Denote as $\mathbf{X}$ the $n \times (p+1)$ matrix with ith row X_i^{T} and as $\mathbf{Y}$ the $n \times 1$ vector with ith element Y_i. The least squares estimator of β is $\hat{\beta} = (\mathbf{X}^{\mathrm{T}}\mathbf{X})^{-1}\mathbf{X}^{\mathrm{T}}\mathbf{Y}$. Show that $\hat{\beta} = (\hat{\beta}_0, \hat{\beta}_{1:p}^{\mathrm{T}})^{\mathrm{T}}$, where

$$\hat{\beta}_0 = \bar{Y} - \bar{X}_{1:p}^{\mathrm{T}}\hat{\beta}_{1:p},$$
$$\hat{\beta}_{1:p} = (\tilde{\mathbf{X}}_{1:p}^{\mathrm{T}}\tilde{\mathbf{X}}_{1:p})^{-1}\tilde{\mathbf{X}}_{1:p}^{\mathrm{T}}\mathbf{Y},$$

where $\bar{Y} = n^{-1}\sum_{i=1}^{n} Y_i$ and $\bar{X}_{1:p} = n^{-1}\sum_{i=1}^{n} X_{i,1:p}$ are the sample averages, and $\tilde{\mathbf{X}}_{1:p}$ is $n \times p$ matrix with ith row $X_{i,1:p} - \bar{X}_{1:p}$.

Ex 1.7. As in Section 1.6, consider a linear regression model:

$$Y_i \sim X_i^{\mathrm{T}}\beta = Z_i\theta + W_i^{\mathrm{T}}\alpha, \quad i = 1, \ldots, n,$$

where $X_i = (Z_i, W_i^{\mathrm{T}})^{\mathrm{T}}$ and $\beta = (\theta, \alpha^{\mathrm{T}})^{\mathrm{T}}$. Denote as $\mathbf{Y}$ the response vector, and partition the covariate matrix $\mathbf{X}$ as $\mathbf{X} = (\mathbf{Z}, \mathbf{W})$. Similarly, partition the LS estimator $\hat{\beta}$ as $\hat{\beta} = (\hat{\theta}, \hat{\alpha}^{\mathrm{T}})^{\mathrm{T}}$. Let R_Y be the residual vector from regressing $\mathbf{Y}$ on $\mathbf{W}$, and R_Z be the residual vector from regressing $\mathbf{Z}$ on $\mathbf{W}$.

(i) Let $\hat{\sigma}^2 = \|\mathbf{Y} - \mathbf{X}\hat{\beta}\|_2^2/(n-p)$ and $\tilde{\sigma}^2 = \|R_Y - R_Z\hat{\theta}\|_2^2/(n-1)$, the application of $\hat{\sigma}^2$ to RonR regression. Show that $\tilde{\sigma}^2 = \frac{n-p}{n-1}\hat{\sigma}^2$.

(ii) Show that $(R_Z^{\mathrm{T}} R_Z)^{-1}$ is identical to the $(1,1)$ element in the model-based variance matrix $(\mathbf{X}^{\mathrm{T}}\mathbf{X})^{-1}$.

(iii) Show that $(R_Z^{\mathrm{T}} R_Z)^{-1}(R_Z^{\mathrm{T}} D R_Z)(R_Z^{\mathrm{T}} R_Z)^{-1}$ is identical to the $(1,1)$ element in the model-robust variance matrix $(\mathbf{X}^{\mathrm{T}}\mathbf{X})^{-1}(\mathbf{X}^{\mathrm{T}} D\mathbf{X})(\mathbf{X}^{\mathrm{T}}\mathbf{X})^{-1}$, where $D = \mathrm{diag}(\hat{r}_1^2, \ldots, \hat{r}_n^2)$ and $\hat{r}_i = Y_i - X_i^{\mathrm{T}}\hat{\beta}$ for $i = 1, \ldots, n$.

Ex 1.8. Simulate a dataset by running the following R code, including `set.seed(0)` for initializing the random number generator:

```
library(MASS)
set.seed(0)

n <- 200
p <- 4

mu <- rep(0,p)
msig <- toeplitz(.8^(1:p-1))
sig <- .5

x <- mvrnorm(n, mu, msig)
y <- x[,1] + pmax(0, apply(x[,2:p],1,sum))^2 /3 + rnorm(n, 0, sig)

cbind(y,x)[1:3,]  # report the first three rows
```

(i) Fit a linear regression model of `y` on `x` (including the intercept), and report all estimated coefficients, model-based and model-robust standard errors. Include degrees of freedom adjustment for model-based standard errors.

(ii) Compute all estimated coefficients, model-based and model-robust standard errors, using residual-on-residual regression. The results should match those from (i).

(iii) In the simulation setting above, can you find the target (or limit) values of all coefficients? Find analytical expressions (as simple as you can) and then compute the values numerically. [Hint: Try to use conditional expectations and reduce the expressions to one-dimensional integration, which can then be evaluated analytically or numerically.]

Ex 1.9. Use the simulation setting described in Exercise 1.8. Run 2000 repeated simulations and report

(i) the Monte Carlo biases and standard deviations of 5 estimated coefficients (including the intercept),

(ii) the square root of the Monte Carlo means of model-based variances (with adjustment of degrees of freedom), and 90% and 95% coverage proportions of associated Wald confidence intervals,

(iii) the square root of the Monte Carlo means of model-robust variances, and 90% and 95% coverage proportions of associated Wald confidence intervals.

For coverage proportions in (ii) and (iii), use the target values found in Exercise 1.8 (iii). Summarize what you learned from this simulation study as much as you can (in one paragraph).

[Note: The quantities mentioned above are defined as follows. Denote by $\hat{\beta}_j^{(k)}$ the estimate of β_j and by $\hat{V}(\hat{\beta}_j^{(k)})$ a variance estimate, model-based or model-robust, from kth simulation, where $j = 1, \ldots, p\,(= 5)$ and $k = 1, \ldots, m\,(= 2000)$. For estimation of β_j, the Monte Carlo mean of point estimates is $\bar{\beta}_j = \frac{1}{m}\sum_{k=1}^{m}\hat{\beta}_j^{(k)}$, the Monte Carlo standard deviation is $\{\frac{1}{m-1}\sum_{k=1}^{m}(\hat{\beta}_j^{(k)} - \bar{\beta}_j)^2\}^{1/2}$, and the Monte Carlo mean of variance estimates is $\frac{1}{m}\sum_{k=1}^{m}\hat{V}(\hat{\beta}_j^{(k)})$. The coverage proportion of $(1-c)$ Wald confidence intervals is $\frac{1}{m}\sum_{k=1}^{m} 1\{\bar{\beta}_j \in [\hat{\beta}_j^{(k)} - z_{c/2}\sqrt{\hat{V}(\hat{\beta}_j^{(k)})},\, \hat{\beta}_j^{(k)} + z_{c/2}\sqrt{\hat{V}(\hat{\beta}_j^{(k)})}\}$, where $\bar{\beta}_j$ is the target value.]

Ex 1.10. Use the simulation setting described in Exercise 1.8, but fix the x-matrix generated (i.e., consider a fixed design).

(i) Find the target (or limit) values of all coefficients? Find analytical expressions (as simple as you can) and then compute the values numerically. [Note: this is similar to Exercise 1.8 (iii), but in a fixed design.]

Run 2000 repeated simulations with the $\mathbf{x}$-matrix fixed, and report

(ii) the Monte Carlo biases and standard deviations of 5 estimated coefficients (including the intercept),

(iii) the square root of the Monte Carlo means of model-based variances (with adjustment of degrees of freedom), and 90% and 95% coverage proportions of associated Wald confidence intervals,

(iv) the square root of the Monte Carlo means of model-robust variances, and 90% and 95% coverage proportions of associated Wald confidence intervals.

For coverage proportions in (ii) and (iii), use the target values found in Question (i). Summarize what you learned from this simulation study as much as you can (in one paragraph).

Ex 1.11. Let $Z \sim \mathrm{N}(\mu, 1)$, where $\mu \in \mathbb{R}$.

(i) For any $\lambda \geq 0$, define the ridge estimator

$$\hat{\mu}_\lambda(z) = \text{argmin}_{t\in\mathbb{R}} \left\{ \frac{1}{2}(t-z)^2 + \frac{1}{2}\lambda t^2 \right\},$$

Compute $\text{MSE}_\lambda(\mu) = E\{(\hat{\mu}_\lambda(Z)-\mu)^2\}$. Plot and compare the function $\text{MSE}_\lambda(\mu)$ over μ for $\lambda = .1, 1, 5, 10$.

(ii) For any $\lambda \geq 0$, define the Lasso estimator

$$\hat{\mu}_\lambda(z) = \text{argmin}_{t\in\mathbb{R}} \left\{ \frac{1}{2}(t-z)^2 + \lambda|t| \right\},$$

Compute $\text{MSE}_\lambda(\mu) = E\{(\hat{\mu}_\lambda(Z)-\mu)^2\}$. Plot and compare the function $\text{MSE}_\lambda(\mu)$ over μ for $\lambda = .1, 1, 5, 10$.

Ex 1.12 (Penalized version of Exercise 1.6). Let $(Y_i, X_{i,1:p}), i = 1, \ldots, n$, be responses and covariates (excluding a constant), where $X_{i,1:p} = (X_{i1}, \ldots, X_{ip})^{\mathrm{T}}$. Let $\tilde{Y}_i = Y_i - \bar{Y}$ and $\tilde{X}_{ij} = X_{ij} - \bar{X}_j$, where $\bar{Y} = n^{-1}\sum_{i=1}^n Y_i$ and $\bar{X}_j = n^{-1}\sum_{i=1}^n X_{ij}$ are the sample averages. Denote $\bar{X}_{1:p} = (\bar{X}_1, \ldots, \bar{X}_p)^{\mathrm{T}}$.

(i) For $\lambda \geq 0$, define the ridge estimator

$$(\hat{\beta}_0, \hat{\beta}_{1:p}) = \text{argmin}_{(\beta_0,\beta_{1:p})} \left\{ \frac{1}{2n}\sum_{i=1}^n (Y_i - \beta_0 - X_{i,1:p}^{\mathrm{T}}\beta_{1:p})^2 + \frac{\lambda}{2}\|\beta_{1:p}\|_2^2 \right\},$$

and

$$\tilde{\beta}_{1:p} = \text{argmin}_{\beta_{1:p}} \left\{ \frac{1}{2n}\sum_{i=1}^n (\tilde{Y}_i - \tilde{X}_{i,1:p}^{\mathrm{T}}\beta_{1:p})^2 + \frac{\lambda}{2}\|\beta_{1:p}\|_2^2 \right\},$$

Show that $\hat{\beta}_{1:p}$ and $\hat{\beta}_0$ can be obtained as $\tilde{\beta}_{1:p}$ and $\bar{Y} - \bar{X}_{1:p}^{\mathrm{T}}\tilde{\beta}_{1:p}$ respectively. In other words, $\tilde{\beta}_{1:p}$ and $\bar{Y} - \bar{X}_{1:p}^{\mathrm{T}}\tilde{\beta}_{1:p}$ are jointly a solution to the first optimization problem. Moreover, show that $\text{tr}(\mathbf{S}_\lambda) = 1 + \text{tr}(\tilde{\mathbf{S}}_\lambda)$. Here $\mathbf{S}_\lambda$ and $\tilde{\mathbf{S}}_\lambda$ are the smoother matrices from the original and "centered" ridge regression, such that

$$\mathbf{X}(\hat{\beta}_0, \hat{\beta}_{1:p}^{\mathrm{T}})^{\mathrm{T}} = \mathbf{S}_\lambda \mathbf{Y}, \quad \tilde{\mathbf{X}}\tilde{\beta}_{1:p} = \tilde{\mathbf{S}}_\lambda \tilde{\mathbf{Y}},$$

where $\mathbf{X} = (\mathbf{1}_p, X_{1,1:p}, \ldots, X_{p,1:p})^{\mathrm{T}}$, with $\mathbf{1}_p$ the $p \times 1$ vector of ones, $\mathbf{Y} = (Y_1, \ldots, Y_n)^{\mathrm{T}}$, $\tilde{\mathbf{X}} = (\tilde{X}_{1,1:p}, \ldots, \tilde{X}_{p,1:p})^{\mathrm{T}}$, and $\tilde{\mathbf{Y}} = (\tilde{Y}_1, \ldots, \tilde{Y}_n)^{\mathrm{T}}$.

(ii) For $\lambda \geq 0$, define the Lasso estimator

$$(\hat{\beta}_0, \hat{\beta}_{1:p}) = \text{argmin}_{(\beta_0,\beta_{1:p})} \left\{ \frac{1}{2n}\sum_{i=1}^n (Y_i - \beta_0 - X_{i,1:p}^{\mathrm{T}}\beta_{1:p})^2 + \lambda\|\beta_{1:p}\|_1 \right\},$$

and

$$\tilde{\beta}_{1:p} = \text{argmin}_{\beta_{1:p}} \left\{ \frac{1}{2n} \sum_{i=1}^{n} (\tilde{Y}_i - \tilde{X}_{i,1:p}^{\mathrm{T}} \beta_{1:p})^2 + \lambda \|\beta_{1:p}\|_1 \right\},$$

Show that $\hat{\beta}_{1:p}$ and $\hat{\beta}_0$ can be obtained as $\tilde{\beta}_{1:p}$ and $\bar{Y} - \bar{X}^{\mathrm{T}} \tilde{\beta}_{1:p}$ respectively. In other words, $\tilde{\beta}_{1:p}$ and $\bar{Y} - \bar{X}^{\mathrm{T}} \tilde{\beta}_{1:p}$ are jointly a solution to the first optimization problem.

Ex 1.13. Let $(Y_i, X_i), i = 1, \ldots, n$, be responses and covariates (possibly including a constant), where $X_i = (X_{i1}, \ldots, X_{ip})^{\mathrm{T}}$. This exercise studies the effect of covariate standardization before fitting penalized regression. Let $\tilde{X}_{ij} = X_{ij} / \|\mathbf{X}_j\|_n$, where $\|\mathbf{X}_j\|_n = (\frac{1}{n} \sum_{i=1}^{n} X_{ij}^2)^{1/2}$.

(i) For $\lambda \geq 0$, define the ridge estimator with the standardized covariates

$$\hat{\gamma} = \text{argmin}_{\gamma = (\gamma_1, \ldots, \gamma_p)^{\mathrm{T}}} \left\{ \frac{1}{2n} \sum_{i=1}^{n} (Y_i - \tilde{X}_i^{\mathrm{T}} \gamma)^2 + \frac{\lambda}{2} \|\gamma\|_2^2 \right\}.$$

Find nonnegative numbers $(c_1, \ldots, c_p)$ such that $\hat{\beta} = (\hat{\gamma}_1 / \|\mathbf{X}_1\|_n, \ldots, \hat{\gamma}_p / \|\mathbf{X}_p\|_n)^{\mathrm{T}}$ is a solution to the optimization problem

$$\min_{\beta = (\beta_1, \ldots, \beta_p)^{\mathrm{T}}} \left\{ \frac{1}{2n} \sum_{i=1}^{n} (Y_i - X_i^{\mathrm{T}} \beta)^2 + \frac{\lambda}{2} \sum_{j=1}^{p} c_j \beta_j^2 \right\},$$

where $\frac{1}{2} \sum_{j=1}^{p} c_j \beta_j^2$ is a weighted ridge penalty.

(ii) For $\lambda \geq 0$, define the Lasso estimator with the standardized covariates

$$\hat{\gamma} = \text{argmin}_{\gamma = (\gamma_1, \ldots, \gamma_p)^{\mathrm{T}}} \left\{ \frac{1}{2n} \sum_{i=1}^{n} (Y_i - \tilde{X}_i^{\mathrm{T}} \gamma)^2 + \lambda \|\gamma\|_1 \right\}.$$

Find nonnegative numbers $(c_1, \ldots, c_p)$ such that $\hat{\beta} = (\hat{\gamma}_1 / \|\mathbf{X}_1\|_n, \ldots, \hat{\gamma}_p / \|\mathbf{X}_p\|_n)^{\mathrm{T}}$ is a solution to the optimization problem

$$\min_{\beta = (\beta_1, \ldots, \beta_p)^{\mathrm{T}}} \left\{ \frac{1}{2n} \sum_{i=1}^{n} (Y_i - X_i^{\mathrm{T}} \beta)^2 + \lambda \sum_{j=1}^{p} c_j |\beta_j| \right\},$$

where $\sum_{j=1}^{p} c_j |\beta_j|$ is a weighted Lasso penalty.

Ex 1.14. Let $(Y_i, X_i), i = 1, \ldots, n$, be responses and covariates (possibly including a constant), where $X_i = (X_{i1}, \ldots, X_{ip})^{\mathrm{T}}$. For $\lambda \geq 0$, define the Lasso estimator

$$\hat{\beta} = \text{argmin}_{\beta = (\beta_1, \ldots, \beta_p)^{\mathrm{T}}} \left\{ \frac{1}{2n} \sum_{i=1}^{n} (Y_i - X_i^{\mathrm{T}} \beta)^2 + \lambda \|\beta\|_1 \right\}.$$

Let $\tilde{X}_{ij} = X_{ij}/\|\mathbf{X}_j\|_n$, where $\|\mathbf{X}_j\|_n = (\frac{1}{n}\sum_{i=1}^n X_{ij}^2)^{1/2}$.

(i) For nonnegative numbers $(\kappa_1, \dots, \kappa_p)$, define

$$\hat{\gamma} = \text{argmin}_{\gamma=(\gamma_1,\dots,\gamma_p)^{\mathrm{T}}} \left\{ \frac{1}{2n} \sum_{i=1}^n (Y_i - \tilde{X}_i^{\mathrm{T}}\gamma)^2 + \lambda \sum_{j=1}^p \kappa_j |\gamma_j| \right\}.$$

Find $(\kappa_1, \dots, \kappa_p)$ such that $\hat{\beta}$ can be obtained as $(\hat{\gamma}_1/\|\mathbf{X}_1\|_n, \dots, \hat{\gamma}_p/\|\mathbf{X}_p\|_n)^{\mathrm{T}}$.

[Note: This question involves similar reasoning as in Exercise 1.13 (ii), but in a different context.]

(ii) Derive the coordinate descent update for computing $\hat{\gamma}$, as an extension of (1.61).

Ex 1.15.

(i) Consider the objective function in Lasso regression,

$$h(\beta) = \frac{1}{2n} \sum_{i=1}^n (Y_i - X_i^{\mathrm{T}}\beta)^2 + \lambda \|\beta\|_1.$$

Suppose that $\tilde{\beta} = (\tilde{\beta}_1, \dots, \tilde{\beta}_p)^{\mathrm{T}}$ satisfies the property that for each $j = 1, \dots, p$, $\tilde{\beta}_j$ minimizes $h(\beta)$ in β_j, with all the other coefficients $\{\tilde{\beta}_k : k \neq j\}$ being fixed, i.e., $\tilde{\beta}_j$ is the coordinate descent update for the jth coefficient at $\tilde{\beta}$. Show that $\tilde{\beta}$ is jointly a minimizer to $h(\beta)$ (i.e., satisfies the stationary condition). Hence coordinate-wise minimization is sufficient for joint minimization of $h(\beta)$.

(ii) Show that the result in (i) holds for an objective function in the form

$$h(\beta) = h_0(\beta) + h_1(\beta),$$

where $\beta = (\beta_1, \dots, \beta_p)^{\mathrm{T}}$, $h_0(\beta)$ is convex and everywhere differentiable, $h_1(\beta) = \sum_{j=1}^p h_{1,j}(\beta_j)$, and $h_{1,j}(\beta_j)$ is convex in β_j but may not be differentiable at some points. In other words, $h(\beta)$ is the sum of convex, differentiable $h_0(\beta)$ and convex, possibly non-differentiable $h_1(\beta)$, and $h_1(\beta)$ is separable in the coordinates of β.

(iii) Provide an example where coordinate-wise minimization does not imply joint minimization for a convex function $h(\beta)$.

Ex 1.16. Consider a modification of the ridge estimator defined as

$$\hat{\beta}_\lambda = \text{argmin}_\beta \left\{ \frac{1}{2n} \sum_{i=1}^n (Y_i - X_i^{\mathrm{T}}\beta)^2 + \lambda \|\beta\|_2 \right\},$$

which is called the root ridge estimator.

(i) Show that a "ball constraint" is satisfied by $\hat{\beta}_\lambda$:

$$\left\| \frac{1}{n} \mathbf{X}^{\mathrm{T}} (\mathbf{Y} - \mathbf{X}\hat{\beta}_\lambda) \right\|_2 \leq \lambda.$$

[Hint: The function $\|\beta\|_2$ is convex and differentiable in β except at $\beta = 0$. Determine the sub-differential of $\|\beta\|_2$ and then use the stationary condition for $\hat{\beta}_\lambda$ as a minimizer of the objective function above.]

(ii) Find a necessary and sufficient condition for $\hat{\beta}_\lambda = 0$.

Ex 1.17. Simulate a dataset (including both training and test data) by running the following R code, including `set.seed(0)`:

```
library(MASS)
set.seed(0)

n <- 200
p <- 20

mu <- rep(0,p)
msig <- toeplitz(.8^(1:p-1))
sig <- .5

x <- mvrnorm(n, mu, msig)
y <- x[,1] + pmax(0, apply(x[,2:4],1,sum))^2 /3 + rnorm(n, 0, sig)

cbind(y,x)[1:3, 1:6]    # report part of the data

### test data
ntest <- 100
xtest <- mvrnorm(ntest, mu, msig)
ytest <- xtest[,1] + pmax(0, apply(xtest[,2:4],1,sum))^2 /3 +
         rnorm(ntest, 0, sig)

cbind(ytest,xtest)[1:3, 1:6]    # report part of the data
```

For all the following questions, please apply all methods to the original dataset, without any standardization on x or y. Try to make your plots informative in selecting the sequence of λ values, etc.

(i) Fit linear regression of y on x (including the intercept), using a ridge penalty on all non-intercept coefficients, on the training data. Produce a path plot of coefficients similar to Figure 1.9 (left).

(ii) Plot (a) training errors, (b) test errors on the test data, (c) 5-fold cross validation errors, and (d) 10-fold cross validation errors, over degrees of freedom, similarly as in Figure 1.9 (right). Report the minimum test error and the test errors obtained using the two selected λ by minimization of cross validation errors and one standard error rule.

(iii) Fit linear regression of y on x (including the intercept), using a Lasso penalty on all non-intercept coefficients, on the training data. Produce a path plot of coefficients similar to Figure 1.10 (left).

(iv) Plot (a) training errors, (b) test errors on the test data, (c) 5-fold cross validation errors, and (d) 10-fold cross validation errors, over the shrinkage fraction, similarly as in Figure 1.10 (right). Report the minimum test error and the test errors obtained using the two selected λ by minimization of cross validation errors and one standard error rule.

(v) Summarize what you learned from this exercise (in one paragraph).

Ex 1.18. Simulate data using the same code as in Exercise 1.17, except with p changed to 100 and 200 respectively. In each case, answer questions (i)–(iv) in Exercise 1.17. Then summarize your findings (in one paragraph).

Ex 1.19. Suppose that $n \times p$ covariate matrix $\mathbf{X} = (X_1, \ldots, X_n)^{\mathrm{T}}$ and $n \times 1$ response vector $\mathbf{Y} = (Y_1, \ldots, Y_n)^{\mathrm{T}}$ are generated as follows: for $i = 1, \ldots, n$,

$$
\begin{aligned}
X_i &= W Z_i + u_i, \\
Y_i &= \theta^{\mathrm{T}} Z_i + v_i,
\end{aligned}
$$

where W is a $p \times d$ constant matrix, θ is a $d \times 1$ constant vector, $Z_1, \ldots, Z_n$ are independent with $Z_i \sim \mathrm{N}(0, I_d)$, $u_1, \ldots, u_n$ are independent with $u_i \sim \mathrm{N}(0, I_p)$, $v_1, \ldots, v_n$ are independent with $v_i \sim \mathrm{N}(0, \sigma_v^2)$, Z_i's, u_i's, and v_i's are also independent of each other, and I_d and I_p are the $d \times d$ and $p \times p$ identity matrix respectively.

Consider linear regression $\mathbf{Y} \sim \mathbf{X}\beta^*$, as indicated by (i) below. We are interested in studying the prediction errors for the ridge estimator of β^*:

$$
\hat{\beta}_\lambda = \operatorname{argmin}_\beta \left\{ \frac{1}{n} \sum_{i=1}^n (Y_i - X_i^{\mathrm{T}} \beta)^2 + \lambda \|\beta\|_2^2 \right\},
$$

where $\lambda \geq 0$ is a tuning parameter. The in-sample and out-of-sample prediction errors conditionally on $\mathbf{X}$ are defined as

$$
\begin{aligned}
\mathrm{Err}_{\mathrm{in}} &= E \left\{ \frac{1}{n} \sum_{i=1}^n (X_i^{\mathrm{T}} \hat{\beta}_\lambda - X_i^{\mathrm{T}} \beta^*)^2 \Big| \mathbf{X} \right\}, \\
\mathrm{Err}_{\mathrm{out}} &= E \left\{ (X_0^{\mathrm{T}} \hat{\beta}_\lambda - X_0^{\mathrm{T}} \beta^*)^2 | \mathbf{X} \right\},
\end{aligned}
$$

where X_0 is a new covariate vector, independent of the training data $(\mathbf{X}, \mathbf{Y})$, and the expectation is taken over $(Y_1, \ldots, Y_n)$ and X_0. The preceding definitions are related to the mean squared errors in predicting new responses, similarly as in Proposition 3.3 in Chapter 3.

(i) Show that (X_i, Y_i), $i = 1, \ldots, n$, satisfy a linear model:

$$Y_i = X_i^{\mathrm{T}} \beta^* + \varepsilon_i,$$

where $\varepsilon_i \sim \mathrm{N}(0, \sigma_\varepsilon^2)$. Show that

$$\beta^* = W(I_d + W^{\mathrm{T}} W)^{-1} \theta, \quad \sigma_\varepsilon^2 = \sigma_v^2 + \theta^{\mathrm{T}} (I_d + W^{\mathrm{T}} W)^{-1} \theta.$$

(ii) Show that $X_i \sim \mathrm{N}(0, \Sigma)$ for $i = 1, \ldots, n$, where

$$\Sigma = I_p + W W^{\mathrm{T}}.$$

(iii) Find explicit expressions for $\mathrm{Err}_{\mathrm{in}}$ and $\mathrm{Err}_{\mathrm{in}}$ in terms of $\beta^*, \sigma_\varepsilon^2$, the (nonzero) eigenvalues and their corresponding eigenvectors of $\hat{\Sigma} = \mathbf{X}^{\mathrm{T}} \mathbf{X}/n$.

In the remaining questions, suppose that (W, θ, σ_v^2) are specified as follows with $p = 2^d$:

$$W = (w_1^{\mathrm{T}}, \ldots, w_p^{\mathrm{T}})^{\mathrm{T}} / \sqrt{d}, \quad \theta = \mathbf{1}_d / \sqrt{d}, \quad \sigma_v^2 = 1,$$

where $w_1, \ldots, w_p$ (with $p = 2^d$) correspond to all possible length-d arrangements of -1 or 1, and $\mathbf{1}_d$ denotes a $d \times 1$ vector of all ones. For example, for $d = 3$ and $p = 8$,

$$\begin{pmatrix} w_1^{\mathrm{T}} \\ \vdots \\ w_p^{\mathrm{T}} \end{pmatrix} = \begin{pmatrix} -1 & -1 & -1 \\ 1 & -1 & -1 \\ -1 & 1 & -1 \\ -1 & -1 & 1 \\ 1 & 1 & -1 \\ 1 & -1 & 1 \\ -1 & 1 & 1 \\ 1 & 1 & 1 \end{pmatrix}.$$

(iv) For the preceding specification, show that the diagonal elements of Σ are all equal to 2, i.e., the marginal variances of the components of X_i are 2. Moreover, show that the eigenvalues of Σ, in the decreasing order, are

$$\lambda_1 = \cdots = \lambda_d = \frac{p}{d} + 1,$$
$$\lambda_{d+1} = \cdots = \lambda_{2^d} = 1.$$

For a simulation study, generate training and test data as follows in R (please run these codes exactly with the random seed):

```
set.seed(1)

### training data
d <- 9
n <- 50
p <- 2^d

theta <- rep(1/sqrt(d), d)
P <- expand.grid(replicate(d, 0:1, simplify = FALSE))
W <- as.matrix((2*P - 1)/sqrt(d))

Z <- matrix(rnorm(n*d), nrow = n, ncol = d)
Y <- Z %*% theta + rnorm(n)
X <- Z %*% t(W) + matrix(rnorm(n*p), nrow = n, ncol = p)
```

For all questions requiring coding, please provide both your R codes and output.

(v) For the following choices of λ,

$$10^{-4}, 10^{-3}, 10^{-2}, 10^{-1}, 1, 2.5, 5, 7.5, 10, 20, 30, 40, 50,$$
$$100, 250, 500, 10^3, 10^4, 10^5, 10^6$$

compute and report Err_{in} and Err_{out} and plot them against $-\log(\lambda)$, given the training data generated above.

(vi) Repeat the experiment for 100 times, i.e., generating training data in the same way as above (except for removing the set.seed line) and computing Err_{in} and Err_{out}. Plot the averages of Err_{in} and Err_{out} against $-\log(\lambda)$.

[Note: This problem is motivated by related work on high-dimensional ridge estimation. See Liang and Tan (2024) and references therein.]

Ex 1.20. Suppose that $X_1, \ldots, X_n$ are IID from a p-variate normal distribution with mean vector 0 and unknown variance matrix Σ (assumed to be non-singular). Given a vector $x^* \in \mathbb{R}^p$, it is of interest to estimate $u^* = \Omega x^*$, where $\Omega = \Sigma^{-1}$ is called the precision matrix. For p much less than n, a simple estimator of u^* is $\hat{u} = \hat{\Sigma}^{-1} x^*$, where $\hat{\Sigma} = \mathbf{X}^{\mathrm{T}}\mathbf{X}/n$, and $\mathbf{X} = (X_1, \ldots, X_n)^{\mathrm{T}}$. Equivalently, $\hat{u}$ can be determined as

$$\hat{u} = \text{argmin}_{u \in \mathbb{R}^p} \left\{ \frac{1}{2} u^{\mathrm{T}} \hat{\Sigma} u - u^{\mathrm{T}} x^* \right\}.$$

For p close to or larger than n, we consider a Lasso penalized estimator

$$\hat{u}_\lambda = \text{argmin}_{u \in \mathbb{R}^p} \left\{ \frac{1}{2} u^{\mathrm{T}} \hat{\Sigma} u - u^{\mathrm{T}} x^* + \lambda \|u\|_1 \right\}, \tag{1.94}$$

where $\lambda > 0$ is a tuning parameter. The Lasso penalized estimator is expected to perform well in the setting where the true value u^* is sparse even with p larger than n. In this problem, you are asked to work out numerical algorithms for computing $\hat{u}_\lambda$ for a pre-specified λ.

(i) For fixed $\lambda > 0$, what is the stationary condition satisfied by $\hat{u}_\lambda$ as a minimizer in (1.94)?

(ii) Derive a coordinate descent algorithm for computing $\hat{u}_\lambda$. Specifically, denote the current iterate as $u^0 = (u_1^0, \ldots, u_p^0)^{\mathrm{T}}$. Provide the formula for updating the jth component of u^0 while fixing the remaining components for any j.

(iii) Derive a proximal gradient algorithm for computing $\hat{u}_\lambda$. Specifically, provide the formula for updating the current iterate u^0 to the next, depending on a step size τ.

For a numerical experiment, generate the following data in R (please run these codes exactly with the random seed):

```
library(MASS)
set.seed(1)

n <- 600
p <- 600

# define Sigma
rho <- 0.3
Sigma <- matrix(rep(0.3, p^2), nrow = p, ncol = p)
diag(Sigma) <- 1

# generate matrix X and vector x_star
X <- mvrnorm(n, rep(0,p), Sigma = Sigma)
x_star <- Sigma %*% c(-1,1,-1,1,1,rep(0, p-5))
```

For all questions requiring coding, please provide both your R codes and output. Set the initial value to 0 (a zero vector). The termination criterion for iterations is when the L_1 norm of the difference between current and updated iterates is less than 10^{-5}.

(iv) Write your R function to implement the coordinate descent algorithm from (ii). Fix the choice of the Lasso tuning parameter, $\lambda = 1.2\sqrt{\frac{\log(p)}{n}}$. Report your results including

- the number of iterations,

- the final iterate (only in the first 10 components), and
- plots of the iterates (only in the jth components, $j = 2, 4, 6, 8, 10$).

Note that an iteration is completed when all the components of u^0 are successively updated once. Check that the stationary condition from (i) is properly satisfied.

(v) Write your R function to implement the proximal gradient algorithm from (iii). Fix the choice of the Lasso tuning parameter, $\lambda = 1.2\sqrt{\frac{\log(p)}{n}}$, and use three choices of the step size, $\tau = 0.01, 0.005, 0.001$. For each step size, report your results as in (iv), and check that the stationary condition from (i) is properly satisfied.

[Note: This problem is motivated by related work on debiased Lasso prediction. See Liang and Tan (2025) and references therein.]

2

Generalized linear regression

In this chapter, we study generalized linear regression, more commonly called generalized linear models, through the following topics. For simplicity, we focus on the random-design setting. First, we discuss numerical algorithms for parameter estimation, including Newton–Raphson and Fisher scoring. Second, we discuss both model-based inference with correct model specification and model-robust inference with model misspecification, by drawing on general, asymptotic theory of M-estimation. Third, we discuss penalized regression, by exploiting quadratic approximation and Majorization-Minimization techniques. We provide numerical demonstration in R, using both existing packages and self-contained code for implementation and application of main methods.

Generalized linear regression is a useful generalization of linear regression to accommodate various types of response data, for example, dichotomous (or binary) responses, polytomous (or multi-valued) responses, or nonnegative responses. A comprehensive reference is McCullagh and Nelder (1989). We use the phrase "generalized linear regression" interchangeably with perhaps the more common "generalized linear models." Similarly as in Chapter 1, we assume understanding of basic material on generalized linear models, and study several topics which are important from the perspective of method development, but may often be left out as technical details in introductory or even intermediate studies.

2.1 Data and model

2.1.1 Data

The form of data is similar as in Chapter 1. Let $\{(X_i, Y_i) : i = 1, \ldots, n\}$ be a collection of n *observations*, where X_i is a *covariate vector* (also called

DOI: 10.1201/9781003680727-2

a regressor vector or a predictor vector) and Y_i is a scalar *response variable* (also called an outcome variable). We write $X_i = (X_{i1}, \ldots, X_{ip})^{\mathrm{T}}$ as a column vector, with X_{ij} denoting the jth component of the ith covariate vector X_i.

2.1.2 Model

Similarly to linear regression in Chapter 1, generalized linear regression can be defined in a fixed or random design, with the corresponding statistical assumptions. The distinction between the fixed and random designs for statistical inference discussed in Chapter 1 can be extended to generalized linear regression. For simplicity, we consider a random design where $(X_1, Y_1), \ldots, (X_n, Y_n)$ are IID as (X, Y). In particular, the IID assumption makes it convenient to apply asymptotic theory of M-estimation for studying sampling properties of estimation and statistical inference (Section 2.3).

Generalized linear regression allows two extensions, which are related to respectively the mean restriction (1.3b) and normality assumption (1.7) in linear regression. First, the conditional mean $E(Y|X)$ can depend nonlinearly on a linear predictor $X^{\mathrm{T}}\beta^*$:

$$E(Y|X) = g^{-}(X^{\mathrm{T}}\beta^*), \tag{2.1}$$

where β^* is an unknown parameter vector, $g(\cdot)$ is a *known*, increasing differentiable function, called the link function, and $g^{-}(\cdot)$ is the inverse function of $g(\cdot)$. For linear regression, the mean restriction (1.3b) corresponds to $g(\cdot)$ being the identity function, $g^{-}(\eta) = \eta$.

Second, the conditional distribution of Y given X can come from an exponential family, with a density function in the form

$$f(y; \theta, \phi) = \exp\left\{ \frac{y\theta - a(\theta)}{\phi} + c(y, \phi) \right\}, \tag{2.2}$$

where θ and $\phi > 0$ are two scalar parameters, $a(\cdot)$ is a *known*, twice differentiable function, and $c(\cdot)$ is a *known* function. [In fact, $a(\cdot)$ needs to be a convex function as discussed after Lemma 2.1.] For a continuous random variable Y, the density function $f(y; \theta, \phi)$ is defined with respect to the Lebesgue measure and corresponds to the usual probability density function. For a discrete random variable Y, $f(y; \theta, \phi)$ is defined with respect to the counting measure, and corresponds to the usual probability mass function.

If ϕ is known, then θ is called the canonical parameter in the exponential family (2.2). By abuse of terminology, even if ϕ is unknown, θ is also called the canonical parameter due to the central role of θ in the exponential family (2.2) as discussed below. To capture the dependency of Y on X, the parameters θ and ϕ are in general two functions of X, denoted as $\theta(X)$ and $\phi(X)$ respectively. In fact, as shown in (2.6), combining (2.1) and (2.2) induces $\theta(X)$ as a specific function of the linear predictor $X^{\mathrm{T}}\beta^*$. As in our later discussion, $\phi(X)$ is often, for simplicity, assumed to be constant in X.

Various common families of univariate distributions can be put in the form (2.2). See Examples 2.1–2.3. However, the exponential family (2.2) has a particular structure. The parameters θ and ϕ are constrained by the normalization condition:

$$\frac{a(\theta)}{\phi} = \log \int \exp\left\{\frac{y\theta}{\phi} + c(y, \phi)\right\} dy. \tag{2.3}$$

There are simple but consequential properties for the exponential family (2.2).

Lemma 2.1 (Mean and variance for exponential family). *For $Y \sim f(y; \theta, \phi)$ in the form (2.2),*

$$E(Y) = a'(\theta), \tag{2.4a}$$
$$\operatorname{var}(Y) = \phi\, a''(\theta), \tag{2.4b}$$

where $a'(\theta)$ and $a''(\theta)$ are the first and second derivatives of $a(\theta)$. Hence $V(\mu) = a''((a')^{-}(\mu))$ is called the variance function, and ϕ is called the dispersion parameter.

Proof. The result can be directly proved by taking derivatives of both sides of (2.3) with respect to θ. This is left as an exercise. For pedagogical purposes, we show that the result can also be deduced by using the following identities:

$$E\left\{\frac{\partial}{\partial\theta} \log f(Y; \theta, \phi)\right\} = 0, \tag{2.5a}$$
$$E\left\{\frac{\partial^2}{\partial\theta^2} \log f(Y; \theta, \phi)\right\} + E\left\{\left(\frac{\partial}{\partial\theta} \log f(Y; \theta, \phi)\right)^2\right\} = 0, \tag{2.5b}$$

which are called first and second Bartlett identities. From the first identity (2.5a),

$$0 = E\left\{\frac{Y - a'(\theta)}{\phi}\right\},$$

which gives $E(Y) = a'(\theta)$. From the second identity (2.5b),

$$0 = E\left\{\frac{-a''(\theta)}{\phi}\right\} + E\left\{\left(\frac{Y - a'(\theta)}{\phi}\right)^2\right\},$$

which gives $\operatorname{var}(Y) = \phi\, a''(\theta)$. □

From Lemma 2.1, several properties of the exponential family (2.2) can be deduced. First, as indicated by (2.4a)–(2.4b), the parameter θ determines both the mean and, up to a scaling factor ϕ, the variance in the exponential family (2.2), through the first and second derivatives of $a(\theta)$ respectively. Furthermore, by (2.4b), $a''(\theta) > 0$ and hence $a(\theta)$ is strictly convex in θ

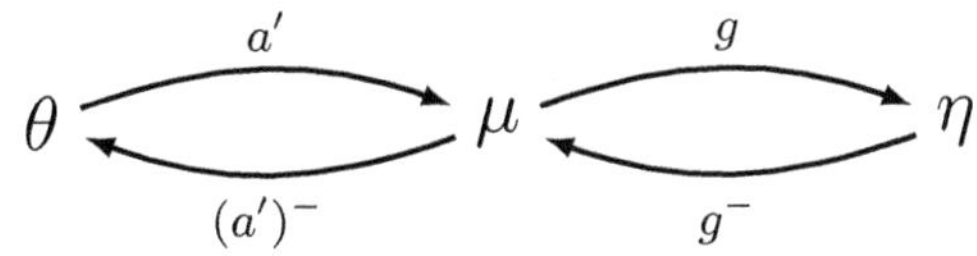

FIGURE 2.1
Relations between canonical parameter $\theta(X)$ in the conditional distribution of Y given X, conditional mean $\mu(X) = E(Y|X)$, and linear predictor $\eta(X) = X^{\mathrm{T}}\beta^*$.

[provided that $f(\cdot;\theta,\phi)$ is a proper density function with $\mathrm{var}(Y) \neq 0$.] Then by (2.4a), $E(Y) = a'(\theta)$ is increasing in θ and hence an invertible function of θ. Therefore, there is a one-to-one mapping between the canonical parameter θ and the mean $\mu = E(Y)$ in the exponential family (2.2). Applying this mapping to Y conditionally on X shows that the canonical parameter $\theta(X)$ in the conditional distribution of Y given X in the form (2.2) is linked one-to-one to the conditional mean $\mu(X) = E(Y|X)$, which in turn is linked one-to-one to the linear predictor $\eta(X) = X^{\mathrm{T}}\beta^*$ by the mean restriction (2.1):

$$\theta(X) = (a')^{-}\{\mu(X)\} = (a')^{-}\{g^{-}(X^{\mathrm{T}}\beta^*)\}. \tag{2.6}$$

See Figure 2.1 for an illustration of the two links, which are associated with the two extensions of generalized linear regression discussed above.

Example 2.1. For linear regression with homoscedastic, normal noises, the conditional distribution of Y given X in (1.7) is from the normal family $\mathrm{N}(\mu, \sigma^2)$, which can be expressed as an exponential family in (2.2):

$$\begin{aligned} f(y;\mu,\sigma^2) &= \frac{1}{\sqrt{2\pi}\sigma}\exp\left\{-\frac{(y-\mu)^2}{2\sigma^2}\right\} \\ &= \exp\left\{\frac{y\mu - \mu^2/2}{\sigma^2} - \frac{y^2}{2\sigma^2} - \frac{1}{2}\log(2\pi\sigma^2)\right\}, \quad y \in \mathbb{R}, \end{aligned}$$

with the definitions

$$\theta = \mu, \quad a(\theta) = \frac{\mu^2}{2}, \quad \phi = \sigma^2.$$

The mean restriction (1.3b), $E(Y|X) = X^{\mathrm{T}}\beta^*$, is obtained by taking $g^{-}(\eta) = \eta$ in (2.1). Hence the canonical parameter $\theta(X)$ in the normal conditional distribution of Y given X, $\mathrm{N}(\mu(X), \sigma^{*2})$, is set to the same as the linear predictor $X^{\mathrm{T}}\beta^*$:

$$\theta(X) = \mu(X) = X^{\mathrm{T}}\beta^*.$$

Example 2.2. For a binary response Y (taking only values 0 or 1), two popular regression models are logistic regression and probit regression, which assume respectively

$$P(Y=1|X) = \frac{e^{X^{\mathrm{T}}\beta^*}}{1+e^{X^{\mathrm{T}}\beta^*}}, \tag{2.7}$$

or

$$P(Y=1|X) = \Phi(X^{\mathrm{T}}\beta^*), \tag{2.8}$$

where β^* is an unknown parameter vector, and $\Phi(\cdot)$ is the cumulative distribution function of $\mathrm{N}(0,1)$. Note that $P(Y=1|X) = E(Y|X)$ for a binary variable Y. The mean restriction (2.7) or (2.8) can be obtained from (2.1) by taking respectively

$$p = g^{-}(\eta) = \frac{e^{\eta}}{1+e^{\eta}} \quad \Longleftrightarrow \quad \eta = g(p) = \log\frac{p}{1-p}, \tag{2.9}$$

or

$$p = g^{-}(\eta) = \Phi(\eta) \quad \Longleftrightarrow \quad \eta = g(p) = \Phi^{-}(p).$$

For binary Y, the conditional distribution of Y given X is Bernoulli, hence fully determined by the mean restriction (2.7) or (2.8). Nevertheless, it is helpful to compare the two different choices of $g(\cdot)$ for the mean restriction from the perspective of viewing the Bernoulli family, Bernoulli(p), as an exponential family in (2.2):

$$\begin{aligned} f(y;p) &= p^y(1-p)^{1-y} = e^{y\log p + (1-y)\log(1-p)} \\ &= \exp\left\{ y\log\frac{p}{1-p} + \log(1-p) \right\}, \quad y = 0,1, \end{aligned}$$

with the definitions

$$\theta = \log\frac{p}{1-p}, \quad a(\theta) = -\log(1-p) = \log(1+e^{\theta}), \quad \phi = 1. \tag{2.10}$$

Combining the relations between θ, p, and η in (2.9) and (2.10) gives $\theta = \log\frac{p}{1-p} = \eta$. Hence for logistic regression, the canonical parameter, $\theta(X)$, in the Bernoulli conditional distribution of Y given X, Bernoulli($p(X)$), is set to the same as the linear predictor $X^{\mathrm{T}}\beta^*$:

$$\theta(X) = \log\frac{p(X)}{1-p(X)} = X^{\mathrm{T}}\beta^*.$$

For probit regression, $\theta(X)$ and $X^{\mathrm{T}}\beta^*$ are related in a more complex manner:

$$\theta(X) = \log\frac{p(X)}{1-p(X)} = \log\frac{\Phi(X^{\mathrm{T}}\beta^*)}{1-\Phi(X^{\mathrm{T}}\beta^*)}.$$

Example 2.3. For a nonnegative response Y (taking nonnegative values), taking $\mu = g^{-}(\eta) = \mathrm{e}^{\eta}$ or $\eta = g(\mu) = \log \mu$ gives the log-linear regression:

$$E(Y|X) = \mathrm{e}^{X^{\mathrm{T}}\beta^*} \quad \Longleftrightarrow \quad \log E(Y|X) = X^{\mathrm{T}}\beta^*, \tag{2.11}$$

where β^* is an unknown parameter vector. If Y is nonnegative, integer-valued, the conditional distribution of Y given X is often assumed to be from the Poisson family, Poisson(μ), which can be expressed as an exponential family in (2.2):

$$\begin{aligned} f(y;\mu) &= \frac{\mu^y \mathrm{e}^{-\mu}}{y!} \\ &= \exp\{y \log \mu - \mu + \log(y!)\}, \quad y = 0, 1, \ldots, \end{aligned}$$

with the definitions

$$\theta = \log \mu, \quad a(\theta) = \mu = \mathrm{e}^{\theta}, \quad \phi = 1.$$

By the preceding choices, the canonical parameter, $\theta(X)$, in the Poisson conditional distribution of Y given X, Poisson($\mu(X)$), is set to the same as the linear predictor $X^{\mathrm{T}}\beta^*$:

$$\theta(X) = \log \mu(X) = X^{\mathrm{T}}\beta^*.$$

The linear regression with normal noises, logistic regression with binary responses, and Poisson log-linear regression are three examples where *canonical links* $g(\cdot)$ are used. In general, for an exponential family (2.2), the canonical link is defined as

$$g(\cdot) = (a')^{-}(\cdot), \tag{2.12}$$

so that by the mean restriction (2.1), the canonical parameter $\theta(X)$ in the conditional distribution of Y given X in the form (2.2) is set to the same as the linear predictor: $\theta(X) = X^{\mathrm{T}}\beta^*$. In other words, if the canonical link $g(\cdot)$ is used, then $\theta = \eta$ in Figure 2.1.

Conceptually, the mean restriction (2.1) and the exponential-family assumption (2.2) are two distinct components of generalized linear regression. As in linear regression, the mean restriction (2.1) alone is sufficient for deriving consistent estimation of β^*, without introducing the distributional assumption (2.2). On the other hand, different choices of the link function $g(\cdot)$ are allowed in the mean restriction (2.1) with the same exponential-family assumption (2.2), as shown by logistic and probit regression in Example 2.2.

2.2 Parameter estimation: Algorithms

We consider maximum likelihood (ML) estimation of β^* in generalized linear regression, defined by the mean restriction (2.1) and exponential-family assumption (2.2), with constant dispersion parameter ϕ^*. As discussed in Section 2.3, the resulting estimator of β^* remains consistent under standard regularity conditions, provided that the mean restriction (2.1) is correctly specified but the exponential-family assumption (2.2) may be dropped. In this section, we introduce the relevant estimators and then discuss computational algorithms.

If ϕ^* is known, the log (conditional) likelihood function of β under (2.1) and (2.2) is

$$\ell_n(\beta) = \sum_{i=1}^n \left\{ \frac{Y_i\theta(X_i;\beta) - a(\theta(X_i;\beta))}{\phi^*} + c(Y_i,\phi^*) \right\},$$

where $\theta(X_i;\beta) = (a')^-(g^-(X_i^{\mathrm{T}}\beta))$ by the relation (2.6). The ML estimator of β^* is defined as $\hat\beta = \operatorname{argmax}_\beta \ell_n(\beta)$. Equivalently, the ML estimator $\hat\beta$ can be obtained as $\hat\beta = \operatorname{argmin}_\beta K_n(\beta)$, independently of ϕ^*, where

$$K_n(\beta) = \frac{-1}{n}\sum_{i=1}^n \{Y_i\theta(X_i;\beta) - a(\theta(X_i;\beta))\}. \tag{2.13}$$

The statistical properties of $\hat\beta$, however, depend on ϕ^* under correct specification of the regression model (see Section 2.3). In the remainder of this section, we mainly study numerical algorithms for computing the ML estimator $\hat\beta$.

If ϕ^* is unknown, then the log-likelihood function of (β,ϕ) is defined as $\ell_n(\beta)$ with ϕ^* replaced by ϕ. Maximum likelihood estimation can be performed jointly for (β^*,ϕ^*), similarly as that for (β^*,σ^{*2}) in linear regression with homoscedastic, normal noises. The ML estimator of β^* is the same as $\hat\beta$ defined above, regardless of estimation of ϕ^*. In general, the ML estimator of ϕ^* may depend on the form of $c(\cdot)$ in a nontrivial manner. Alternatively, a tractable, moment estimator of ϕ^* based on (2.4b) is

$$\hat\phi = \frac{1}{n-p}\sum_{i=1}^n \frac{(Y_i - \hat\mu_i)^2}{\hat V_i}, \tag{2.14}$$

where $\hat\mu_i = g^-(X_i^{\mathrm{T}}\hat\beta)$ and $\hat V_i = V(\hat\mu_i) = a''((a')^-(\hat\mu_i))$ by the definition of $V(\cdot)$ in Lemma 2.1. For linear regression with homoscedastic, normal noises and $\phi^* = \sigma^{*2}$ as in Example 2.1, the estimator $\hat\phi$ recovers the common estimator $\hat\sigma^2$ for σ^{*2} in (1.10).

The dispersion estimator $\hat\phi$ in (2.14) is related to the Pearson's statistic

$$S_X = \sum_{i=1}^n \frac{(Y_i - \hat\mu_i)^2}{\hat V_i}.$$

For linear regression with homoscedastic, normal noises, $\frac{S_X}{\phi^*}$ has an exact χ^2_{n-p} distribution (χ^2 distribution with $n-p$ degrees of freedom), and $\hat\phi$ is an unbiased estimator for $\phi^* = \sigma^{*2}$. In general, if the model assumptions (2.1) and

(2.2) are correctly specified, then $\frac{S_X}{\phi^*}$ can be expected to be approximately distributed as χ^2_{n-p} or simply χ^2_n in the classical setting ($n \to \infty$ and p is fixed). [Question: How can this statement be made precise?] Then the estimator $\hat{\phi}$ is a consistent estimator for ϕ^* because $\chi^2_n/n \xrightarrow{P} 1$ as $n \to \infty$.

Another statistic for goodness of fit is the residual deviance (also called the unrescaled deviance)

$$D_X = -2 \sum_{i=1}^{n} \left\{ Y_i \hat{\theta}_i - a(\hat{\theta}_i) \right\}, \tag{2.15}$$

where $\hat{\theta}_i = \theta(X_i; \hat{\beta})$. The scaled deviance, $\frac{D_X}{\phi^*}$, is twice the fitted negative log-likelihood, $-2\ell_n(\hat{\beta})$, up to an additive constant. The residual deviance is useful for comparing two nested models, with covariate vectors $\mathring{X} \subset X$ of dimensions $\mathring{p} < p$. If the model assumptions (2.1) and (2.2) are correctly specified with $\mathring{X}$ (and hence also with X), then

$$\frac{D_{\mathring{X}} - D_X}{\phi^*} \tag{2.16}$$

is the log-likelihood ratio statistic, which converges in distribution to $\chi^2_{p-\mathring{p}}$ in the classical setting ($n \to \infty$ and p is fixed). The null deviance, $D_{\mathring{X}}$, is defined by taking $\mathring{X}$ to be the constant 1 (hence a regression model with the intercept only). Then the log-likelihood ratio statistic (2.16) can be used to test the null hypothesis that the non-intercept coefficients are all zero, assuming that the regression model with covariate vector X is correctly specified.

2.2.1 Newton–Raphson algorithm

For a twice differentiable, scalar function $h(\beta)$, consider the minimization problem

$$\min_{\beta} \; h(\beta).$$

The Newton–Raphson algorithm proceeds as follows.

- Pick an initial value β^0.
- Compute for $k = 1, 2, \cdots$,

$$\beta^k = \beta^{k-1} - \left(\frac{\partial^2 h}{\partial \beta \partial \beta^{\mathrm{T}}} \right)^{-1} \frac{\partial h}{\partial \beta} \Big|_{\beta = \beta^{k-1}}, \tag{2.17}$$

  until some termination criterion is met.

To solve $\max_\beta \; h(\beta)$, replacing h by $-h$ leads to seemingly the same algorithm. It is informative to examine how the Newton–Raphson algorithm is derived. See Figure 2.2, and compare with Figure 1.5 for the MM algorithm.

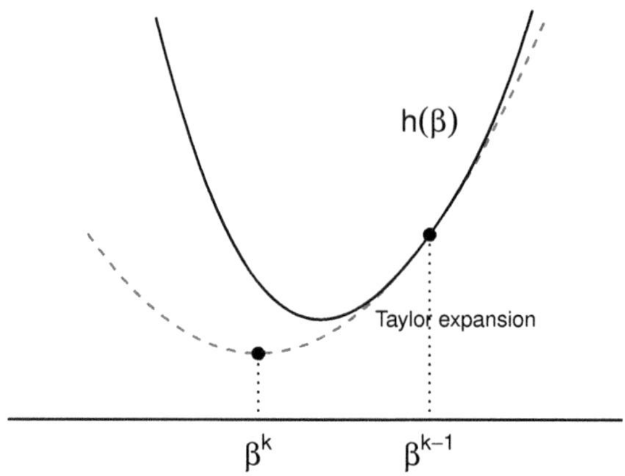

FIGURE 2.2
An iteration of the Newton–Raphson algorithm.

First, a quadratic approximation is obtained from the Taylor expansion:

$$h(\beta) \approx h(\beta^{k-1}) + \frac{\partial h}{\partial \beta^{\mathrm{T}}}(\beta^{k-1})(\beta - \beta^{k-1}) \\ + \frac{1}{2}(\beta - \beta^{k-1})^{\mathrm{T}} \frac{\partial^2 h}{\partial \beta \partial \beta^{\mathrm{T}}}(\beta^{k-1})(\beta - \beta^{k-1}).$$

Then the quadratic approximation (on the right-hand side of the above display) is minimized by taking the gradient and setting to 0:

$$0 = \frac{\partial h}{\partial \beta}(\beta^{k-1}) + \frac{\partial^2 h}{\partial \beta \partial \beta^{\mathrm{T}}}(\beta^{k-1})(\beta - \beta^{k-1}),$$

which yields the Newton–Raphson update (2.17). Compared with the derivation of the gradient descent (1.65), the main difference is that the quadratic approximation is minimized with exactly the Hessian, $(\partial^2 h/\partial\beta\partial\beta^{\mathrm{T}})(\beta^{k-1})$, at the current iterate β^{k-1}, instead of a constant upper bound $\tau^{-1}I$. The Newton–Raphson algorithm is known to converge fast when iterating near a minimizer, but otherwise may be unstable.

To improve the stability of the Newton–Raphson algorithm, a backtracking line search can be incorporated at each iteration k as follows.

- Compute β^k as in (2.17).
- If $h(\beta^k) < h(\beta^{k-1})$, then continue to the $(k+1)$th iteration. Otherwise, find $0 < \eta < 1$ such that $\tilde{\beta}^k = \beta^{k-1} + \eta(\beta^k - \beta^{k-1})$ satisfies $h(\tilde{\beta}^k) < h(\beta^{k-1})$ and then reset $\beta^k = \tilde{\beta}^k$. In practice, this can be implemented by searching over $\eta = \frac{1}{2^i}$ for $i = 1, 2, \ldots$.

We show that all sufficiently small $\eta > 0$ satisfies the desired property under mild conditions. This is similar to the fact that the GD update, $\beta^{k-1} - \tau \frac{\partial h}{\partial \beta}(\beta^{k-1})$, always leads to a smaller value of h than $h(\beta^{k-1})$ for all sufficiently small $\tau > 0$.

Lemma 2.2 (Justification for backtracking). *In the preceding setting, assume that* $\frac{\partial^2 h}{\partial \beta \partial \beta^{\mathrm{T}}}(\beta)$ *is continuous for* β *in a neighborhood of* β^{k-1}, $\frac{\partial^2 h}{\partial \beta \partial \beta^{\mathrm{T}}}(\beta^{k-1})$ *is positive definite, and* $\frac{\partial h}{\partial \beta}(\beta^{k-1}) \neq 0$. *Then there exists* $0 < \eta_0 < 1$ *such that* $\tilde{\beta}^k = \beta^{k-1} + \eta(\beta^k - \beta^{k-1})$ *satisfies* $h(\tilde{\beta}^k) < h(\beta^{k-1})$ *for any* $\eta \in (0, \eta_0]$.

Proof. Let $\Delta(\eta) = h(\tilde{\beta}^k) - h(\beta^{k-1}) = h(\beta^{k-1} + \eta(\beta^k - \beta^{k-1})) - h(\beta^{k-1})$. Then $\Delta(0) = 0$ and, by the chain rule,

$$\begin{aligned}\frac{\partial \Delta}{\partial \eta}(0) &= \frac{\partial h}{\partial \beta^{\mathrm{T}}}(\beta^{k-1})(\beta^k - \beta^{k-1}) \\ &= -\frac{\partial h}{\partial \beta^{\mathrm{T}}}(\beta^{k-1}) \left\{ \frac{\partial^2 h}{\partial \beta \partial \beta^{\mathrm{T}}}(\beta^{k-1}) \right\}^{-1} \frac{\partial h}{\partial \beta}(\beta^{k-1}) < 0,\end{aligned}$$

because $\frac{\partial^2 h}{\partial \beta \partial \beta^{\mathrm{T}}}(\beta^{k-1})$ is positive definite, and $\frac{\partial h}{\partial \beta}(\beta^{k-1}) \neq 0$. Moreover, $\frac{\partial \Delta}{\partial \eta}(\eta)$ is continuous for η in a neighborhood of 0, by the continuity of $\frac{\partial^2 h}{\partial \beta \partial \beta^{\mathrm{T}}}(\beta)$ in a neighborhood of β^{k-1}. Hence there exists $\eta_0 > 0$ such that $\frac{\partial \Delta}{\partial \eta}(\eta) < 0$ for any $\eta \in [0, \eta_0]$. By the mean value theorem, for any $\eta \in (0, \eta_0]$,

$$\Delta(\eta) = \Delta(\eta) - \Delta(0) = \eta \left\{ \frac{\partial \Delta}{\partial \eta}(\dot{\eta}) \right\} < 0,$$

where $\dot{\eta}$ lies between 0 and η and hence $\dot{\eta} \in (0, \eta_0]$. □

2.2.2 Fisher scoring and iteratively reweighted least squares

We apply the Newton–Raphson algorithm to maximize the log-likelihood function $\ell_n(\beta)$ in generalized linear regression. For convenience, we use the following notation. For $i = 1, \ldots, n$, let

$$\eta_i = X_i^{\mathrm{T}} \beta, \quad \mu_i = g^{-}(\eta_i), \quad \theta_i = (a')^{-}(\mu_i), \quad V_i = a''(\theta_i),$$

where the dependency on β is suppressed in the notation. From the middle two expressions, the following relations also hold: $\eta_i = g(\mu_i)$ and $\mu_i = a'(\theta_i)$. Then the average negative log-likelihood function (also called the likelihood loss) can be rewritten as

$$K_n(\beta) = \frac{-1}{n} \ell_n(\theta) = \frac{-1}{n} \sum_{i=1}^{n} \left\{ \frac{Y_i \theta_i - a(\theta_i)}{\phi^*} + c(Y_i, \phi^*) \right\}. \tag{2.18}$$

[Equivalently, for computing $\hat{\beta}$, the objective function can be taken as in (2.13), by dropping the constant ϕ^*. The above function (2.18), precisely the average negative log-likelihood, facilitates the later discussion of the second Bartlett identity in model-based inference.] The gradient of $K_n(\theta)$, also called the score function, can be directly calculated:

$$\begin{aligned}\frac{\partial K_n}{\partial \beta}(\beta) &= \frac{-1}{n\phi^*}\sum_{i=1}^{n}(Y_i-\mu_i)\frac{\partial \theta_i}{\partial \beta} \\ &= \frac{-1}{n\phi^*}\sum_{i=1}^{n}(Y_i-\mu_i)V_i^{-1}\frac{\partial \mu_i}{\partial \eta_i}X_i, \end{aligned} \tag{2.19}$$

where

$$\frac{\partial \theta_i}{\partial \beta} = \frac{\partial \theta_i}{\partial \mu_i}\frac{\partial \mu_i}{\partial \eta_i}\frac{\partial \eta_i}{\partial \beta} = V_i^{-1}\frac{\partial \mu_i}{\partial \eta_i}X_i.$$

The first step follows from the chain rule and the second step follows from the differentiation rule of inverse functions, $\frac{\partial \theta_i}{\partial \mu_i} = (\frac{\partial \mu_i}{\partial \theta_i})^{-1} = (a''(\theta_i))^{-1} = V_i^{-1}$, with $\mu_i = a'(\theta_i)$. The Hessian of $K_n(\theta)$ can also be calculated:

$$\begin{aligned}&\frac{\partial^2 K_n}{\partial \beta \partial \beta^{\mathrm{T}}}(\beta) \\ &= \frac{-1}{n\phi^*}\sum_{i=1}^{n}\left\{V_i^{-1}\frac{\partial \mu_i}{\partial \eta_i}X_i\frac{\partial}{\partial \beta^{\mathrm{T}}}(Y_i-\mu_i) + (Y_i-\mu_i)\frac{\partial}{\partial \beta^{\mathrm{T}}}\left(V_i^{-1}\frac{\partial \mu_i}{\partial \eta_i}X_i\right)\right\} \\ &= \underbrace{\frac{1}{n\phi^*}\sum_{i=1}^{n}V_i^{-1}\left(\frac{\partial \mu_i}{\partial \eta_i}\right)^2 X_i X_i^{\mathrm{T}}}_{\mathbf{I}} + \underbrace{\frac{-1}{n\phi^*}\sum_{i=1}^{n}(Y_i-\mu_i)\frac{\partial^2 \theta_i}{\partial \beta \partial \beta^{\mathrm{T}}}}_{\mathbf{II}}. \end{aligned} \tag{2.20}$$

The Hessian $\frac{\partial^2 K_n}{\partial \beta \partial \beta^{\mathrm{T}}}(\beta)$ is the sum of two terms. The first term **I** is positive definite unless in degenerate cases, whereas the second term **II** may be neither positive definite nor negative definite. Hence the Hessian may fail to be positive definite, which presents a source of instability for the Newton–Raphson algorithm. Without the positive-definiteness of the Hessian, even the guarantee provided in Lemma 2.2 for backtracking is no longer valid.

In relation with the foregoing issue, we discuss two distinct scenarios. In the first scenario, suppose that a canonical link is used, $g(\cdot) = (a')^{-}(\cdot)$. Then for each i,

$$\theta_i = (a')^{-}(g^{-}(\eta_i)) = \eta_i,$$

and hence $\frac{\partial \theta_i}{\partial \beta} = \frac{\partial \eta_i}{\partial \beta} = X_i$, independent of β, and $\frac{\partial^2 \theta_i}{\partial \beta \partial \beta^{\mathrm{T}}} = 0$. As a result, several interesting properties are obtained. First, the average negative log-likelihood function in (2.18) becomes

$$K_n(\beta) = \frac{-1}{n}\sum_{i=1}^{n}\frac{Y_i X_i^{\mathrm{T}}\beta - a(X_i^{\mathrm{T}}\beta)}{\phi^*} + \text{constant},$$

and hence is convex in β by the convexity of $a(\cdot)$ discussed after Lemma 2.1. Second, the gradient becomes

$$\frac{\partial K_n}{\partial \beta}(\beta) = \frac{-1}{n\phi^*}\sum_{i=1}^{n}(Y_i - \mu_i)X_i,$$

which mimics the gradient of the least-squares objective, $\frac{\partial}{\partial\beta}\sum_{i=1}^n (Y_i - X_i^{\mathrm{T}}\beta)^2/2 = -\sum_{i=1}^n (Y_i - X_i^{\mathrm{T}}\beta)X_i$, even though μ_i is nonlinear in β. Third, the second term **II** of the Hessian vanishes, **II** $= 0$, and the Hessian reduces to the first term **I** only:

$$\begin{aligned}
\frac{\partial^2 K_n}{\partial\beta\partial\beta^{\mathrm{T}}}(\beta) &= \frac{1}{n\phi^*}\sum_{i=1}^{n} V_i^{-1}\left(\frac{\partial\mu_i}{\partial\eta_i}\right)^2 X_i X_i^{\mathrm{T}} \\
&= \frac{1}{n\phi^*}\sum_{i=1}^{n}\frac{\partial\mu_i}{\partial\eta_i}X_i X_i^{\mathrm{T}}, \qquad (2.21)
\end{aligned}$$

which is always nonnegative definite. [The second equality follows because $\frac{\partial\mu_i}{\partial\eta_i} = \frac{\partial\mu_i}{\partial\theta_i} = a''(\theta_i) = V_i$ with the canonical link.] This reaffirms that the average negative log-likelihood function, $K_n(\beta)$, is convex in β. The Newton–Raphson update (2.17) reduces to

$$\beta^k = \beta^{k-1} + \left\{\sum_{i=1}^{n}\frac{\partial\mu_i}{\partial\eta_i}X_i X_i^{\mathrm{T}}\right\}^{-1}\sum_{i=1}^{n}(Y_i - \mu_i)X_i\Big|_{\beta=\beta^{k-1}}. \qquad (2.22)$$

Due to the convexity of $K_n(\beta)$, the Newton–Raphson algorithm is expected to be well-behaved, and the backtracking line search can be justified with the guarantee in Lemma 2.2.

In the second scenario, suppose that a non-canonical link is used. Then as discussed above, the Hessian $\frac{\partial^2 K_n}{\partial\beta\partial\beta^{\mathrm{T}}}(\beta)$ may not be positive definite, in the presence of the second term **II**. In this case, the Newton–Raphson algorithm can be modified by exploiting the following observation. The model expectation of the second term **II** in the Hessian vanishes, $E_\beta(\mathbf{II}|X_1,\ldots,X_n) = 0$, and the expected Hessian reduces to the first term **I** only and hence is always nonnegative definite:

$$E_\beta\left\{\frac{\partial^2 K_n}{\partial\beta\partial\beta^{\mathrm{T}}}(\beta)\Big| X_1,\ldots,X_n\right\} = \mathbf{I},$$

where $E_\beta(\cdot|X_1,\ldots,X_n)$ denotes the expectation with respect to the conditional distribution of $(Y_1,\ldots,Y_n)$ given $(X_1,\ldots,X_n)$ under model assumption (2.1) with parameter β^* replaced by β. By replacing the Hessian $\frac{\partial^2 K_n}{\partial\beta\partial\beta^{\mathrm{T}}}(\beta)$ with its model expectation in the Newton–Raphson update (2.17), Fisher scoring yields the following update:

$$\beta^k = \beta^{k-1} + \left\{\sum_{i=1}^{n} V_i^{-1}\left(\frac{\partial\mu_i}{\partial\eta_i}\right)^2 X_i X_i^{\mathrm{T}}\right\}^{-1}\sum_{i=1}^{n}(Y_i - \mu_i)V_i^{-1}\frac{\partial\mu_i}{\partial\eta_i}X_i\Big|_{\beta=\beta^{k-1}}. \qquad (2.23)$$

The Fisher-scoring algorithm reduces to the Newton–Raphson algorithm (2.22) if a canonical link is used, but provides a desirable modification if a non-canonical link is used, for example, probit regression. The update direction, $\beta^k - \beta^{k-1}$, in (2.23) can be proved to be a descent direction for minimizing $K_n(\beta)$ similarly as in Lemma 2.2. Hence the backtracking line search can be safely incorporated to improve stability of the algorithm.

Another advantage of Fisher scoring is that the update (2.23) can be implemented by performing iteratively reweighted least squares. For each i, define the weight and working response

$$W_i = V_i^{-1}\left(\frac{\partial \mu_i}{\partial \eta_i}\right)^2, \quad \tilde{Y}_i = X_i^{\mathrm{T}}\beta + (Y_i - \mu_i)\frac{\partial \eta_i}{\partial \mu_i},$$

depending on β, although the dependency is suppressed in the notation. The Fisher-scoring update (2.23) can be rewritten as

$$\begin{aligned}\beta^k &= \beta^{k-1} + \left\{\sum_{i=1}^n W_i X_i X_i^{\mathrm{T}}\right\}^{-1} \sum_{i=1}^n (Y_i - \mu_i) W_i \frac{\partial \eta_i}{\partial \mu_i} X_i \bigg|_{\beta=\beta^{k-1}} \\ &= \left\{\sum_{i=1}^n W_i X_i X_i^{\mathrm{T}}\right\}^{-1} \sum_{i=1}^n W_i X_i \tilde{Y}_i \bigg|_{\beta=\beta^{k-1}}, \end{aligned} \tag{2.24}$$

which is a weighted least-squares estimator with covariate vector X_i, weight W_i, and working response $\tilde{Y}_i$ for $i = 1, \ldots, n$. Both W_i and $\tilde{Y}_i$ need to be evaluated at the current iterate β^{k-1} in each iteration. The underlying quadratic approximation to $K_n(\beta)$ can be rewritten as a weighted least-squares objective:

$$K_{n,Q}(\beta; \beta^{k-1}) = \frac{1}{2n\phi^*} \sum_{i=1}^n W_i(\beta^{k-1}) \left\{\tilde{Y}_i(\beta^{k-1}) - X_i^{\mathrm{T}}\beta\right\}^2 + \text{constant},$$

where the factor $\frac{1}{\phi^*}$ is needed to match the definition of $K_n(\beta)$ in (2.18).

By Lemma 2.1, it can be shown that

$$W_i^{-1} = V_i \left(\frac{\partial \eta_i}{\partial \mu_i}\right)^2 = \frac{1}{\phi^*}\mathrm{var}_\beta\left(X_i^{\mathrm{T}}\beta + (Y_i - \mu_i)\frac{\partial \eta_i}{\partial \mu_i}\Big| X_i\right),$$

where $\mathrm{var}_\beta(\cdot|X_i)$ denotes the expectation under model assumption (2.1) with parameter β^* replaced by β. Hence W_i is an inverse variance weight, being inversely proportional to the model variance of the working response $\tilde{Y}_i$ given X_i. If a canonical link is used, then the weight and working response can be simplified to

$$W_i = \frac{\partial \mu_i}{\partial \eta_i} = V_i, \quad \tilde{Y}_i = X_i^{\mathrm{T}}\beta + (Y_i - \mu_i)\frac{\partial \eta_i}{\partial \mu_i} = X_i^{\mathrm{T}}\beta + V_i^{-1}(Y_i - \mu_i), \tag{2.25}$$

because in the case, $\frac{\partial \mu_i}{\partial \eta_i} = V_i$ as mentioned earlier. The weight $W_i = V_i$ is inversely proportional to the conditional variance of $V_i^{-1}(Y_i - \mu_i)$ given X_i. Typically, inverse variance weighting is associated with efficient estimation (i.e., asymptotic minimum-variance estimation), which is expected for the ML estimator $\hat{\beta}$ under correct model specification. See Exercise 2.1 for an example where inverse variance weighting leads to minimum-variance estimation.

Example 2.4. For logistic regression (2.7), a canonical link, $g(p) = \log \frac{p}{1-p}$ in (2.9), is used. The average negative log-likelihood function is

$$K_n(\beta) = \frac{-1}{n}\ell_n(\beta) = \frac{-1}{n}\sum_{i=1}^{n}\left\{Y_i X_i^{\mathrm{T}}\beta - \log(1 + \mathrm{e}^{X_i^{\mathrm{T}}\beta})\right\}.$$

The quadratic approximation from the Taylor expansion of $K_n(\beta)$ at β^{k-1} is

$$\begin{aligned} K_{n,Q}(\beta;\beta^{k-1}) = K_n(\beta^{k-1}) &- \frac{1}{n}\sum_{i=1}^{n}(Y_i - \mu_i)\Big|_{\beta=\beta^{k-1}} X_i^{\mathrm{T}}(\beta - \beta^{k-1}) \\ &+ \frac{1}{2}(\beta - \beta^{k-1})^{\mathrm{T}}\frac{1}{n}\sum_{i=1}^{n}\mu_i(1-\mu_i)\Big|_{\beta=\beta^{k-1}} X_i X_i^{\mathrm{T}}(\beta - \beta^{k-1}), \end{aligned}$$

where $\mu_i = g^{-}(X_i^{\mathrm{T}}\beta) = \mathrm{e}^{X_i^{\mathrm{T}}\beta}/(1+\mathrm{e}^{X_i^{\mathrm{T}}\beta})$. The Newton–Raphson update (2.22) becomes

$$\begin{aligned} \beta^k &= \mathrm{argmin}_\beta\, K_{n,Q}(\beta;\beta^{k-1}) \\ &= \beta^{k-1} + \left\{\sum_{i=1}^{n}\mu_i(1-\mu_i)X_i X_i^{\mathrm{T}}\right\}^{-1}\sum_{i=1}^{n}(Y_i - \mu_i)X_i\Big|_{\beta=\beta^{k-1}}. \end{aligned}$$

The quadratic approximation can be rewritten as

$$K_{n,Q}(\beta;\beta^{k-1}) = \frac{1}{2n}\sum_{i=1}^{n}W_i(\beta^{k-1})\left\{\tilde{Y}_i(\beta^{k-1}) - X_i^{\mathrm{T}}\beta\right\}^2 + \text{constant},$$

and the Newton–Raphson update can be rewritten as the weighted least-squares estimator (2.24),

$$\beta^k = \left\{\sum_{i=1}^{n}W_i(\beta^{k-1})X_i X_i^{\mathrm{T}}\right\}^{-1}\sum_{i=1}^{n}W_i(\beta^{k-1})X_i\tilde{Y}_i(\beta^{k-1}),$$

where the weight and working response are

$$W_i(\beta) = \mu_i(1-\mu_i), \quad \tilde{Y}_i(\beta) = X_i^{\mathrm{T}}\beta + \frac{Y_i - \mu_i}{\mu_i(1-\mu_i)},$$

with $\frac{\partial \mu_i}{\partial \eta_i} = V_i = \mu_i(1-\mu_i)$ for logistic regression.

The weighted least-squares estimator (2.24) needs to be computed iteratively. A major computational cost is to repeatedly invert the matrix $\sum_{i=1}^n W_i(\beta^{k-1})X_iX_i^{\mathrm{T}}$, with the weights W_i depending on the current iterate β^{k-1}. For logistic regression, to reduce computational cost per iteration, there is a useful application of the MM algorithm discussed in Section 1.8. In fact, a simple upper bound of the Hessian of the objective function $K_n(\beta)$ is

$$\frac{\partial^2 K_n}{\partial\beta\partial\beta^{\mathrm{T}}}(\beta) = \frac{1}{n}\sum_{i=1}^n \mu_i(1-\mu_i)X_iX_i^{\mathrm{T}} \leq \frac{1}{n}\sum_{i=1}^n \frac{1}{4}X_iX_i^{\mathrm{T}}.$$

Hence a majorizing, quadratic function for $K_n(\beta)$ at β^{k-1} is

$$\begin{aligned} K_{n,\mathrm{M1}}(\beta;\beta^{k-1}) &= K_n(\beta^{k-1}) - \frac{1}{n}\sum_{i=1}^n (Y_i-\mu_i)\Big|_{\beta=\beta^{k-1}} X_i^{\mathrm{T}}(\beta-\beta^{k-1}) \\ &\quad + \frac{1}{2}(\beta-\beta^{k-1})^{\mathrm{T}}\frac{1}{n}\sum_{i=1}^n \frac{1}{4}X_iX_i^{\mathrm{T}}(\beta-\beta^{k-1}) \qquad (2.26) \\ &= \frac{1}{2n}\sum_{i=1}^n \frac{1}{4}\left\{\tilde{Y}_i(\beta^{k-1}) - X_i^{\mathrm{T}}\beta\right\}^2 + \text{constant}, \end{aligned}$$

where the working response is

$$\tilde{Y}_i(\beta) = X_i^{\mathrm{T}}\beta + \frac{Y_i-\mu_i}{1/4}.$$

The resulting MM update is a (unweighted) least-squares estimator

$$\begin{aligned} \beta^k &= \mathrm{argmin}_\beta\, K_{n,\mathrm{M1}}(\beta;\beta^{k-1}) \\ &= \left(\sum_{i=1}^n X_iX_i^{\mathrm{T}}\right)^{-1}\sum_{i=1}^n X_i\tilde{Y}_i(\beta^{k-1}). \end{aligned}$$

The matrix $\sum_{i=1}^n X_iX_i^{\mathrm{T}}$ only needs to be inverted once (at the beginning of the iterations). Compared with the Newton–Raphson algorithm, the modified algorithm is not only computationally less costly per iteration but also more stable due to the monotonicity of the MM algorithm, even though it may converge at a slower theoretical rate.

2.3 Model-based and model-robust inferences

2.3.1 Asymptotic theory of M-estimation

Asymptotic theory of M-estimation provides a general, useful collection of asymptotic properties for M-estimators, which are defined by minimizing (or maximizing) certain sample objective functions. We provide a high-level account in the classical setting where the number of parameters is fixed and the sample size tends to infinity. See Tan (2025), Chapter 2, for associated technical details.

Suppose that $Z_1, \ldots, Z_n$ are IID variables as Z. An M-estimator $\hat\beta$ is defined in the form $\hat\beta = \operatorname{argmin}_\beta K_n(\beta)$ with a sample objective function

$$K_n(\beta) = \frac{1}{n}\sum_{i=1}^n \rho(Z_i;\beta),$$

where β is a parameter vector of dimension p and $\rho(\cdot;\beta)$ is a scalar, objective function for an individual datum. The target value (or the limit value as justified later) $\bar\beta$ for $\hat\beta$ is defined as $\bar\beta = \operatorname{argmin}_\beta K(\beta)$ with the population objective function

$$K(\beta) = E\{\rho(Z;\beta)\}.$$

Given that $\rho(z;\beta)$ is differentiable in β for any fixed z (among other regularity conditions), $\hat\beta$ is a solution to

$$0 = \frac{1}{n}\sum_{i=1}^n \frac{\partial\rho}{\partial\beta}(Z_i;\beta), \tag{2.27}$$

and $\bar\beta$ is a solution to

$$0 = E\left\{\frac{\partial\rho}{\partial\beta}(Z;\beta)\right\}. \tag{2.28}$$

Hence $\hat\beta$ is also a Z-estimator, defined as a solution to a system of estimating equations (i.e., a system of equations used for estimation purposes). For identifiability, we assume throughout that $\bar\beta$ is a unique minimizer of $K_n(\beta)$ and a unique solution to equation (2.28). Asymptotic theory of Z-estimation may appear to subsume that of M-estimation described below, although regularity conditions required for theory of M-estimation is usually weaker than that of Z-estimation. See Section 2.7.1 for a description of asymptotic theory of Z-estimation.

The estimator $\hat\beta$ is often motivated by certain modeling assumptions, but its asymptotic properties can be well characterized, independently of those modeling assumptions, under suitable regularity conditions in the setting of $n \to \infty$ and fixed p. Various sets of regularity conditions can be formulated about smoothness and integrability of functions related to $\rho(Z;\beta)$ and others, but the details are not discussed here. Assume that $\rho(z;\beta)$ is twice differentiable in β for any fixed z. Let

$$\Sigma(\beta) = H^{-1}(\beta)G(\beta)H^{-1}(\beta),$$
$$H(\beta) = E\left\{\frac{\partial^2\rho}{\partial\beta\partial\beta^{\mathrm{T}}}(Z;\beta)\right\}, \quad G(\beta) = E\left\{\left(\frac{\partial\rho}{\partial\beta}(Z;\beta)\right)^{\otimes 2}\right\},$$

and similarly

$$\hat{\Sigma}(\beta) = \hat{H}^{-1}(\beta)\hat{G}(\beta)\hat{H}^{-1}(\beta),$$
$$\hat{H}(\beta) = \tilde{E}\left\{\frac{\partial^2 \rho}{\partial\beta\partial\beta^{\mathrm{T}}}(Z;\beta)\right\}, \quad \hat{G}(\beta) = \tilde{E}\left\{(\frac{\partial\rho}{\partial\beta}(Z;\beta))^{\otimes 2}\right\},$$

where $\tilde{E}(\cdot)$ denotes the sample average as before, and $u^{\otimes 2} = uu^{\mathrm{T}}$ for a column vector u. By the stationary conditions (2.27)–(2.28), $G(\bar{\beta})$ and $\hat{G}(\hat{\beta})$ are equivalent to $G(\bar{\beta}) = \mathrm{var}\{\frac{\partial\rho}{\partial\beta}(Z;\bar{\beta})\}$ and $\hat{G}(\hat{\beta}) = \widetilde{\mathrm{var}}\{\frac{\partial\rho}{\partial\beta}(Z;\hat{\beta})\}$, where $\widetilde{\mathrm{var}}(\cdot)$ denotes the sample variance, for simplicity, without the $1/(n-1)$ adjustment.

Proposition 2.1 (Asymptotic theory of M-estimation). *In the preceding setting, the following results hold under suitable regularity conditions. As $n \to \infty$ and p is fixed, $\hat{\beta} \underset{\mathcal{P}}{\to} \bar{\beta}$ and*

$$\sqrt{n}(\hat{\beta} - \bar{\beta}) \underset{\mathcal{D}}{\to} \mathrm{N}(0, \Sigma(\bar{\beta})),$$
$$\hat{\Sigma}(\hat{\beta}) \underset{\mathcal{P}}{\to} \Sigma(\bar{\beta}).$$

Either $n^{-1}\hat{\Sigma}(\hat{\beta})$ or $\hat{\Sigma}(\hat{\beta})$ is called a sandwich variance estimator. From Proposition 2.1, valid Wald confidence intervals can be obtained for individual elements of $\bar{\beta}$, similarly as discussed in Section 1.4. In fact, Proposition 1.6 about the LS estimator under possible misspecification of linear regression (1.3b) in a random design can be deduced from Proposition 2.1 by taking $Z = (X^{\mathrm{T}}, Y)^{\mathrm{T}}$ and

$$\rho(Z;\beta) = \frac{1}{2}(Y - X^{\mathrm{T}}\beta)^2.$$

The matrices $H(\beta)$ and $\hat{H}(\beta)$ become constant in β, H and $\hat{H}$, because $\rho(z;\beta)$ is quadratic in β. If linear regression (1.3b) is correctly specified, then the rescaled asymptotic variance of $\hat{\beta}$, $H^{-1}G(\bar{\beta})H^{-1}$, in Proposition 1.6 remains valid with $\bar{\beta}$ identical to β^* under general heteroscedastic noises, and reduces to $\sigma^{*2}H^{-1}$ under homoscedastic noises with constant variance σ^{*2}. Similarly, Proposition 2.1 can also be applied to the ML estimator $\hat{\beta}$ in generalized linear regression, even though $\hat{\beta}$ does not admit a closed form.

Bibliographical note. Asymptotic theory of M-estimation (as well as Z-estimation) has been discussed in various books, for example, Manski (1988) and van der Vaart (1998). As mentioned in Section 1.4, early motivations for the development of M-estimation include robust estimation (Huber, 1967) and ML estimation with misspecified models (White, 1982).

2.3.2 Model-based evaluation and inference

The ML estimator $\hat{\beta}$ is an M-estimator, defined as a maximizer of the log-likelihood function $\ell_n(\beta)$ in generalized linear regression. The asymptotic variance of $\hat{\beta}$ can be derived from Proposition 2.1, whether or not generalized linear regression is correctly specified. This leads to model-robust evaluation and inference discussed later. However, under correct model specification, the asymptotic variance of $\hat{\beta}$ can be simplified to recover standard results in likelihood theory.

There are two modeling assumptions involved in generalized linear regression. Given that β^* is the parameter of interest, the primary assumption is the mean restriction (2.1), and the secondary, distributional assumption is the exponential-family assumption (2.2) with constant dispersion parameter ϕ^*. We study model-based evaluation in mainly two distinct cases. In the first, narrower case, we assume that both mean restriction (2.1) and the exponential-family assumption (2.2) with constant ϕ^* are valid (see Proposition 2.2). In a variation of the first case, the exponential-family assumption (2.2) is relaxed to the variance restriction (2.31) below. In the second, broader case, we assume that the mean restriction (2.1) is valid, while the variance restriction (2.31) may be misspecified (see Proposition 2.3).

We introduce the following notation:

$$\eta(X;\beta) = X^{\mathrm{T}}\beta, \quad \mu(X;\beta) = g^{-}(\eta(X;\beta)), \quad \theta(X;\beta) = (a')^{-}(\mu(X;\beta)). \tag{2.29}$$

Then the mean restriction (2.1) states

$$E(Y|X) = \mu(X;\beta^*). \tag{2.30}$$

By Lemma 2.1, the exponential-family assumption (2.2) implies

$$\mathrm{var}(Y|X) = \phi^* V(E(Y|X)), \tag{2.31}$$

where $V(\mu) = a''((a')^{-}(\mu))$, called the variance function. The estimator $\hat{\beta}$ is defined as $\hat{\beta} = \mathrm{argmin}_\beta K_n(\beta)$, where $K_n(\beta) = -n^{-1}\ell_n(\beta) = \tilde{E}\{\rho(X,Y;\beta)\}$ with

$$\rho(X,Y;\beta) = \frac{-1}{\phi^*}\{Y\theta(X;\beta) - a(\theta(X;\beta))\}.$$

[The factor $\frac{1}{\phi^*}$ can be removed from the definition of ρ, without affecting the convergence properties in Proposition 2.2. But then the second Bartlett identity below needs to be modified as $G(\beta^*) = \phi^* H(\beta^*)$.] By the previous expressions (2.19) and (2.20), the gradient and Hessian of $\rho(X,Y;\beta)$ in β are

$$\frac{\partial\rho}{\partial\beta} = \frac{-1}{\phi^*}(Y-\mu)V^{-1}(\mu)\frac{\partial\mu}{\partial\eta}X, \tag{2.32a}$$

$$\frac{\partial^2\rho}{\partial\beta\partial\beta^{\mathrm{T}}} = \frac{1}{\phi^*}V^{-1}(\mu)\left(\frac{\partial\mu}{\partial\eta}\right)^2 XX^{\mathrm{T}} + \frac{-1}{\phi^*}(Y-\mu)\frac{\partial^2\theta}{\partial\beta\partial\beta^{\mathrm{T}}}, \tag{2.32b}$$

where the abbreviations are used, $\eta = \eta(X;\beta)$, $\mu = \mu(X;\beta)$, $\theta = \theta(X;\beta)$, and $V(\mu) = V(\mu(X;\beta))$. By using both (2.30) and (2.31), it can be calculated that

$$0 = E\left\{\frac{\partial \rho}{\partial \beta}(X,Y;\beta^*)\right\}, \tag{2.33a}$$

$$H(\beta^*) = E\left\{\frac{\partial^2 \rho}{\partial \beta \partial \beta^{\mathrm{T}}}(X,Y;\beta^*)\right\} = \frac{1}{\phi^*} E\left\{V^{-1}(\mu)\left(\frac{\partial \mu}{\partial \eta}\right)^2 XX^{\mathrm{T}}\right\}\Big|_{\beta=\beta^*}, \tag{2.33b}$$

$$G(\beta^*) = \mathrm{var}\left\{\frac{\partial \rho}{\partial \beta}(X,Y;\beta^*)\right\} = \frac{1}{\phi^{*2}} E\left\{(Y-\mu)^2 V^{-2}(\mu)\left(\frac{\partial \mu}{\partial \eta}\right)^2 XX^{\mathrm{T}}\right\}\Big|_{\beta=\beta^*} \tag{2.33c}$$

$$= \frac{1}{\phi^*} E\left\{V^{-1}(\mu)\left(\frac{\partial \mu}{\partial \eta}\right)^2 XX^{\mathrm{T}}\right\}\Big|_{\beta=\beta^*}. \tag{2.33d}$$

Equation (2.33a), called the first Bartlett identity, indicates that β^* satisfies the stationary condition (2.28), and hence $\bar{\beta} = \beta^*$ by the uniqueness of $\bar{\beta}$ satisfying (2.28). Equations (2.33b) and (2.33d) imply the second Bartlett identity, $H(\beta^*) = G(\beta^*)$, i.e.,

$$E\left\{\frac{\partial^2 \rho}{\partial \beta \partial \beta^{\mathrm{T}}}(X,Y;\beta^*)\right\} = \mathrm{var}\left\{\frac{\partial \rho}{\partial \beta}(X,Y;\beta^*)\right\}.$$

Hence the asymptotic variance $\Sigma(\beta^*)$ reduces to $\Sigma(\beta^*) = H^{-1}(\beta^*)G(\beta^*)H^{-1}(\beta^*) = I^{-1}(\beta^*)$, which can be consistently estimated by the sample version $\hat{I}^{-1}(\hat{\beta})$ in addition to $\hat{\Sigma}(\hat{\beta})$ under suitable regularity conditions, where

$$I(\beta) = \frac{1}{\phi^*} E\left\{V^{-1}(\mu)\left(\frac{\partial \mu}{\partial \eta}\right)^2 XX^{\mathrm{T}}\right\},$$

$$\hat{I}(\beta) = \frac{1}{\phi^*} \tilde{E}\left\{V^{-1}(\mu)\left(\frac{\partial \mu}{\partial \eta}\right)^2 XX^{\mathrm{T}}\right\}.$$

Note that $I(\beta^*) = H(\beta^*) = G(\beta^*)$ is called the (expected) Fisher information, and $\Sigma(\beta^*) = I^{-1}(\beta^*)$ is the inverse Fisher information. To summarize, the preceding discussion leads to the following asymptotic properties for $\hat{\beta}$ and $\hat{I}(\hat{\beta})$.

Proposition 2.2 (Generalized linear regression with correct mean restriction and exponential-family assumption). *Suppose that generalized linear regression is correctly specified, including the mean restriction (2.1) and the exponential-family assumption (2.2) with constant dispersion parameter ϕ^*. Then the following results hold under suitable regularity conditions. As $n \to \infty$ and p is fixed, $\hat{\beta} \underset{\mathcal{P}}{\to} \beta^*$ and*

$$\sqrt{n}(\hat{\beta} - \beta^*) \underset{\mathcal{D}}{\to} \mathrm{N}(0, I^{-1}(\beta^*)), \tag{2.34a}$$

$$\hat{I}(\hat{\beta}) \underset{\mathcal{P}}{\to} I(\beta^*). \tag{2.34b}$$

We discuss several extensions of the preceding theory. First, Proposition 2.2 can be extended if the mean restriction (2.1) is retained but the exponential-family assumption (2.2) is relaxed to the variance restriction (2.31), where $V(\cdot)$ is a known function which can be directly specified regardless of $a(\cdot)$. In this case, the estimator $\hat{\beta}$ is redefined as a solution to

$$0 = \tilde{E}\left\{\frac{-1}{\phi^*}(Y-\mu)V^{-1}(\mu)\frac{\partial\mu}{\partial\eta}X\right\}, \tag{2.35}$$

where $\eta = \eta(X;\beta)$ and $\mu = \mu(X;\beta)$ are defined as in (2.29) and $V(\mu) = V(\mu(X;\beta))$. Equivalently, $\hat{\beta}$ can be defined as $\hat{\beta} = \text{argmin}_\beta K_n(\beta)$, where $K_n(\beta) = \tilde{E}\{\rho(X,Y;\beta)\}$ with

$$\rho(X,Y;\beta) = \frac{-1}{\phi^*}\int_Y^\mu V^{-1}(t)(Y-t)\,\mathrm{d}t, \tag{2.36}$$

provided that the integral exists. [Question: What happens if the integral does not exist?] The function $-\rho(X,Y;\beta)$ is known as the log quasi-likelihood function of β for an individual datum (X,Y). Hence $\hat{\beta}$ is called the maximum quasi-likelihood (MQL) estimator. The gradient of ρ remains the same as (2.32a), but the Hessian of ρ becomes

$$\frac{\partial^2\rho}{\partial\beta\partial\beta^{\mathrm{T}}} = \underbrace{\frac{1}{\phi^*}V^{-1}(\mu)\left(\frac{\partial\mu}{\partial\eta}\right)^2 XX^{\mathrm{T}}}_{\mathbf{I}} + \underbrace{\frac{-1}{\phi^*}(Y-\mu)\frac{\partial}{\partial\eta}\left(V^{-1}(\mu)\frac{\partial\mu}{\partial\eta}\right)XX^{\mathrm{T}}}_{\mathbf{II}}. \tag{2.37}$$

With these modifications, the discussion leading to Proposition 2.2 remains valid and hence the asymptotic properties (2.34a)–(2.34b) hold for $\hat{\beta}$ and $\hat{I}(\hat{\beta})$. In addition, the Newton–Raphson and Fisher scoring algorithms can be similarly applied as in Section 2.2. Given the exponential-family assumption (2.2) is relaxed, the condition (2.12) for a canonical link $g(\cdot)$ can be extended to be $\frac{\partial\mu}{\partial\eta} = V(\mu)$ or equivalently

$$g'(\cdot) = V^{-1}(\cdot). \tag{2.38}$$

If a canonical link is used, then the gradient of ρ reduces to $\frac{\partial\rho}{\partial\beta} = \frac{-1}{\phi^*}(Y-\mu)X$, and the Hessian of ρ reduces to the first term **I** only, $\frac{\partial^2\rho}{\partial\beta\partial\beta^{\mathrm{T}}} = \frac{1}{\phi^*}\frac{\partial\mu}{\partial\eta}XX^{\mathrm{T}}$, so that $K_n(\beta)$ is convex and the Newton–Raphson algorithm is expected to be well-behaved.

Second, as mentioned in Section 2.2, although the estimator $\hat{\beta}$ is defined, independently of ϕ^*, the asymptotic variance of $\hat{\beta}$ does depend on ϕ^*. If the exponential-family assumption (2.2) is relaxed to variance restriction (2.31), the moment estimator $\hat{\phi}$ in (2.14) can be defined with $\hat{V}_i = V(\hat{\mu}_i)$ directly based on $V(\cdot)$. Under the variance restriction (2.31), $\hat{\phi}$ can be shown to be consistent for ϕ^* (whether or not ϕ^* is known):

$$\hat{\phi} \underset{\mathcal{P}}{\rightarrow} \phi^*.$$

If ϕ^* is unknown, then $\hat{I}(\hat{\beta})$ with ϕ^* replaced by $\hat{\phi}$ is a consistent estimator for $I(\beta^*)$.

Third, the reduction of the asymptotic variance $\Sigma(\beta^*)$ to $I^{-1}(\beta^*)$ in Proposition 2.2 relies on the variance restriction (2.31). Without this assumption, the simplification of $G(\beta^*)$ from (2.33c) to (2.33d) does not in general hold. Nevertheless, Proposition 2.1 remains applicable to $\hat{\beta}$, with the asymptotic variance $\Sigma(\beta^*) = H^{-1}(\beta^*)G(\beta^*)H^{-1}(\beta^*)$ based on $G(\beta^*)$ in (2.33c), even if the variance restriction (2.31) is relaxed, i.e., the conditional variance $\text{var}(Y|X)$ is left to be unrestricted. This leads to the following result, which extends Proposition 2.2 in a similar manner as Proposition 1.4 accommodates linear regression with heteroscedastic noises (under a correctly specified mean restriction). As the dispersion parameter ϕ^* is dropped, the estimator $\hat{\phi}$ is no longer needed in variance estimation by $\hat{\Sigma}(\hat{\beta})$.

Proposition 2.3 (Generalized linear regression with correct mean restriction and misspecified variance restriction). *Suppose that the mean restriction (2.1) is correctly specified, but the variance restriction (2.31) may be misspecified. Then the following results hold under suitable regularity conditions. As $n \rightarrow \infty$ and p is fixed, $\hat{\beta} \underset{\mathcal{P}}{\rightarrow} \beta^*$ and*

$$\sqrt{n}(\hat{\beta} - \beta^*) \underset{\mathcal{D}}{\rightarrow} \mathrm{N}(0, \Sigma(\beta^*)), \tag{2.39a}$$

$$\hat{\Sigma}(\hat{\beta}) \underset{\mathcal{P}}{\rightarrow} \Sigma(\beta^*), \tag{2.39b}$$

where $\Sigma(\beta^*) = H^{-1}(\beta^*)G(\beta^*)H^{-1}(\beta^*)$ with $H(\beta^*)$ and $G(\beta^*)$ in (2.33b) and (2.33c) but with the factors $\frac{1}{\phi^*}$ and $\frac{1}{\phi^{*2}}$ removed, $\hat{\Sigma}(\hat{\beta}) = \hat{H}^{-1}(\hat{\beta})\hat{G}(\beta)\hat{H}^{-1}(\beta)$, and $\hat{H}(\hat{\beta})$ and $\hat{G}(\hat{\beta})$ are defined as the sample versions of $H(\beta^*)$ and $G(\beta^*)$, with $E(\cdot)$ replaced by $\tilde{E}(\cdot)$ and β^* replaced by $\hat{\beta}$.

2.3.3 Model-robust evaluation and inference

As mentioned earlier, Proposition 2.1 can be applied to $\hat{\beta}$ even if the primary modeling assumption, the mean restriction (2.1), may be misspecified in generalized linear regression. In addition to the mean restriction, the secondary, variance restriction (2.31) may also be misspecified. The variance function $V(\cdot)$ can be directly introduced, independently of the exponential-family assumption (2.2).

Let $\hat{\beta}$ be the MQL estimator, defined as a solution to estimating equation (2.35) or as a minimizer of the objective function $K_n(\beta) = \tilde{E}\{\rho(X, Y; \beta)\}$ with ρ in (2.36). The target value $\bar{\beta}$ is defined as a solution, assumed to be unique, to the population estimating equation,

$$0 = E\left\{(Y-\mu)V^{-1}(\mu)\frac{\partial\mu}{\partial\eta}X\right\}, \tag{2.40}$$

or equivalently as a unique minimizer of $K(\beta) = E\{\rho(X,Y;\beta)\}$, where ρ is defined as in (2.36) but with the factor $\frac{1}{\phi^*}$ removed. Let

$$H(\beta) = E\left\{V^{-1}(\mu)\left(\frac{\partial\mu}{\partial\eta}\right)^2 XX^{\mathrm{T}} + (-1)(Y-\mu)\frac{\partial}{\partial\eta}\left(V^{-1}(\mu)\frac{\partial\mu}{\partial\eta}\right)XX^{\mathrm{T}}\right\}, \tag{2.41a}$$

$$G(\beta) = E\left\{(Y-\mu)^2V^{-2}(\mu)\left(\frac{\partial\mu}{\partial\eta}\right)^2 XX^{\mathrm{T}}\right\}. \tag{2.41b}$$

Application of Proposition 2.1 leads to the following result.

Proposition 2.4 (Generalized linear regression with misspecified mean or variance restriction)**.** *Suppose that the mean restriction (2.1) and the variance restriction (2.31) may be misspecified. Then the following results hold under suitable regularity conditions. As $n\to\infty$ and p is fixed, $\hat\beta \underset{\mathcal{P}}{\longrightarrow} \bar\beta$ and*

$$\sqrt{n}(\hat\beta-\bar\beta) \underset{\mathcal{D}}{\longrightarrow} \mathrm{N}(0,\Sigma(\bar\beta)), \tag{2.42a}$$

$$\hat\Sigma(\hat\beta) \underset{\mathcal{P}}{\longrightarrow} \Sigma(\bar\beta), \tag{2.42b}$$

where $\Sigma(\beta) = H^{-1}(\beta)G(\beta)H^{-1}(\beta)$ with $H(\beta)$ and $G(\beta)$ defined in (2.41a) and (2.41b), $\hat\Sigma(\beta) = \hat H^{-1}(\beta)\hat G(\beta)\hat H^{-1}(\beta)$, and $\hat H(\beta)$ and $\hat G(\beta)$ are defined as the sample versions of $H(\beta)$ and $G(\beta)$, with $E(\cdot)$ replaced by $\tilde E(\cdot)$.

The asymptotic properties (2.42a)–(2.42b) appear similar to (2.39a)–(2.39b) in Proposition 2.3, with the true value β^* replaced by the target value $\bar\beta$. However, there are subtle differences in whether the asymptotic variances can be simplified. The asymptotic variances $\Sigma(\bar\beta)$ and $\hat\Sigma(\hat\beta)$ in (2.42a)–(2.42b) cannot in general be put in the form of $\Sigma(\beta^*)$ and $\hat\Sigma(\hat\beta)$ in (2.39a)–(2.39b), which are simplified by invoking the mean restriction (2.1). [Proposition 2.3 remains valid if $\Sigma(\beta^*)$ and $\hat\Sigma(\hat\beta)$ are defined instead in the form of $\Sigma(\beta)$ and $\hat\Sigma(\beta)$ in Proposition 2.4.] Except for the removal of $\frac{1}{\phi^*}$ or $\frac{1}{\phi^{*2}}$ (due to the removal of $\frac{1}{\phi^*}$ from ρ), $G(\bar\beta)$ is defined as $G(\beta^*)$ in (2.33c) with β^* replaced by $\bar\beta$, but $H(\bar\beta)$ may not reduce to (2.33b) with β^* replaced by $\bar\beta$, because the second term **II** in the Hessian of ρ, (2.37), may not be eliminated with $E(Y|X) \neq \mu(X;\bar\beta)$. Nevertheless, in the case of a canonical link $g(\cdot)$ satisfying (2.38), the second term **II** in (2.37) vanishes due to $V^{-1}(\mu)\frac{\partial\mu}{\partial\eta} \equiv 1$. In this case, $H(\bar\beta)$ does reduce to (2.33b) with β^* replaced by $\bar\beta$ and the factor $\frac{1}{\phi^*}$ removed, and hence $\Sigma(\bar\beta)$ and $\hat\Sigma(\hat\beta)$ in (2.42a)–(2.42b) are of the same form as $\Sigma(\beta^*)$ and $\hat\Sigma(\hat\beta)$ in (2.39a)–(2.39b).

TABLE 2.1
Consistency of MQL estimator and variance estimators (after rescaled by n)

	Correct mean restriction (2.1)		Misspecified mean restriction (2.1)
	Correct variance restriction (2.31)	Misspecified variance restriction (2.31)	
MQL estimator $\hat{\beta}$	consistent for β^* in (2.1)	consistent for β^* in (2.1)	consistent for $\bar{\beta}$ solving (2.40)
Variance estimator $\hat{I}^{-1}(\hat{\beta})$ in (2.34b)	consistent for $I^{-1}(\beta^*)$ in (2.34a)	inconsistent for $\Sigma(\beta^*)$ in (2.39a)	inconsistent for $\Sigma(\bar{\beta})$ in (2.42a)
Variance estimator $\hat{\Sigma}(\hat{\beta})$ in (2.39b) and (2.42b)	consistent for $I^{-1}(\beta^*)$ in (2.34a)	consistent for $\Sigma(\beta^*)$ in (2.39a)	consistent for $\Sigma(\bar{\beta})$ in (2.42a)

Note: Assume that a canonical link $g(\cdot)$ satisfying (2.38) is used, so that the variance estimators $\hat{\Sigma}(\hat{\beta})$ in (2.39b) and (2.42b) are the same.

Table 2.1 summarizes consistency (or lack of) for the MQL estimator $\hat{\beta}$ and the variance estimators $\hat{I}^{-1}(\hat{\beta})$ and $\hat{\Sigma}(\hat{\beta})$ for $n\,\mathrm{var}(\hat{\beta})$, in the convenient case where a canonical link $g(\cdot)$ satisfying (2.38) is used. See Table 1.1 for a related summary in linear regression. The variance estimator $\hat{I}^{-1}(\hat{\beta})$, with ϕ^* estimated by $\hat{\phi}$ if needed, is consistent if both the mean restriction (2.1) and the variance restriction (2.31) are correctly specified, in a similar manner as $\hat{\sigma}^2\hat{H}^{-1}$ is consistent in linear regression with homoscedastic noises. The variance estimator $\hat{\Sigma}(\hat{\beta})$ remains consistent if the mean restriction (2.1) and the variance restriction (2.31) may be misspecified, in a similar manner as in linear regression with heteroscedastic noises or misspecified mean restriction. Therefore, $\hat{I}^{-1}(\hat{\beta})$ is a model-based variance estimator, and $\hat{\Sigma}(\hat{\beta})$ is a model-robust variance estimator. Wald confidence intervals using $\hat{I}^{-1}(\hat{\beta})$ provide model-based inference about β^*, whereas those using $\hat{\Sigma}(\hat{\beta})$ enable model-robust inference about $\bar{\beta}$, which coincides with β^* if the mean restriction is correctly specified.

Example 2.5. Consider logistic regression (2.7), where a canonical link is used. The exponential-family assumption (2.2) is automatically satisfied by the Bernoulli family with $\phi^* = 1$. Hence the variance restriction (2.31) is satisfied with $V(\mu) = \mu(1-\mu)$. The ML estimator $\hat{\beta}$ is an M-estimator, $\hat{\beta} = \mathrm{argmin}_{\beta} n^{-1}\sum_{i=1}^n \rho(X_i, Y_i; \beta)$, with

$$\rho(X, Y; \beta) = -\left\{ Y X^{\mathrm{T}}\beta - \log(1 + \mathrm{e}^{X^{\mathrm{T}}\beta}) \right\}.$$

The gradient and Hessian of $\rho(X, Y; \beta)$ in β are

$$\frac{\partial \rho}{\partial \beta}(X, Y; \beta) = -(Y - \mu)X,$$
$$\frac{\partial^2 \rho}{\partial \beta \partial \beta^{\mathrm{T}}}(X, Y; \beta) = \mu(1-\mu)XX^{\mathrm{T}},$$

where $\mu = \mathrm{e}^{X^{\mathrm{T}}\beta}/(1 + \mathrm{e}^{X^{\mathrm{T}}\beta})$, with the dependency on X and β suppressed. Hence $\hat{\beta}$ is also a Z-estimator as a solution to the estimating equation ("score equation")

$$0 = \frac{1}{n}\sum_{i=1}^{n}(Y_i - \mathrm{expit}(X_i^{\mathrm{T}}\beta))X,$$

where $\mathrm{expit}(u) = \mathrm{e}^u/(1 + \mathrm{e}^u)$. Let

$$H(\beta) = E\left\{ \frac{\partial^2 \rho}{\partial \beta \partial \beta^{\mathrm{T}}}(X, Y; \beta) \right\} = E\left\{ \mu(1-\mu)XX^{\mathrm{T}} \right\},$$
$$G(\beta) = E\left\{ \left(\frac{\partial \rho}{\partial \beta}(X, Y; \beta) \right)^{\otimes 2} \right\} = E\left\{ (Y-\mu)^2 XX^{\mathrm{T}} \right\},$$

and accordingly

$$\hat{H}(\beta) = \tilde{E}\{\mu(1-\mu)XX^{\mathrm{T}}\},$$
$$\hat{G}(\beta) = \tilde{E}\{(Y-\mu)^2 XX^{\mathrm{T}}\}.$$

The model-based theory from Proposition 2.2 can be stated as follows. If logistic regression (2.7), serving as the mean restriction (2.1), is correctly specified, then $H(\beta^*) = G(\beta^*)$ and hence $\sqrt{n}(\hat{\beta} - \beta^*) \longrightarrow \mathrm{N}(0, \Sigma(\beta^*))$, where $\Sigma(\beta^*) = H^{-1}(\beta^*)G(\beta^*)H^{-1}(\beta^*)$ reduces to $\Sigma(\beta^*) = H^{-1}(\beta^*)$. Moreover, a consistent estimator for the asymptotic variance of $\hat{\beta}$, after rescaled by n, is $\hat{H}^{-1}(\hat{\beta})$. [Another estimator $\hat{G}^{-1}(\hat{\beta})$ is also consistent, but less commonly used.] The model-robust theory from Proposition 2.4 can be stated as follows. If logistic regression (2.7) may be misspecified, then $\sqrt{n}(\hat{\beta} - \bar{\beta}) \longrightarrow \mathrm{N}(0, \Sigma(\bar{\beta}))$, where $\bar{\beta}$ is defined as a unique minimizer of $K(\beta) = E\{\rho(X, Y; \beta)\}$, and $\Sigma(\bar{\beta}) = H^{-1}(\bar{\beta})G(\bar{\beta})H^{-1}(\bar{\beta})$. Moreover, a consistent estimator for the asymptotic variance of $\hat{\beta}$, after rescaled by n, is $\hat{\Sigma}(\hat{\beta}) = H^{-1}(\hat{\beta})G(\hat{\beta})H^{-1}(\hat{\beta})$.

2.4 Numerical demonstration: Logistic regression

We demonstrate implementation of logistic regression in R, mainly ML estimation and model-based and model-robust variance estimation. Unlike in Section 1.5 for linear regression, we use the built-in R function `glm()` to find the coefficient estimator $\hat{\beta}$, and leave to readers the exercises of writing self-contained code to implement the Newton–Raphson and Fisher scoring algorithms while calling only basic mathematical functions.

Consider the data (`SAheart.data`) obtained from a larger survey dataset of Rousseauw et al. (1983) and used in Hastie et al. (2009), Section 4.4.2. The data include a binary response variable, the presence or absence of coronary heart disease (`chd`), and several covariates, systolic blood pressure (`sbp`), smoking level (`tobacco`), low density lipoprotein cholesterol (`ldl`), `adiposity`, family history of heart disease (`famhist`), type-A behavior (`typea`), `obesity`, alcohol consumption level (`alcohol`), and `age` in 462 white males in South Africa. To facilitate the evaluation of prediction performance later (Section 2.6), the original dataset is randomly split into a training set of size 308 and a test set of size 154. For safe replication, the dataset with the additional variable `train` is saved as an accompanying data file.

```
> heart <- read.table("SAheart.data", header=TRUE, sep=",",
+                     row.names=1)  #first col is row names
> heart$famhist <- ifelse(heart$famhist=="Present", 1, 0)
>
```

```
> set.seed(12)
> heart$train <- rep(TRUE, 462)
> heart$train[sample(1:462, size=154)] <- FALSE
>
> heart[1:5,]
  sbp tobacco  ldl adiposity famhist typea obesity alcohol age
1 160   12.00 5.73     23.11       1    49   25.30   97.20  52
2 144    0.01 4.41     28.61       0    55   28.87    2.06  63
3 118    0.08 3.48     32.28       1    52   29.14    3.81  46
4 170    7.50 6.41     38.03       1    51   31.99   24.26  58
5 134   13.60 3.50     27.78       1    60   25.99   57.34  49
  chd train
1   1 FALSE
2   1 FALSE
3   0 FALSE
4   1  TRUE
5   1  TRUE
>
> train <- heart[heart$train,]  # size 308
> test <- heart[!heart$train,]  # size 154
```

We use the built-in R function `glm()` to fit logistic regression of the response variable `chd` on the 7 covariates in addition to the intercept as in Hastie et al. (2009), but in the training data. From `summary(fit.glm)` below, the column `Estimate` give the individual components of the ML estimator $\hat{\beta}$, whereas the column `Std.Error` gives the model-based standard errors, which are the square roots of the diagonal elements of the model-based variance estimator $\hat{H}^{-1}(\hat{\beta})/n$. See Example 2.5 for specific formulas.

```
> y <- train[,10]
> x <- as.matrix(train[,c(1:3,5,7:9)])
> # 4th and 6th column, "adiposity" and "typea", not used
>
> fit.glm <- glm(y ~ x, family=binomial(link="logit"))
> summary(fit.glm)
......
Coefficients:
             Estimate Std. Error z value Pr(>|z|)
(Intercept) -3.251967   1.163971  -2.794 0.005208 **
xsbp        -0.001529   0.007176  -0.213 0.831313
xtobacco     0.059427   0.030630   1.940 0.052358 .
```

```
xldl         0.256945   0.074138   3.466 0.000529 ***
xfamhist     1.031406   0.274779   3.754 0.000174 ***
xobesity    -0.034551   0.035474  -0.974 0.330056
xalcohol     0.005344   0.005291   1.010 0.312521
xage         0.038682   0.012628   3.063 0.002190 **
---
Signif. codes:  0 '***' 0.001 '**' 0.01 '*' 0.05 '.' 0.1 ' ' 1
......
    Null deviance: 401.48  on 307  degrees of freedom
Residual deviance: 325.17  on 300  degrees of freedom
......
```

To confirm the correctness of $\hat{\beta}$ found by `glm()`, the following shows that the score equation is numerically satisfied, i.e., the gradient of the average negative log-likelihood is 0 at $\hat{\beta}$.

```
> expit <- function(x)
+   1/(1+exp(-x))
>
> yfit <- as.vector(expit(cbind(1,x)%*%fit.glm$coef))
> apply((y-yfit) *cbind(1,x), 2, mean)
                          sbp       tobacco           ldl
-5.356977e-11 -6.930466e-09 -2.787409e-11 -1.868562e-10
      famhist       obesity       alcohol           age
-7.167458e-12 -1.268335e-09 -4.661069e-10 -1.258020e-09
```

To verify the preceding statement about `Std.Error` from `glm()` and incorporate model-robust inference, we write the following function, `my.logit()`, which returns the model-based and model-robust SEs (`se` and `r.se`) as described in Example 2.5, together with the point estimates (`coef`) taken from `glm()`.

```
my.logit <- function(y,x, intercept=TRUE) {
  if (intercept) {
    x <- cbind(1,x)
  } else {
    x <- cbind(x)
  }

  n <- dim(x)[1]
  fit.glm <- glm(y ~ -1 + x, family=binomial(link="logit"))
  #"-1" to exclude default intercept
```

```
  bet <- fit.glm $coef
  yfit <- as.vector(expit(x%*%bet))
  H <- t(x)%*%((yfit*(1-yfit)) *x)/n

  # model-based variance
  evar <- solve(H) /n

  # robust variance
  res <- y-yfit
  r.evar <- solve(H, t(x)%*%(res^2*x)/n) %*% solve(H) /n

  out <- cbind(bet, sqrt(diag(evar)), sqrt(diag(r.evar)))
  if (intercept)  rownames(out)[1] <- "(inter)"
  colnames(out) <- c("coef", "se", "r.se")
  out
}
```

The following is the output from using the R function `my.logit()` to fit logistic regression of `chd` on the 7 covariates in addition to the intercept. By comparison, the model-based SEs (`se`) match the reported SEs (`Std.Error`) from `glm()`. The model-robust SEs (`r.se`) shows various differences from the model-based SEs, ranging from being 12% smaller for `tobacco` to 4% greater for the intercept and 3% greater for `obesity`.

```
> my.logit(y,x)
                 coef          se        r.se
(inter)  -3.251967465 1.163980990 1.214653252
xsbp     -0.001528644 0.007176153 0.007289761
xtobacco  0.059427375 0.030629905 0.027028848
xldl      0.256945112 0.074138777 0.068447679
xfamhist  1.031405730 0.274780588 0.272467520
xobesity -0.034551246 0.035473741 0.036479432
xalcohol  0.005343626 0.005291029 0.005394052
xage      0.038682116 0.012628098 0.011681107
```

2.5 Penalized regression: Algorithms

We study penalized estimation for generalized linear regression. The discussion is mainly about computational algorithms, without dealing with associated statistical theory.

Consider generalized linear regression with mean restriction (2.1) and exponential-family assumption (2.2) with constant dispersion parameter ϕ^*. For simplicity, assume that a canonical link $g(\cdot)$ is used, satisfying $g = (a')^-$ as in (2.12). From (2.18) with $\theta_i = \eta_i = X_i^{\mathrm{T}}\beta$, the average negative log-likelihood function (or the likelihood loss) is, up to a factor $\frac{1}{\phi^*}$,

$$K_n(\beta) = \frac{-1}{n}\sum_{i=1}^n \{Y_i X_i^{\mathrm{T}}\beta - a(X_i^{\mathrm{T}}\beta)\} + \text{constant},$$

which is convex in β by the convexity of $a(\cdot)$ discussed after Lemma 2.1. A penalized maximum likelihood estimator is defined in the form

$$\hat{\beta}_\lambda = \operatorname{argmin}_\beta \{K_n(\beta) + \lambda\, \mathrm{pen}(\beta)\},$$

where $\lambda \geq 0$ is a tuning parameter and $\mathrm{pen}(\beta)$ is a penalty function in β. Two popular choices are the ridge penalty, $\mathrm{pen}(\beta) = \frac{1}{2}\|\beta\|_2^2$, and the Lasso penalty, $\mathrm{pen}(\beta) = \|\beta\|_1$. The corresponding estimator is denoted as $\hat{\beta}_{\lambda,L_2}$ or $\hat{\beta}_{\lambda,L_1}$, and called the ridge-penalized or Lasso-penalized ML estimator of β or, for short, the ridge or Lasso estimator of β.

For Lasso-penalized ML estimation, the stationary condition for $\hat{\beta}_\lambda$ is

$$\frac{1}{n}\sum_{i=1}^n (Y_i - a'(X_i^{\mathrm{T}}\hat{\beta}_\lambda))X_{ij} \begin{cases} = \lambda, & \text{if } \hat{\beta}_{\lambda,j} > 0, \\ \in [-\lambda, \lambda], & \text{if } \hat{\beta}_{\lambda,j} = 0, \\ = -\lambda, & \text{if } \hat{\beta}_{\lambda,j} < 0, \end{cases}$$

which reduces to (1.59) for linear regression with $a(u) = u^2/2$. Unlike Lasso-penalized linear regression, there is in general no closed-form expression for $\hat{\beta}_\lambda$ even in the univariate case ($p = 1$), so that a direct application of coordinate descent is not effective for computing $\hat{\beta}_\lambda$.

To compute the penalized estimator $\hat{\beta}_\lambda$, a simple but effective strategy is to iteratively form a quadratic approximation to the loss function $K_n(\beta)$, assumed to be twice differentiable, and then solve a penalized weighted least-squares problem including the possibly non-smooth penalty term. As discussed in Section 2.2, the Newton–Raphson algorithm iteratively optimizes a quadratic approximation by the Taylor expansion when the objective function is twice differentiable. In the following discussion, it is helpful (but not necessary) to take $\mathrm{pen}(\beta)$ to be the Lasso penalty, so that the objective function, $K_n(\beta) + \lambda\, \mathrm{pen}(\beta)$, is non-smooth and the Newton–Raphson algorithm is not applicable. In the case of the ridge penalty, the update (2.43) below is equivalent to applying the Newton–Raphson algorithm directly to the objective function, $K_n(\beta) + \frac{\lambda}{2}\|\beta\|_2^2$. [Exercise: Verify this statement.]

By the gradient and Hessian calculation underlying the Newton–Raphson update (2.22), the quadratic approximation from the Taylor expansion of $K_n(\beta)$ about β^{k-1} is

$$\begin{aligned} & K_{n,Q}(\beta;\beta^{k-1}) \\ &= K_n(\beta^{k-1}) - \frac{1}{n}\sum_{i=1}^{n}(Y_i - \mu_i)\Big|_{\beta=\beta^{k-1}} X_i^{\mathrm{T}}(\beta - \beta^{k-1}) \\ &\quad + \frac{1}{2n}(\beta - \beta^{k-1})^{\mathrm{T}} \sum_{i=1}^{n} \frac{\partial \mu_i}{\partial \eta_i}\Big|_{\beta=\beta^{k-1}} X_i X_i^{\mathrm{T}}(\beta - \beta^{k-1}), \end{aligned}$$

where $\eta_i = X_i^{\mathrm{T}}\beta$ and $\mu_i = a'(\eta_i) = g^{-}(\eta_i)$. Equivalently, the quadratic approximation can be rewritten as a weighted least-squares objective:

$$K_{n,Q}(\beta;\beta^{k-1}) = \frac{1}{2n}\sum_{i=1}^{n} W_i(\beta^{k-1})\left\{\tilde{Y}_i(\beta^{k-1}) - X_i^{\mathrm{T}}\beta\right\}^2 + \text{constant},$$

where the weight and working response are from (2.25):

$$W_i(\beta) = \frac{\partial \mu_i}{\partial \eta_i}, \quad \tilde{Y}_i(\beta) = X_i^{\mathrm{T}}\beta + \left(\frac{\partial \mu_i}{\partial \eta_i}\right)^{-1}(Y_i - \mu_i),$$

Replacing $K_n(\beta)$ with $K_{n,Q}(\beta, \beta^{k-1})$ and minimizing the approximate objective $K_n(\beta) + \lambda\,\mathrm{pen}(\beta)$ leads to the following update:

$$\beta^k = \mathrm{argmin}_\beta \left\{K_{n,Q}(\beta;\beta^{k-1}) + \lambda \mathrm{pen}(\beta)\right\}. \tag{2.43}$$

In the case of the Lasso penalty, $\mathrm{pen}(\beta) = \|\beta\|_1$, the preceding optimization corresponds to a Lasso-penalized weighted least-squares problem, and can be solved (up to some termination criterion) by properly extending the iterative algorithms for Lasso (unweighted) linear regression in Section 1.8. The details are left as an exercise.

The quadratic approximation $K_{n,Q}(\beta;\beta^{k-1})$ can be modified by exploiting the Majorization-Minimization (MM) technique. For logistic regression, a majorizing, quadratic function for $K_n(\beta)$ at β^{k-1} is obtained as $K_{n,\mathrm{M1}}(\beta;\beta^{k-1})$ in (2.26):

$$\begin{aligned} K_{n,\mathrm{M1}}(\beta;\beta^{k-1}) &= K_n(\beta^{k-1}) - \frac{1}{n}\sum_{i=1}^{n}(Y_i - \mu_i)\Big|_{\beta=\beta^{k-1}} X_i^{\mathrm{T}}(\beta - \beta^{k-1}) \\ &\quad + \frac{1}{2}(\beta - \beta^{k-1})^{\mathrm{T}} \frac{1}{n}\sum_{i=1}^{n} \frac{1}{4} X_i X_i^{\mathrm{T}}(\beta - \beta^{k-1}). \end{aligned}$$

Then $K_{n,\text{M1}}(\beta;\beta^{k-1}) + \lambda\text{pen}(\beta)$ is also a majorizing function for the full objective $K_n(\beta) + \lambda\text{pen}(\beta)$. The resulting MM update,

$$\beta^k = \text{argmin}_\beta \left\{ K_{n,\text{M1}}(\beta;\beta^{k-1}) + \lambda\text{pen}(\beta) \right\}, \tag{2.44}$$

satisfies the monotonicity property, whereas the update (2.43) may not. For the Lasso penalty, $\text{pen}(\beta) = \|\beta\|_1$, the preceding optimization corresponds to a Lasso problem with unweighted least squares, and can be solved by applying the algorithms (coordinate descent, active-set descent, or proximal gradient) for linear regression in Section 1.8.

In addition, suppose that $\frac{1}{n}\sum_{i=1}^n X_i X_i^{\text{T}} \le \tau^{-1} I$ for a constant $\tau > 0$. Then a further majorizing function for $K_n(\beta)$ at β^{k-1} is

$$K_{n,\text{M2}}(\beta;\beta^{k-1}) = K_n(\beta^{k-1}) - \frac{1}{n}\sum_{i=1}^n (Y_i - \mu_i)\Big|_{\beta=\beta^{k-1}} X_i^{\text{T}}(\beta - \beta^{k-1}) + \frac{1}{2}\frac{1}{4\tau}\|\beta - \beta^{k-1}\|_2^2.$$

The resulting MM update can be shown to be

$$\begin{aligned} \beta^k &= \text{argmin}_\beta \left\{ K_{n,\text{M2}}(\beta;\beta^{k-1}) + \lambda\text{pen}(\beta) \right\} \\ &= S_{4\tau\lambda}\left\{ \beta^{k-1} + 4\tau\frac{1}{n}\sum_{i=1}^n (Y_i - \mu_i) X_i \Big|_{\beta=\beta^{k-1}} \right\}, \end{aligned} \tag{2.45}$$

which is reminiscent of the proximal gradient update (1.68) for Lasso linear regression. [Exercise: Show that the update (2.45) can also be obtained by applying just one iteration of the proximal gradient algorithm with the initial value β^{k-1} to the Lasso problem (2.44).]

Bibliographical note. For penalized ML estimation, the strategy of forming a quadratic approximation and then solving a penalized least-squares problem has been advocated by Friedman et al. (2007) and implemented by the R package `glmnet` (Friedman et al., 2021), where coordinate descent is used for the latter purpose. This strategy has been extended by Tan (2020b) to tackle penalized non-ML estimation for logistic regression. Lasso-penalized ML and non-ML estimation for logistic regression is provided by the R package `RCAL` (Tan & Sun, 2020), where active-set descent is used for solving a Lasso least-squares problem.

2.6 Numerical demonstration: Penalized logistic regression

We demonstrate numerical application of penalized logistic regression in R. Unlike in Section 1.9 for linear regression, we use the R package `glmnet`

(Friedman et al., 2021) and leave to readers the exercises of writing self-contained code to implement the algorithms in Section 2.5. Only Lasso estimation is discussed, and ridge estimation can be similarly treated. As shown in the accompanying R code file, the R package RCAL (Tan & Sun, 2020) can also be used to fit logistic regression by Lasso-penalized ML estimation.

We use the heart disease data and consider logistic regression of the binary response chd on the 7 covariates (in addition to the intercept), as in Section 2.4. The original dataset is split into a training set of size 308 and a test set of size 154. Given a regression model, the training set is used to fit the model (i.e., decide the coefficient estimates), whereas the test set is used to evaluate the prediction performance of fitted models.

```
> train <- heart[heart$train,]  # size 308
> test <- heart[!heart$train,]  # size 154
>
> y <- train[,10]
> x <- as.matrix(train[,c(1:3,5,7:9)])
> # 4th and 6th column, "adiposity" and "typea", not used
>
> y.test <- test[,10]
> x.test <- as.matrix(test[,c(1:3,5,7:9)])
```

2.6.1 Lasso logistic regression

In practice, there are often two modifications from the general discussion in Section 2.5. First, Lasso estimation with an intercept term is typically fitted with the intercept non-penalized, i.e., the Lasso objective is modified as

$$K_{1,n}(\beta) = \frac{1}{n}\sum_{i=1}^{n} \rho_0(Y_i, \beta_0 + X_{i,1:p}^{\mathrm{T}}\beta_{1:p}) + \lambda\|\beta_{1:p}\|_1,$$

where $\rho_0(y, \eta) = -y\eta + \log(1 + \mathrm{e}^{\eta})$, $\beta = (\beta_0, \beta_{1:p}^{\mathrm{T}})^{\mathrm{T}}$, $X_{i,1:p} = (X_{i1}, \dots, X_{ip})^{\mathrm{T}}$, and $\lambda \ge 0$ is a tuning parameter. Second, particularly when the covariates are of considerably different scales, covariate standardization is applied before fitting and then the coefficient estimates are returned to the original scale. Similarly as in Exercise 1.13 for linear regression, this amounts to modifying the Lasso objective as

$$K_{1,n}(\beta) = \frac{1}{n}\sum_{i=1}^{n} \rho_0(Y_i, \beta_0 + X_{i,1:p}^{\mathrm{T}}\beta_{1:p}) + \lambda\sum_{j=1}^{p} \mathrm{SD}_j|\beta_j|, \tag{2.46}$$

where SD_j is the sample standard deviation of the jth covariate $(X_{1j}, \dots, X_{nj})$ (using the $1/n$ formula), and $\sum_{j=1}^{p} \mathrm{SD}_j|\beta_j|$ is a weighted L_1 norm. The Lasso estimator of β, defined as a minimizer of $K_{1,n}(\beta)$ in (2.46), can be computed by similar ideas as discussed in Section 2.5, i.e., iteratively forming a quadratic approximation (or a quadratic majorization) and solving a Lasso-penalized least-squares problem with a non-penalized intercept after covariate standardization. The details are left as an exercise.

The following output using **glmnet** with **standarsize=T** gives the Lasso estimates corresponding to the minimizer of the objective function (2.46) with the choices of λ in an arithmetic sequence from 0 to 0.2 with increment 0.04.

```
> fit.lasso <- glmnet(x, y, family="binomial", alpha=1, standardize=T,
+                      lambda=seq(.2,0,by=-.04))
> # use "6:1", for increasing lambda
> coef.mat <- rbind(fit.lasso $a0, fit.lasso $bet) [,6:1]
> coef.mat
......
                  s5          s4          s3          s2
        -3.252059976 -2.89626202 -2.04534710 -1.29667221
sbp     -0.001528242  .           .           .
tobacco  0.059434026  0.03671846  0.01174427  .
ldl      0.256948504  0.14387181  0.07846868  0.01989521
famhist  1.031418673  0.67006916  0.36740537  0.09191060
obesity -0.034548886  .           .           .
alcohol  0.005343476  .           .           .
age      0.038680463  0.02672759  0.02028311  0.01344495
                  s1         s0
        -0.697395234 -0.5877867
sbp      .            .
tobacco  .            .
ldl      .            .
famhist  .            .
obesity  .            .
alcohol  .            .
age      0.002580932  .
```

To confirm that glmnet with standardize=T minimizes (2.46), we verify that the stationary condition is satisfied (up to machine precision), for example, at $\lambda = 0.04$, by examining the empirical inner products of the residual vector and the covariate columns. [Exercise: What is the stationary condition for minimizing the Lasso objective (2.46)?] The coefficient estimates associated with tobacco, idl, famhist, and age are positive, and the empirical inner products are 0.04 times the covariate standard deviations. The coefficient estimates associated with the remaining three covariates are 0, and the empirical inner products fall in the interval $[-0.04, 0.04]$ after divided by the covariate standard deviations.

```
> j <- 2
> coef.mat[,j]
                    sbp     tobacco         ldl     famhist
-2.89626202  0.00000000  0.03671846  0.14387181  0.67006916
    obesity     alcohol         age
 0.00000000  0.00000000  0.02672759
>
> xbar <- apply(x,2,mean)
> sd.x <- sqrt(apply((x-rep(xbar,each=n))^2, 2, mean))
>
> apply(x*as.vector(y-expit(cbind(1,x)%*%coef.mat[,j])), 2, mean) /sd.x
         sbp      tobacco          ldl      famhist      obesity
 0.010626132  0.039990687  0.039995918  0.040000815 -0.005688244
     alcohol          age
 0.028934015  0.040006795
```

Figure 2.3 shows the rescaled Lasso estimates and the training and test errors versus the shrinkage fraction of L_1 norms, computed from a more dense sequence of λ choices. The estimates are rescaled mainly to improve visualization. The training or test errors are defined in terms of the likelihood loss in the training or test set as later in Section 2.6.2. See the accompanying R code file for computation details. As λ decreases (and hence the fraction of L_1 norms increases), the number of nonzero coefficient estimates increases from 0 to 7, reflecting the sparsity of Lasso estimation. In addition, as λ decreases, the training error keeps decreasing, whereas the test error exhibits a U-shaped pattern, similarly as in Figure 1.7.

2.6.2 Cross validation

To select the tuning parameter in penalized estimation for generalized linear regression, it is common to use (one-time) validation or cross validation similarly as in Section 1.9.3 for linear regression. For completeness, we formally describe cross validation for penalized ML estimation in logistic regression.

Suppose that the available dataset of size N is randomly split into a training set of size n and a test set of size $N-n$, denoted respectively as $\{(X_i, Y_i) : i = 1, \ldots, n\}$ and $\{(X_i, Y_i) : i = n+1, \ldots, N\}$. For logistic regression, a penalized ML estimator is defined as

$$\hat{\beta}_\lambda = \text{argmin}_\beta \left\{ \frac{1}{n} \sum_{i=1}^n \rho_0(Y_i, X_i^{\mathrm{T}}\beta) + \lambda \, \text{pen}(\beta) \right\},$$

where $\rho_0(y, \eta) = -y\eta + \log(1 + \mathrm{e}^\eta)$ as in (2.46), $\lambda \geq 0$ is a tuning parameter, and $\text{pen}(\beta)$ is a penalty function in β. The function $\rho_0(y, \eta)$ is the likelihood loss (i.e., the negative log-likelihood) for a response y and a

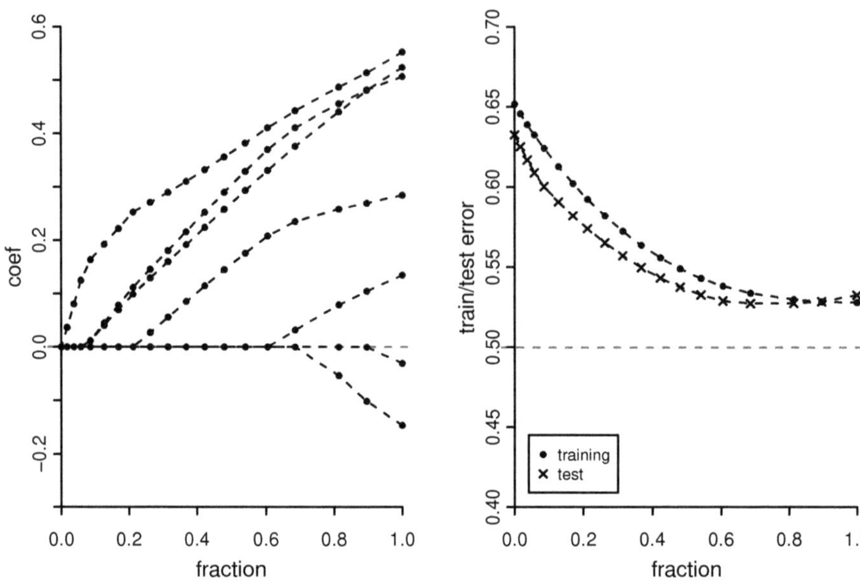

FIGURE 2.3
Plots of the rescaled Lasso estimates $\hat{\gamma}_j = \mathrm{SD}_j \cdot \hat{\beta}_j$, $j = 1, \ldots, p$, versus the fraction of L_1 norms (left) and the training and test errors versus the shrinkage fraction (right), as the λ choice is varied, in the heart disease example. The shrinkage fraction is $\frac{\|\hat{\gamma}_{1:p}(\lambda)\|_1}{\|\hat{\gamma}_{1:p}(0)\|_1}$, where $\hat{\gamma}_{1:p}(\lambda)$ is the rescaled Lasso estimator given λ. A horizontal line is also placed at the test error for the (infeasible) ML estimator trained on the training and test data combined.

linear predictor η. Possible dependency of $\mathrm{pen}(\beta)$ on the training set [such as $(\mathrm{SD}_1, \ldots, \mathrm{SD}_p)$ in (2.46)] is suppressed in the notation. Given λ and $\hat{\beta}_\lambda$, the training error is defined as the mean likelihood loss in the training set, $\mathrm{err}_\lambda = n^{-1} \sum_{i=1}^n \rho_0(Y_i, X_i^{\mathrm{T}} \hat{\beta}_\lambda)$, and the test error is defined as that in the test set, $\widehat{\mathrm{Err}}_\lambda = (N-n)^{-1} \sum_{i=n+1}^N \rho_0(Y_i, X_i^{\mathrm{T}} \hat{\beta}_\lambda)$. The training or test error may also be rescaled by 2, to give the mean of the residual deviance in the training or test set by the definition (2.15).

Given a choice of K, typically 5 or 10, K-hold cross validation proceeds as follows.

- Randomly split the training set into K approximately equal-sized parts, i.e., split $\{1, 2, \ldots, n\}$ into K approximately equal-sized subsets $\mathcal{I}_1, \ldots, \mathcal{I}_K$.
- For each λ and $k = 1, \ldots, K$, compute

$$\tilde{\beta}_\lambda^{(k)} = \mathrm{argmin}_\beta \left\{ \frac{1}{|\mathcal{I}_k^c|} \sum_{i \in \mathcal{I}_k^c} \rho_0(Y_i, X_i^{\mathrm{T}} \beta) + \lambda \, \mathrm{pen}(\beta) \right\},$$

and then $\widetilde{\text{Err}}_\lambda^{(k)} = \frac{1}{|\mathcal{I}_k|}\sum_{i\in\mathcal{I}_k}\rho_0(Y_i, X_i^{\text{T}}\tilde{\beta}_\lambda^{(k)})$. The term pen($\beta$) may depend on the training subset $\{(X_i, Y_i) : i \in \mathcal{I}_k^c\}$ as in (2.46). Compute the CV error

$$\text{CV}(\lambda) = \frac{1}{K}\sum_{k=1}^{K}\widetilde{\text{Err}}_\lambda^{(k)}.$$

- Select λ by minimizing CV(λ) or, in practice, select λ with the smallest CV(λ) over a grid of values $\lambda_1, \ldots, \lambda_m$, i.e., $\hat{\lambda} = \text{argmin}_{\lambda_1,\ldots,\lambda_m}\text{CV}(\lambda)$.
- For the selected λ, compute $\hat{\beta}_\lambda$ and report err_λ and $\widehat{\text{Err}}_\lambda$.

An alternative approach, called "one standard error rule" (Breiman et al., 1984; Hastie et al., 2009), is to select the largest $\lambda \in \{\lambda_1, \ldots, \lambda_m\}$ such that

$$\text{CV}(\lambda) \le \text{CV}(\hat{\lambda}) + \frac{\text{SD}(\hat{\lambda})}{\sqrt{K}},$$

where $\text{SD}(\lambda) = [\frac{1}{K-1}\sum_{k=1}^{K}\{\widetilde{\text{Err}}_\lambda^{(k)} - \text{CV}(\lambda)\}^2]^{1/2}$, the sample standard deviation of the validation errors. The term $\frac{\text{SD}(\lambda)}{\sqrt{K}}$ is referred to as the SE of CV(λ).

We write the following function, `my.logit.lasso.cv()`, to implement cross validation (with K coded as `fold`) for Lasso logistic regression with a sequence of λ choices. The function is similar to `my.ridge.cv()` and `my.lasso.cv()` for linear regression as described in Section 1.9.3, except that `glmnet` is used instead of our own written functions for implementing penalized estimation with fixed λ. See the accompanying R code file for details.

```
my.logit.lasso.cv <- function(fold, y,x, lam.seq, permut=NULL) {
  x <- cbind(x)
  n <- dim(x)[1]
  if (is.null(permut))
    permut <- sample(1:n)

  fold.id <- split(n, fold)
  fold.id <- fold.id[permut]

  m <- length(lam.seq)
  err <- matrix(NA, nrow=m, ncol=fold)
  for (j in 1:fold) {
    test <- which(fold.id==j)
    y.test <- y[test]
    x.test <- x[test, , drop=F]
```

```
    y.train <- y[-test]
    x.train <- x[-test, , drop=F]
    out <- glmnet(x.train, y.train, family="binomial", alpha=1,
           standardize=T, lambda=lam.seq[m:1])   # decreasing lambda
    coef.mat <- rbind(out $a0, out $bet) [,m:1]
    coef.mat <- as.matrix(coef.mat)
    for (k in 1:m)
      err[k,j] <- logit.loss(y.test, cbind(1,x.test) %*% coef.mat[,k])
  }
  err.ave <- apply(err, 1, mean)
  err.se <- apply(err, 1, sd)/sqrt(fold)

  sel <- which.min(err.ave)
  sel <- c(sel, max(which(err.ave <= err.ave[sel] + err.se[sel])))
  out.all <- glmnet(x, y, family="binomial", alpha=1, standardize=T,
                    lambda=lam.seq[sel[2:1]])   # decreasing lambda

  list(out=out.all, sel=sel, lam=lam.seq[sel],
       err.ave=err.ave, err.se=err.se)
}
```

The following output using `my.logit.lasso.cv()` gives the results of 10-fold cross validation for Lasso regression in the heart disease example. Figure 2.4, extending Figure 2.3, shows the CV errors and their SEs and the two selected λ values, $\lambda = 0$ (no Lasso penalty, selected by minimum CV error) or $\lambda = .15$ (selected by the one standard error rule).

```
> lam.seq <- c(0, .005, seq(.01, .2, by=.01))
> out.lasso.cv <- my.logit.lasso.cv(fold=10, y,x, lam.seq, permut=1:n)
> out.lasso.cv
......
$lam
[1] 0.02 0.07

$err.ave
 [1] 0.5559682 0.5531367 0.5511980 0.5490028 0.5492963 0.5527143
 [7] 0.5582213 0.5651226 0.5730347 0.5819265 0.5912342 0.6006340
[13] 0.6105794 0.6206601 0.6300005 0.6372016 0.6435714 0.6503242
[19] 0.6544721 0.6551449 0.6551449 0.6551449
```

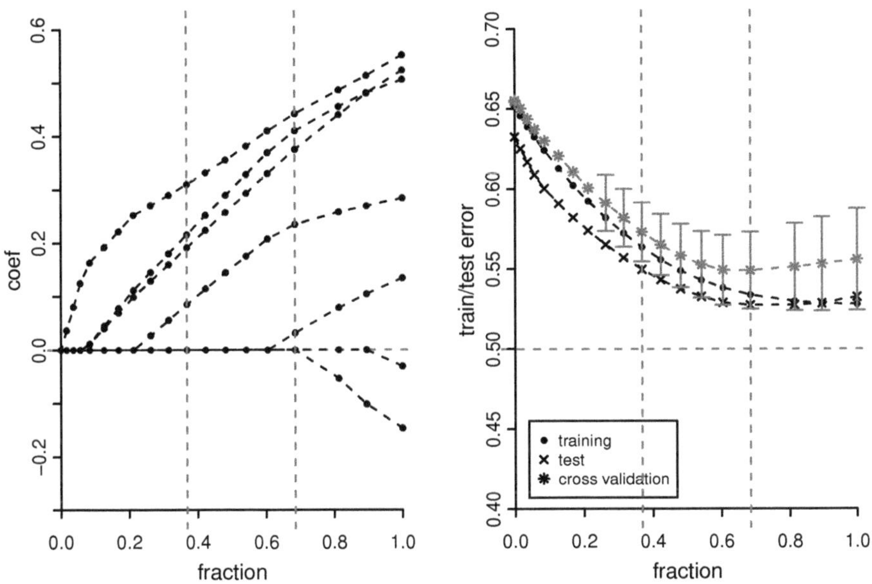

FIGURE 2.4
Extension of Figure 2.3, showing the 10-fold CV errors and their SEs versus the shrinkage fraction (right). Two vertical lines are also placed at the shrinkage fraction corresponding to the two selected λ values.

```
$err.se
 [1] 0.03166639 0.02932053 0.02720181 0.02403738 0.02179087
 [6] 0.02064400 0.01980230 0.01907041 0.01848630 0.01805370
[11] 0.01768450 0.01733279 0.01711790 0.01697544 0.01667194
[16] 0.01649210 0.01627897 0.01610919 0.01637941 0.01640522
[21] 0.01640522 0.01640522
```

2.7 Appendices

2.7.1 Asymptotic theory of Z-estimation

To complement Section 2.3.1 on M-estimation, we describe asymptotic theory of Z-estimation in the classical setting.

A Z-estimator (with Z indicating "zero") $\hat{\beta}_n$ is defined as a solution to $T_n(\beta) = 0$ with a sample estimating function

$$T_n(\beta) = \frac{1}{n} \sum_{i=1}^{n} \psi(Z_i; \beta), \tag{2.47}$$

where $\psi(\cdot;\beta)$ is a $p \times 1$ estimating function (of the same dimension, p, as β) for an individual datum. The target value $\bar{\beta}$ for $\hat{\beta}_n$ is defined as a solution to $T(\beta) = 0$ with the population estimating function

$$T(\beta) = E\{\psi(Z;\beta)\}. \tag{2.48}$$

The asymptotic properties of $\hat{\beta}_n$ can be well characterized under suitable regularity conditions in the classical setting of $n \to \infty$ and fixed p. Details of such regularity conditions are not discussed here. Assume that $\psi(z;\beta)$ is differentiable in β for any fixed z. Let

$$\Sigma(\beta) = H^{-1}(\beta)G(\beta)H^{-\mathrm{T}}(\beta),$$
$$H(\beta) = E\left\{\frac{\partial \psi}{\partial \beta^{\mathrm{T}}}(Z;\beta)\right\}, \quad G(\beta) = E\left\{\psi(Z;\beta)^{\otimes 2}\right\},$$

and similarly

$$\hat{\Sigma}(\beta) = \hat{H}^{-1}(\beta)\hat{G}(\beta)\hat{H}^{-\mathrm{T}}(\beta),$$
$$\hat{H}(\beta) = \tilde{E}\left\{\frac{\partial \psi}{\partial \beta^{\mathrm{T}}}(Z;\beta)\right\}, \quad \hat{G}(\beta) = \tilde{E}\left\{\psi(Z;\beta)^{\otimes 2}\right\},$$

where $B^{-\mathrm{T}} = (B^{-1})^{\mathrm{T}}$ for a matrix B and $u^{\otimes 2} = uu^{\mathrm{T}}$ for a column vector u. See Exercise 1.2 for the interpretation of $\frac{\partial \psi}{\partial \beta^{\mathrm{T}}}$, with a vector-valued ψ and a vector-valued β.

Proposition 2.5 (Asymptotic theory of Z-estimation). *In the preceding setting, the following results hold under suitable regularity conditions. As $n \to \infty$ and p is fixed, $\hat{\beta} \xrightarrow[\mathcal{P}]{} \bar{\beta}$ and*

$$\sqrt{n}(\hat{\beta} - \bar{\beta}) \xrightarrow[\mathcal{D}]{} \mathrm{N}(0, \Sigma(\bar{\beta})),$$
$$\hat{\Sigma}(\hat{\beta}) \xrightarrow[\mathcal{P}]{} \Sigma(\bar{\beta}).$$

For an M-estimator $\hat{\beta}$ with (twice-differentiable) objective function $\rho(z;\beta)$, which is then a Z-estimator with $\psi(z;\beta) = \frac{\partial}{\partial \beta}\rho(z;\beta)$, Proposition 2.5 recovers Proposition 2.1. In this case, $H(\beta)$ and $\hat{H}(\beta)$ reduce to

$$H(\beta) = E\left\{\frac{\partial^2 \rho}{\partial \beta \partial \beta^{\mathrm{T}}}(Z;\beta)\right\}, \quad \hat{H}(\beta) = \tilde{E}\left\{\frac{\partial^2 \rho}{\partial \beta \partial \beta^{\mathrm{T}}}(Z;\beta)\right\},$$

and hence are symmetric matrices, because $\frac{\partial \psi}{\partial \beta^{\mathrm{T}}}(z;\beta) = \frac{\partial^2 \rho}{\partial \beta \partial \beta^{\mathrm{T}}}(z;\beta)$ is symmetric (see Exercise 1.2). On the other hand, Proposition 2.5 is also applicable to a Z-estimator $\hat{\beta}$ which cannot be determined as an M-estimator from any objective function. As a simple rule, such a situation occurs for $\hat{\beta}$ with an estimating function $\psi(z;\beta)$ such that $\frac{\partial \psi}{\partial \beta^{\mathrm{T}}}(z;\beta)$ is an asymmetric matrix. [Exercise: Explain why.]

Exercises

Ex 2.1. Let $Z_1 \sim \mathrm{N}(\mu, \sigma_1^2)$ and, independently, $Z_2 \sim \mathrm{N}(\mu, \sigma_2^2)$, where μ is unknown and (σ_1^2, σ_2^2) are known. [More generally, (σ_1^2, σ_2^2) are allowed to be known up to a constant of proportionality, i.e., $(\sigma_1^2, \sigma_2^2) = (c_1\sigma^2, c_2\sigma^2)$ with known (c_1, c_2) but possibly unknown σ^2.] Show that the optimal choice of c in minimizing the variance of $(1-a)Z_1 + aZ_2$, an unbiased estimator of μ, is determined such that

$$(a, 1-a) \propto (\sigma_1^{-2}, \sigma_2^{-2}).$$

Ex 2.2. Present *in analytical details* the Newton–Raphson and the Fisher scoring algorithms for fitting the probit regression.

Ex 2.3. Use the dataset `birthwt` discussed in Venables and Ripley (2002), Section 7.2, and consider the logit and probit regression models in Exercise 2.4. The dataset is available in the R package `MASS` (Ripley et al., 2020) as well as the accompanying data files.

Write your own R programs to

(i) fit logit regression using Newton–Raphson,

(ii) fit logit regression using the MM algorithm with a quadratic majorizing function as described after Example 2.4,

(iii) fit probit regression using Newton–Raphson,

(iv) fit probit regression using Fisher scoring.

For *each* of your programs (a)–(d), report the number of iterations until convergence and the coefficient estimates obtained with the following choices. Summarize your findings.

- starting value β^0: set to be 0 or the least-squares estimates,
- convergence declared at iteration k when $\|\beta^k - \beta^{k-1}\|_2 \le 10^{-3}$ or $\|\beta^k - \beta^{k-1}\|_2 \le 10^{-6}$.
- convergence declared at iteration k when $|K_n(\beta^k) - K_n(\beta^{k-1})| \le 10^{-3}$ or $|K_n(\beta^k) - K_n(\beta^{k-1})| \le 10^{-6}$, where $K_n(\cdot)$ denotes the average log-likelihood function (that is, log-likelihood divided by n).

Ex 2.4. For the dataset `birthwt`, write your own R code to replicate the following results obtained from the built-in R function `glm()`, including

(i) (point) Estimate,

(ii) (model-based) Std. Error,

(iii) Null deviance,

(iv) Residual deviance.

Report the starting value β^0 and the convergence criterion used.

```
> library(MASS)
> attach(birthwt)
>
> ### data pre-processing
> race <- factor(race, label=c("white", "black", "other"))
> ptd <- factor(ptl>0)
> ftv <- factor(ftv)
> levels(ftv)[-(1:2)] <- "2+"
>
> ### logit
> bwt.glm <- glm(low ~ age + lwt + race + smoke + ptd + ht + ui + ftv,
family=binomial)
>
> summary(bwt.glm, cor=F)
......
Coefficients:
            Estimate Std. Error z value Pr(>|z|)
(Intercept)  0.82302    1.24471   0.661  0.50848
age         -0.03723    0.03870  -0.962  0.33602
lwt         -0.01565    0.00708  -2.211  0.02705 *
raceblack    1.19241    0.53597   2.225  0.02609 *
raceother    0.74069    0.46174   1.604  0.10869
smoke        0.75553    0.42502   1.778  0.07546 .
ptdTRUE      1.34376    0.48062   2.796  0.00518 **
ht           1.91317    0.72074   2.654  0.00794 **
ui           0.68019    0.46434   1.465  0.14296
ftv1        -0.43638    0.47939  -0.910  0.36268
ftv2+        0.17901    0.45638   0.392  0.69488
---
Signif. codes:  0 ‘***’ 0.001 ‘**’ 0.01 ‘*’ 0.05 ‘.’ 0.1 ‘ ’ 1
......
    Null deviance: 234.67  on 188  degrees of freedom
Residual deviance: 195.48  on 178  degrees of freedom
......

>
```

```
> ### probit
> bwt.glm2 <- glm(low ~ age + lwt + race + smoke + ptd + ht + ui + ftv,
family=binomial(link="probit"))
>
> summary(bwt.glm2, cor=F)
......
Coefficients:
             Estimate Std. Error z value Pr(>|z|)
(Intercept)  0.461585   0.722771   0.639  0.52306
age         -0.023300   0.022514  -1.035  0.30071
lwt         -0.008925   0.004056  -2.200  0.02779 *
raceblack    0.704876   0.318383   2.214  0.02683 *
raceother    0.441680   0.266840   1.655  0.09788 .
smoke        0.464624   0.246865   1.882  0.05982 .
ptdTRUE      0.816927   0.288368   2.833  0.00461 **
ht           1.132272   0.426167   2.657  0.00789 **
ui           0.414494   0.281152   1.474  0.14041
ftv1        -0.279556   0.279582  -1.000  0.31736
ftv2+        0.086412   0.269392   0.321  0.74839
---
Signif. codes:  0 '***' 0.001 '**' 0.01 '*' 0.05 '.' 0.1 ' ' 1
......
    Null deviance: 234.67  on 188  degrees of freedom
Residual deviance: 194.99  on 178  degrees of freedom
......
```

Ex 2.5. For the dataset `birthwt`, write R code to compute model-robust standard errors for the point estimates of coefficients from logit regression and probit regression.

Ex 2.6. Present *in analytical details* the coordinate descent algorithm for fitting logistic regression with a Lasso penalty (on non-intercept coefficients only), depending on how a quadratic approximation or majorization is performed to the likelihood loss function.

(i) Use the quadratic approximation directly from the Taylor expansion,

(ii) Use the quadratic majorizing function as described in Section 2.5.

For analytical details, please show the formulas for updating a single coordinate by weighted or unweighted least squares with a Lasso penalty. Please also show the details for updating the intercept which is non-penalized.

Ex 2.7. For the dataset `birthwt`, write your own R code based on Exercise 2.6 to replicate the results of Lasso penalized logistic regression from the R package `glmnet`.

Note: Please create the covariate matrix according to logistic regression in Exercise 2.4. Please demonstrate your replication of results from `glmnet` with `standardize=T` given a sufficiently dense sequence of lambda values for the Lasso penalty. Produce a path plot of coefficients similar to Figure 2.3 (left) for linear regression.

3

Smoothing methods: Splines and kernels

In this chapter, we study smoothing methods, mainly univariate curve fitting, through the following topics. First, we discuss ridge penalized splines and cubic smoothing splines, which are two popular smoothing methods using spline basis functions. Second, we discuss nonparametric smoothing based on reproducing kernel Hilbert spaces (RKHS), which provides a useful general framework for studying both smoothing splines and other methods such as Gaussian kernel regression. Third, we discuss Lasso penalized splines and total variation splines, which are Lasso counterparts to ridge penalized splines and smoothing splines. Fourth, we briefly discuss generalized nonparametric regression by combining smoothing with generalized linear regression. We provide numerical demonstration in R, using both existing packages and self-contained code for implementation and application of main methods.

Smoothing methods involve fitting smooth curves (or surfaces) to capture the dependency of responses on univariate (or multivariate) covariates. There has been an extensive and growing literature on this topic. For example, textbook accounts (which are closely related to our discussion) include Hastie et al. (2009), Chapter 5, or Ruppert et al. (2003), Chapter 3. For smoothing splines and related RKHSs, a classical monograph is Wahba (1990), and a comprehensive treatment is provided by Gu (2013) in the framework of functional ANOVA models for multivariate smoothing. Total variation splines are formally studied by Mammen and van de Geer (1997).

We mainly discuss univariate smoothing, i.e., fitting a smooth curve for a scalar response against a scalar covariate, although RKHS regression in Section 3.4 can be applied to both univariate and multivariate smoothing. We discuss kernel methods in the sense of RKHS regression with reproducing kernels. We do not discuss kernel smoothing known as local regression, where weighting kernels are used for localization as in Nadaraya–Waston smoothing. See Hastie et al. (2009), Chapter 6, and references therein.

DOI: 10.1201/9781003680727-3

3.1 Ridge penalized splines

3.1.1 Data and model

Let $\{(Z_i, Y_i) : i = 1, \ldots, n\}$ be a training dataset, where Z_i is a scalar *covariate* and Y_i is a scalar *response variable.* A *linear* smoothing (or nonparametric regression) model postulates that

$$Y_i = f^*(Z_i) + \varepsilon_i, \quad i = 1, \ldots, n, \tag{3.1}$$

where $f^*(\cdot)$ is an unknown function, assumed to be smooth from some function class $\mathcal{F}$, and ε_i is a noise variable, assumed to satisfy $E(\varepsilon_i) = 0$ in a fixed design or $E(\varepsilon_i|Z_i) = 0$ in a random design for $i = 1, \ldots, n$. The word "linear" is referred to the fact that $E(Y_i)$ or $E(Y_i|Z_i)$ is linear in (more precisely, identical to) $f^*(Z_i)$. We give a list of examples to illustrate how various classes of piecewise polynomials can be specified for $f^*(\cdot)$.

- *Linear regression.* The class $\mathcal{F}$ consists of functions in the form

  $$f(z) = b_0 + b_1 z.$$

  Then (3.1) reduces to linear regression

  $$Y_i = b_0^* + b_1^* Z_i + \varepsilon_i, \quad i = 1, \ldots, n,$$

  where (b_0^*, b_1^*) are the parameters such that $f^*(z) = b_0^* + b_1^* z$.

- *Linear spline with 1 knot.* The class $\mathcal{F}$ consists of functions in the form

  $$\begin{aligned} f(z) &= b_0 + b_1 z + c_1 (z - \xi_1)_+ \\ &= \begin{cases} b_0 + b_1 z, & z \le \xi_1, \\ b_0 + b_1 z + c_1 (z - \xi_1), & z > \xi_1, \end{cases} \end{aligned}$$

  where $\xi_1 \in \mathbb{R}$ is a constant, called a knot, and $a_+ = 0$ if $a \le 0$ or a if $a > 0$. Formally, with a fixed knot ξ_1, the class $\mathcal{F}$ is the set of all functions $f(z)$ in the preceding form for $b_0, b_1 \in \mathbb{R}$. A different choice of ξ_1 leads to a different class $\mathcal{F}$. The function $f(z)$ is continuous everywhere and it is piecewise linear: linear with slope b_1 if $z \le \xi_1$ and linear with slope $b_1 + c_1$ if $z > \xi_1$. Hence the coefficient c_1 indicates the change of the slope at the knot. See Figure 3.1.

- *Linear spline with k knots.* The class $\mathcal{F}$ consists of functions in the form

  $$f(z) = b_0 + b_1 z + \sum_{j=1}^{k} c_j (z - \xi_j)_+,$$

  where $\xi_1 < \cdots < \xi_k$ are fixed knots. The boundary knots can be set to $\xi_0 = -\infty$ and $\xi_{k+1} = \infty$. The function $f(z)$ is continuous everywhere

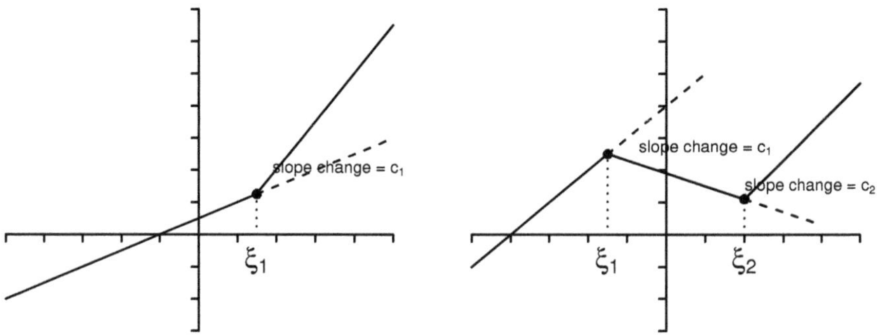

FIGURE 3.1
A linear spline with 1 knot (left) or with 2 knots (right).

and piecewise linear (hence called a linear spline): it is a linear function between two consecutive knots, but the slope varies across each knot. The coefficient c_j indicates the change of the slope at the knot ξ_j. See Figure 3.1. The smoothing model (3.1) corresponds to linear regression with the "covariates" (or more appropriately here, the regressors) for Y_i taken to be $1, Z_i, (Z_i - \xi_1)_+, \ldots, (Z_i - \xi_k)_+$, i.e., the following *basis functions* evaluated at Z_i:

$$1, z, (z - \xi_1)_+, \ldots, (z - \xi_k)_+,$$

which are called truncated linear basis functions.

- *Quadratic spline with k knots.* The class $\mathcal{F}$ consists of functions in the form

$$f(z) = b_0 + b_1 z + b_2 z^2 + \sum_{j=1}^{k} c_j \{(z - \xi_j)_+\}^2,$$

where $\xi_1 < \cdots < \xi_k$ are fixed knots. The function $f(z)$ is differentiable everywhere and piecewise quadratic (hence called a quadratic spline): it is a quadratic function between two consecutive knots. [Hence a quadratic spline is more than being continuous and piecewise quadratic.] The slopes from the left and right of each knot are the same, but the curvatures (i.e., second derivatives) differ on the two sides of the knot. The coefficient c_j indicates the change of the curvature at the knot ξ_j. See Figure 3.2. The corresponding basis functions, called truncated quadratic bases, are

$$1, z, z^2, \{(z - \xi_1)_+\}^2, \ldots, \{(z - \xi_k)_+\}^2.$$

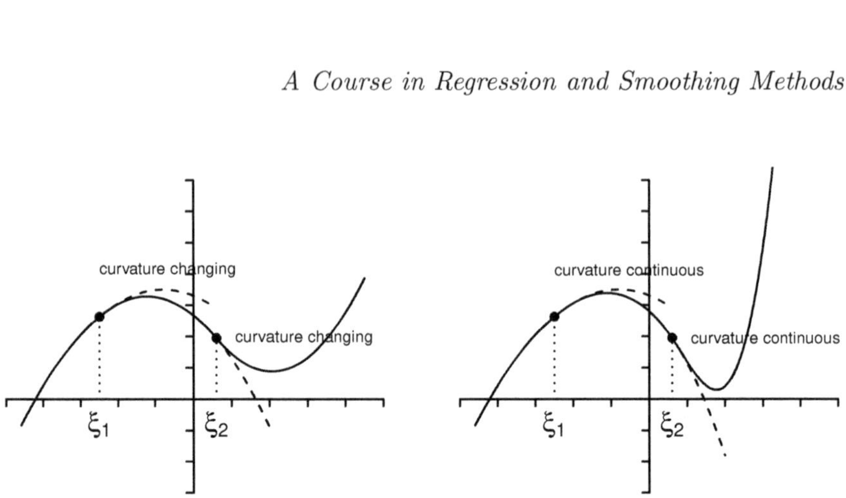

FIGURE 3.2
A quadratic spline with 2 knots (left) and a cubic spline with 2 knots (right). The two functions are the same to the left of ξ_1 and pass the same two points at ξ_1 and ξ_2. A dashed line to the right of a knot indicates the same polynomial function as to the near left.

- *Cubic spline with k knots.* The class $\mathcal{F}$ consists of functions in the form

$$f(z) = b_0 + b_1 z + b_2 z^2 + b_3 z^3 + \sum_{j=1}^{k} c_j \{(z - \xi_j)_+\}^3,$$

where $\xi_1 < \cdots < \xi_k$ are fixed knots. The function $f(z)$ is twice differentiable everywhere and piecewise cubic (hence called a cubic spline): it is a cubic polynomial between two consecutive knots. The slopes and curvatures from the left and right of each knot are the same, but the third derivatives differ on the two sides of the knot. See Figure 3.2. The corresponding basis functions, called truncated cubic bases, are

$$1, z, z^2, z^3, \{(z - \xi_1)_+\}^3, \ldots, \{(z - \xi_k)_+\}^3.$$

- *Polynomial spline with degree q and k knots.* The class $\mathcal{F}$ consists of functions in the form

$$f(z) = b_0 + b_1 z + \cdots + b_q z^q + \sum_{j=1}^{k} c_j \{(z - \xi_j)_+\}^q,$$

where $\xi_1 < \cdots < \xi_k$ are fixed knots. The corresponding basis functions, called truncated power bases, can be collected into a column vector

$$\phi_{\mathrm{T}}(z) = (1, z, \ldots, z^q, \{(z - \xi_1)_+\}^q, \ldots, \{(z - \xi_k)_+\}^q)^{\mathrm{T}}. \qquad (3.2)$$

In practice, polynomial splines with degree $q > 3$ are rarely used.

3.1.2 Estimation

The smoothing model (3.1) with f^* specified as a polynomial spline with degree q and k knots can be put as linear regression:

$$\begin{aligned} Y_i &= X_i^{\mathrm{T}}\beta^* + \varepsilon_i \\ &= b_0^* + b_1^* Z_i + \cdots + b_q^* Z_i^q + \sum_{j=1}^{k} c_j^* \{(Z_i - \xi_j)_+\}^q + \varepsilon_i, \quad i = 1, \ldots, n, \end{aligned} \tag{3.3}$$

where $\beta^* = (b_0^*, \ldots, b_q^*, c_1^*, \ldots, c_k^*)^{\mathrm{T}}$ is a vector of $p = 1 + q + k$ unknown coefficients and

$$X_i = \phi_{\mathrm{T}}(Z_i) = (1, Z_i, \ldots, Z_i^q, \{(Z_i - \xi_1)_+\}^q, \ldots, \{(Z_i - \xi_k)_+\}^q)^{\mathrm{T}}$$

is a vector of basis functions evaluated at Z_i, hence called a basis vector. In the matrix notation, the linear regression is $\mathbf{Y} \sim \mathbf{X}_{\mathrm{T}}\beta^*$, where

$$\mathbf{Y} = \begin{pmatrix} Y_1 \\ \vdots \\ Y_n \end{pmatrix},$$

$$\mathbf{X}_{\mathrm{T}} = \begin{pmatrix} 1 & Z_1 & \ldots & Z_1^q & \{(Z_1 - \xi_1)_+\}^q & \ldots & \{(Z_1 - \xi_k)_+\}^q \\ \vdots & \vdots & \vdots & \vdots & \vdots & \vdots & \vdots \\ 1 & Z_n & \ldots & Z_n^q & \{(Z_n - \xi_1)_+\}^q & \ldots & \{(Z_n - \xi_k)_+\}^q \end{pmatrix}.$$

For convenience, $\mathbf{X}_{\mathrm{T}}$ is called a basis matrix. As in Chapter 1, the least-squares estimator $\hat{\beta} = (\hat{b}_0, \ldots, \hat{b}_q, \hat{c}_1, \ldots, \hat{c}_q)^{\mathrm{T}}$ for β^* is

$$\begin{aligned} \hat{\beta} &= \operatorname{argmin}_\beta \frac{1}{n} \sum_{i=1}^{n} (Y_i - X_i^{\mathrm{T}}\beta)^2 \\ &= \left(\sum_{i=1}^{n} X_i X_i^{\mathrm{T}} \right)^{-1} \left(\sum_{i=1}^{n} X_i Y_i \right) \\ &= (\mathbf{X}_{\mathrm{T}}^{\mathrm{T}} \mathbf{X}_{\mathrm{T}})^{-1} (\mathbf{X}_{\mathrm{T}}^{\mathrm{T}} \mathbf{Y}). \end{aligned}$$

Then the fitted function for f^* is

$$\hat{f}(z) = \phi_{\mathrm{T}}(z)^{\mathrm{T}} \hat{\beta} = \hat{b}_0 + \hat{b}_1 z + \cdots + \hat{b}_q z^q + \sum_{j=1}^{k} \hat{c}_j \{(z - \xi_j)_+\}^q,$$

which is called a (non-penalized) regression spline.

The regression spline can be made to fit the training data arbitrarily well, by using an increasing number of knots. For example, for a linear spline with $k = n - 2$ knots (hence $k + 2 = n$ regressors), the training data with distinct $Z_1, \ldots, Z_n$ can be fitted perfectly by least squares, i.e., $\hat{f}(Z_i) = Y_i$ for $i =$

$1, \dots, n$. But then the fitted function $\hat{f}(\cdot)$ may be too wiggly (or rough) to provide useful predictions at new data points (outside the training data). To control roughness and promote smoothness of the fitted function, it is natural to employ penalized estimation as in Section 1.8. In the following, we discuss ridge estimation which is conventionally more popular than Lasso estimation, to be discussed later in Section 3.6.

As a modification from ridge estimation in Section 1.8, it is customary to use the generalized ridge estimator

$$\begin{aligned}\hat{\beta}_\lambda &= \mathrm{argmin}_\beta \left\{ \frac{1}{2n} \sum_{i=1}^n (Y_i - X_i^{\mathrm{T}} \beta)^2 + \frac{\lambda}{2} \sum_{j=1}^k c_j^2 \right\} \\ &= \mathrm{argmin}_\beta \left\{ \frac{1}{2n} \sum_{i=1}^n (Y_i - X_i^{\mathrm{T}} \beta)^2 + \frac{\lambda}{2} \beta^{\mathrm{T}} D \beta \right\},\end{aligned} \tag{3.4}$$

where $\lambda > 0$ is a tuning parameter, and D, called the ridge weight matrix, is the diagonal matrix

$$D = \begin{pmatrix} 0_{(q+1)\times(q+1)} & 0_{(q+1)\times k} \\ 0_{k\times(q+1)} & I_{k\times k} \end{pmatrix}.$$

As a roughness measure, the ridge penalty is placed only on the coefficients $(c_1, \dots, c_k)$, which are associated with the truncated basis functions and encode the changes of the qth derivatives across the knots. Setting the gradient of the penalized objective to 0,

$$0 = -\frac{1}{n} \sum_{i=1}^n (Y_i - X_i^{\mathrm{T}} \beta) X_i + \lambda D \beta,$$

and solving for β gives a closed-form expression

$$\begin{aligned}\hat{\beta}_\lambda &= \left(\frac{1}{n} \sum_{i=1}^n X_i X_i^{\mathrm{T}} + \lambda D \right)^{-1} \left(\frac{1}{n} \sum_{i=1}^n X_i Y_i \right) \\ &= (\mathbf{X}_{\mathrm{T}}^{\mathrm{T}} \mathbf{X}_{\mathrm{T}} + n\lambda D)^{-1} (\mathbf{X}_{\mathrm{T}}^{\mathrm{T}} \mathbf{Y}).\end{aligned} \tag{3.5}$$

Then the fitted function for f^*,

$$\hat{f}_\lambda(z) = \phi_{\mathrm{T}}(z)^{\mathrm{T}} \hat{\beta}_\lambda = \hat{b}_{\lambda,0} + \hat{b}_{\lambda,1} z + \cdots + \hat{b}_{\lambda,q} z^q + \sum_{j=1}^k \hat{c}_{\lambda,j} \{(z - \xi_j)_+\}^q,$$

is called a (ridge) penalized spline.

In practice, regression splines or penalized splines are often used with two different ways of specifying the knots. The first approach is to specify a small number of knots as sample quantiles of the observed covariates $Z_1, \dots, Z_n$, for example, 3 knots at the 25%, 50%, 75% sample quantiles or 9 knots at

10%, ..., 90% sample quantiles. In this approach, regression splines are typically used. The second approach is to specify a large number of knots or even the maximum possible number of knots, for example, $k = n - 2$ knots for a linear spline, and then employ penalized estimation in fitting the spline. A prominent method related to the second approach is smoothing splines, which will be discussed in Section 3.3.

The truncated power basis functions are conceptually simple, facilitating a direct interpretation of the coefficients c_j for the truncated basis functions and hence a motivation for the particular ridge penalty in (3.4). Alternatively, there is an equivalent set of basis functions, called B-spline basis functions, which can be numerically more appealing due to their local supports. See Hastie et al. (2009), Chapter 5 Appendix. In the following, we describe how the fitting process needs to be modified to achieve equivalent fitting results when the truncated power basis functions are changed to the B-spline ones.

For polynomial splines with degree q and k knots, the B-spline basis functions, collected in a column vector $\phi_{\mathrm{B}}(z) = (\phi_{\mathrm{B},1}(z), \ldots, \phi_{\mathrm{B},p}(z))^{\mathrm{T}}$, are related to the truncated power basis functions in (3.2) via a linear transformation: $\phi_{\mathrm{B}}(z)^{\mathrm{T}} = \phi_{\mathrm{T}}(z)^{\mathrm{T}} L$ or

$$(\phi_{\mathrm{B},1}(z), \ldots, \phi_{\mathrm{B},p}(z)) = (1, z, \ldots, z^q, \{(z - \xi_1)_+\}^q, \ldots, \{(z - \xi_k)_+\}^q) L,$$

where L is a $p \times p$ invertible matrix, with $p = 1 + q + k$. The basis matrix obtained by evaluating the B-spline basis functions at the data points is

$$\mathbf{X}_{\mathrm{B}} = \begin{pmatrix} \phi_{\mathrm{B},1}(Z_1) & \ldots & \phi_{\mathrm{B},p}(Z_1) \\ \vdots & \vdots & \vdots \\ \phi_{\mathrm{B},1}(Z_n) & \ldots & \phi_{\mathrm{B},p}(Z_n) \end{pmatrix}.$$

Then $\mathbf{X}_{\mathrm{B}}$ is also related to the basis matrix $\mathbf{X}_{\mathrm{T}}$ as

$$\mathbf{X}_{\mathrm{B}} = \mathbf{X}_{\mathrm{T}} L.$$

The linear regression, $\mathbf{Y} \sim \mathbf{X}_{\mathrm{T}} \beta^*$, can be put as $\mathbf{Y} \sim \mathbf{X}_{\mathrm{B}} \theta^*$, where θ^* is related to β^* as $\theta^* = L^{-1} \beta^*$. By replacing $\mathbf{X}_{\mathrm{T}}$ in $\hat{\beta}$ with $\mathbf{X}_{\mathrm{B}}$, the least-squares estimator for θ^* is

$$\hat{\theta} = (\mathbf{X}_{\mathrm{B}}^{\mathrm{T}} \mathbf{X}_{\mathrm{B}})^{-1} (\mathbf{X}_{\mathrm{B}}^{\mathrm{T}} \mathbf{Y}),$$

which can be easily shown to satisfy $\hat{\theta} = L^{-1} \hat{\beta}$. Hence the fitted values match on the training data, $\mathbf{X}_{\mathrm{B}} \hat{\theta} = \mathbf{X}_{\mathrm{T}} \hat{\beta}$, or more generally the fitted functions also match, $\phi_{\mathrm{B}}(\cdot)^{\mathrm{T}} \hat{\theta} = \phi_{\mathrm{T}}(\cdot)^{\mathrm{T}} \hat{\beta}$. However, to match the fitting results from penalized estimation, the ridge estimator for θ^* is defined with a new ridge weight matrix $\tilde{D}$ as

$$\begin{aligned} \hat{\theta}_\lambda &= \operatorname{argmin}_\theta \left\{ \frac{1}{2n} \|\mathbf{Y} - \mathbf{X}_{\mathrm{B}} \theta\|_2^2 + \frac{\lambda}{2} \theta^{\mathrm{T}} \tilde{D} \theta \right\} \\ &= (\mathbf{X}_{\mathrm{B}}^{\mathrm{T}} \mathbf{X}_{\mathrm{B}} + n\lambda \tilde{D})^{-1} (\mathbf{X}_{\mathrm{B}}^{\mathrm{T}} \mathbf{Y}). \end{aligned} \tag{3.6}$$

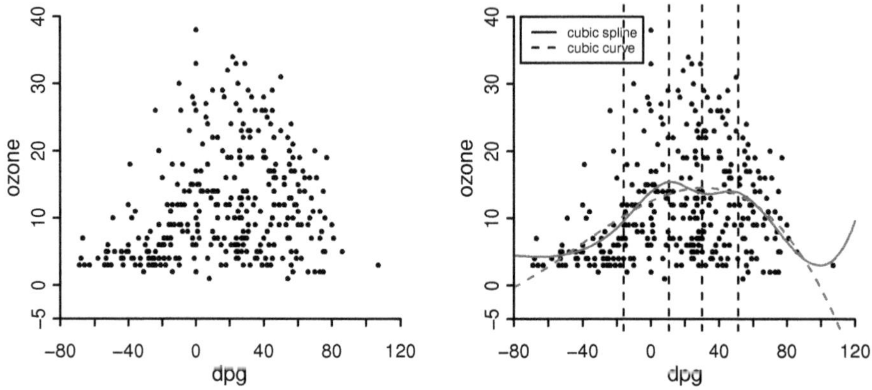

FIGURE 3.3
Scatter plot of ozone versus dpg in the ozone dataset (left) and the fitted (non-penalized) cubic spline as well as the fitted cubic curve (i.e., a polynomial of degree 3) by least squares (right). Vertical lines are placed at the four knots used for the cubic spline.

To match the penalized objective functions in (3.4) and (3.6), we require

$$\mathbf{X}_{\mathrm{T}}\beta = \mathbf{X}_{\mathrm{B}}\theta, \quad \beta^{\mathrm{T}} D\beta = \theta^{\mathrm{T}}\tilde{D}\theta,$$

which are satisfied by setting $\theta = L^{-1}\beta$ and $\tilde{D} = L^{\mathrm{T}}DL$. Therefore, if $\tilde{D} = L^{\mathrm{T}}DL$, then $\hat{\theta}_\lambda = \hat{\beta}_\lambda L^{-1}$ so that the fitted functions match, $\phi_{\mathrm{B}}(\cdot)\hat{\theta}_\lambda = \phi_{\mathrm{T}}(\cdot)\hat{\beta}_\lambda$. Although D is diagonal, the new weight matrix $\tilde{D}$ is no longer diagonal. Given the basis matrix $\mathbf{X}_{\mathrm{B}}$, the transformation matrix L is still required to compute $\tilde{D}$ and the ridge estimator $\hat{\theta}_\lambda$, even though the detail of L can be avoided for computing the least-squares estimator $\hat{\theta}$.

3.2 Numerical demonstration: Penalized splines

We demonstrate implementation of ridge penalized splines using truncated power or B-spline basis functions. As shown below, consider the ozone dataset used in Breiman and Friedman (1985), which contains daily measurements of ozone concentration (ozone) and several meteorological quantities in the Los Angeles basin in 1976. For illustration, we study univariate smoothing (or curve fitting) of ozone against the Dagget pressure gradient (dpg). Figure 3.3 (left) gives the scatter plot of the two variables.

```
> data <- read.table(file="LAozone.data", header=T, sep=",")
> data[1:5,]
```

```
  ozone   vh wind humidity temp  ibh dpg ibt vis doy
1     3 5710    4       28   40 2693 -25  87 250   3
2     5 5700    3       37   45  590 -24 128 100   4
3     5 5760    3       51   54 1450  25 139  60   5
4     6 5720    4       69   35 1568  15 121  60   6
5     4 5790    6       19   45 2631 -33 123 100   7
>
> attach(data)
```

We write the following function, `tr.pow()`, to compute the truncated power basis functions (excluding the constant) at a set of points (supplied as a vector `x`), depending on the degree and knots (supplied as an integer `degree` and a vector `knots` respectively). To prevent potentially widely different scales over the range of covariate data, the basis functions are computed after rescaling the covariate data to [0, 1], provided that the data points are contained between the two points specified by `ends`. (The knots remain to be specified in the original scale.) The function returns a basis matrix with the number of rows equal to the length of `x` and the number of columns equal to `degree` plus the length of `knots`.

```
tr.pow <- function(x, degree=3, knots, ends=c(0,1)) {
  n <- length(x)
  k <- length(knots)
  out <- matrix(0, n, degree+k)   #degree + number of knots

  x <- (x-ends[1])/(ends[2]-ends[1])          # rescaling
  knots <- (knots-ends[1])/(ends[2]-ends[1])

  for (j in 1:degree)
    out[,j] <- x^j
  for (j in 1:k)
    out[, degree+j] <- ifelse(x> knots[j], (x-knots[j])^degree, 0)
  out
}
```

For a sequence of points `x` in $(-80, 120)$, which includes the range of `dpg` $(-69, 107)$, we apply the function `tr.pow()` to obtain the truncated cubic basis matrix `bX1`, where 4 knots are used at the 20%, 40%, 60%, 80% quantiles of the covariate data `dpg`. [The elements of `x` are called grid points, whereas those of `dpg` are called data points.] For comparison, we also apply the built-in function `bs()`, available from the R package `splines`, to obtain the cubic B-spline basis matrix `bx2`. The returned basis matrix from `bs()` depends on both the arguments `knots` indicating the interior knots and `Boundary.knots`

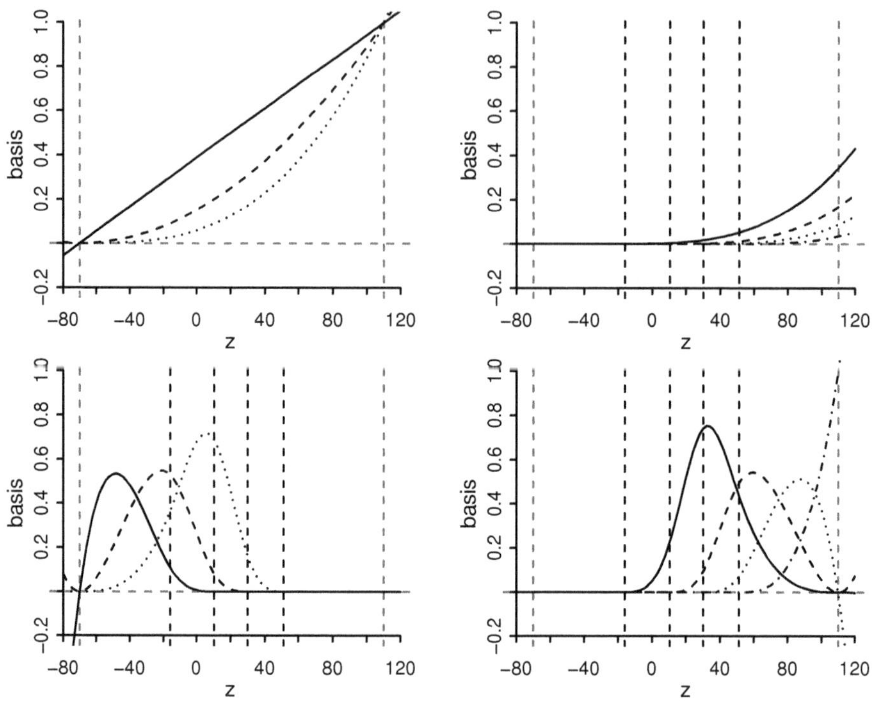

FIGURE 3.4
Plots of the truncated cubic basis functions (first row) and cubic B-spline basis functions (second row), excluding the constant. Vertical lines are placed at the four (interior) knots as well as `ends` for truncated cubic basis or `Boundary.knots` for B-spline basis.

needed to anchor the B-spline basis functions. The warning message from `bs()` can be ignored, because the data points in `dpg` are inside the range of `Boundary.knots`, even though some points in `x` are outside the range. See `help(bs)` and the accompanying R code file for more information.

Figure 3.4 shows the basis functions represented as the columns of `bX1` and `bX2`. As mentioned above, the truncated cubic basis functions are computed after rescaling based on `ends`: (z, z^2, z^3) are plotted after being shifted to $z = -70$ and then divided by $(180, 180^2, 180^3)$ respectively, and the functions, $\{(z-\xi_j)_+\}^3$, are plotted after being divided by 180^3. The B-spline basis functions are, by design, of restrained scales over the range defined by `Boundary.knots`, but may behave in an uncontrolled manner outside the range.

```
> knots <- quantile(dpg, c(.2,.4,.6,.8))
> ends <- c(-70, 110)
> x <- seq(-80,120,2)
```

```
>
> bX1 <- tr.pow(x, degree=3, knots=knots, ends=ends)
> dim(bX1)
[1] 101   7
>
> library(splines)
> bX2 <- bs(x, degree=3, knots=knots, Boundary.knots=c(-70,110))
Warning message:
......
  some 'x' values beyond boundary knots may cause ill-conditioned bases
> dim(bX2)
[1] 101   7
```

We fit a non-penalized cubic spline (as well as a cubic curve) by least squares using the R function `lm()` with the response `ozone` and the basis matrix, `X1` or `X2` (i.e., $\mathbf{X}_{\mathrm{T}}$ or $\mathbf{X}_{\mathrm{B}}$ in Section 3.1). The basis matrix `X1` or `X2` is obtained by evaluating the truncated power or B-spline basis functions at the *data points* in `dpg`. Hence `X1` and `X2` differ from `bX1` and `bX2`, which give the associated basis functions at the *grid points* in `x`. Nevertheless, the fitted spline functions, represented as the vectors `ffit.tp` and `ffit.bs`, are determined by multiplying `bX1` and `bX2` with the coefficient vectors computed from `lm()`. As indicated by the comparison of `ffit.tp` and `ffit.bs` below, the two fitted functions, based on the truncated power and B-spline basis, are numerically identical to each other.

Figure 3.3 (right) shows the fitted (non-penalized) cubic spline as well as the cubic curve, i.e., `ffit.tp` or `ffit.bs` and `ffit.cc` at the grid points `x`. The cubic spline appears to fit the data more satisfactorily than the cubic curve. Both curves, however, exhibit some limitations at the right end of `dpg`: a wiggly turn by the cubic spline and a downward bend by the cubic curve. The wiggly turn in the cubic spline would be missing if the function is plotted *only* at the data points in `dpg`. See the accompanying R code file for details.

```
> X1 <- tr.pow(dpg, degree=3, knots=knots, ends=ends)
> X2 <- bs(dpg, degree=3, knots=knots, Boundary.knots=c(-70,110))
>
> ffit.tp <- cbind(1,bX1)%*%lm(ozone ~ X1)$coef
> ffit.bs <- cbind(1,bX2)%*%lm(ozone ~ X2)$coef
> ffit.cc <- cbind(1,bX1[,1:3])%*%lm(ozone ~ X1[,1:3])$coef
>
> cbind(x, ffit.tp, ffit.bs, ffit.cc)[1:10,]
       x
 [1,] -80 4.461655 4.461655 -0.25179047
 [2,] -78 4.414351 4.414351  0.01938234
```

```
 [3,] -76 4.370981 4.370981  0.29964721
 [4,] -74 4.332483 4.332483  0.58848465
 [5,] -72 4.299797 4.299797  0.88537512
 [6,] -70 4.273862 4.273862  1.18979912
 [7,] -68 4.255618 4.255618  1.50123713
 [8,] -66 4.246002 4.246002  1.81916963
 [9,] -64 4.245955 4.245955  2.14307711
[10,] -62 4.256415 4.256415  2.47244005
```

We write the following functions, `pen.spline0()` and `pen.spline()`, to implement ridge penalized splines using truncated power basis functions and the ridge estimator of the coefficient vector in (3.4) and (3.5). The tuning parameter `lam` corresponds to $n\lambda$ in the notation of Section 3.1. The function `pen.spline0()` requires, as an input argument, the basis matrix `X` at the data points. The function `pen.spline()` combines `tr.pow()` and `pen.spline0()` and serves as a higher-level interface, with the covariate data points `x` and the covariate knots, `knots`, as input arguments. Each of the two functions returns a list including the coefficient estimates (`coef`), the fitted values at the covariate data points (`fit`), the smoother matrix (`S`), and the degrees of freedom (`df`), which is discussed in Section 3.3.

```
pen.spline0 <- function(y, X, degree=3, lam) {
  X <- cbind(1, X)   #truncated power basis matrix
  n <- dim(X)[1]
  k <- dim(X)[2]

  D <- diag(k)
  D[1:(1+degree), 1:(1+degree)] <- 0
  coef <- solve( t(X)%*%X/n+lam/n*D, t(X)%*%y/n )
  fit <- X%*% coef

  S <- X%*% solve( t(X)%*%X/n+lam/n*D, t(X)/n )
  df <- sum(diag(S))

  list(coef=c(coef), fit=c(fit), S=S, df=df)
}

pen.spline <- function(y, x, degree=3, knots, ends, lam) {
  X <- tr.pow(x, degree, knots, ends)
  out <- pen.spline0(y, X, degree, lam)
  list(coef=out$coef, fit=out$fit, S=out$S, df=out$df)
}
```

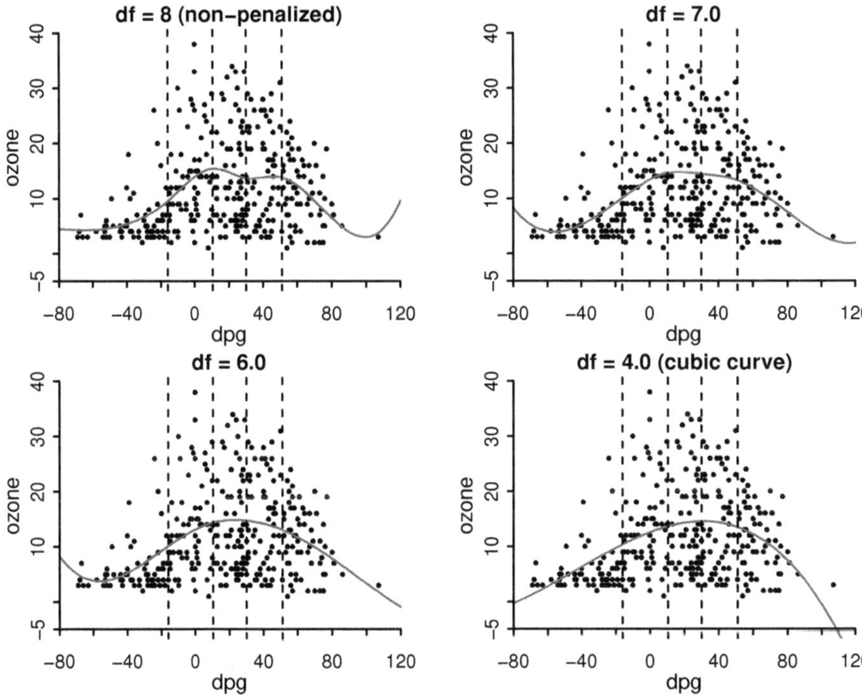

FIGURE 3.5
Plots of ridge-penalized cubic splines for `ozone` against `dpg` in the ozone dataset, with several choices of tuning parameter λ. The corresponding degrees of freedom are stated in the title of each plot. Vertical lines are placed at the four knots which are quantiles of `dpg`.

Figure 3.5 shows the ridge-penalized cubic splines obtained using `pen.spline0()` with several choices of tuning parameter λ, including two extreme cases leading to the non-penalized cubic spline or the cubic curve as in Figure 3.3. The following R code illustrates how the (1,2) plot with `df = 7.0` can be produced. The fitted spline function is evaluated at the grid points `x`, not just the data points `dpg`, and then the function values, `ffit.pen1`, are plotted. As the tuning parameter λ varies, the cubic spline takes different shapes, especially near the two ends of the data points `dpg`.

```
> out1 <- pen.spline0(ozone, X1, degree=3, lam=10.7e-6)
> out1$df
[1] 7.00523
> ffit.pen1 <- cbind(1,bX1)%*%out1$coef
> plot(dpg, ozone, xlim=c(-80,120), ylim=c(-5,40))
> lines(x, ffit.pen1, col="blue")
> abline(v=knots, lty=2)
```

3.3 Cubic smoothing splines

The ridge-penalized splines in Section 3.1 are derived by taking the function class $\mathcal{F}$ to be a finite-dimensional linear space of splines with specific basis functions and then converting the smoothing problem to fitting linear regression with the regressors defined from the basis functions. For penalized estimation, the ridge penalty is introduced directly on the coefficients associated with the basis functions in the spline function. This approach can be said to be basis driven. In this section, we describe a function-driven approach where $\mathcal{F}$ is an infinite-dimensional space of smooth functions and a roughness penalty is defined on any function in $\mathcal{F}$ directly through its derivatives.

For preparation, we introduce a subclass of cubic splines, called natural cubic splines, which are constrained to be linear beyond the boundary knots. As stated in Section 3.1, a cubic spline with knots $\xi_1 < \cdots < \xi_k$ can be represented as

$$f(z) = b_0 + b_1 z + b_2 z^2 + b_3 z^3 + \sum_{j=1}^{k} c_j\{(z-\xi_j)_+\}^3, \tag{3.7}$$

in terms of $k+4$ basis functions and associated coefficients. With k fixed knots, the set of cubic splines is $(k+4)$ dimensional, hence over-parameterized for interpolating the function values at the k knots. Given a cubic spline $f(z)$ defined as (3.7), it can be shown that for any $\tilde{c}_1, \tilde{c}_k \in \mathbb{R}$, the function $\tilde{f}(z) = f(z) + \tilde{c}_1\{(\xi_1 - z)_+\}^3 + \tilde{c}_k(\{(z-\xi_k)_+\}^3$ is also a cubic spline with the same knots $\xi_1, \ldots, \xi_k$, and $\tilde{f}$ coincides with f in the interval $[\xi_1, \xi_k]$ but may wildly deviate from f outside the interval [Exercise]. See Figure 3.6. There also exist infinite choices of cubic splines, each of which admits the same knots and associated values as f but it may differ from f both inside and outside the interval $[\xi_1, \xi_k]$.

A natural cubic spline with knots $\xi_1 < \cdots < \xi_k$ (including interior knots $\xi_2, \ldots, \xi_{k-1}$ and boundary knots ξ_1, ξ_k) is defined by requiring that $f(z)$ is linear beyond the two boundary knots ξ_1 and ξ_k, i.e.,

$$f^{(2)}(z) = f^{(3)}(z-) = 0 \quad \text{for} \quad z \le \xi_1, \tag{3.8a}$$

$$f^{(2)}(z) = f^{(3)}(z+) = 0 \quad \text{for} \quad z \ge \xi_k, \tag{3.8b}$$

where $f^{(j)}(z)$ denotes the jth derivative, and $f^{(j)}(z\mp)$ denotes the jth derivative from the left or right. These boundary constraints lead to 4 linear constraints on the $k+4$ coefficients in (3.7), so that a natural cubic spline can be represented in terms of k basis functions and associated coefficients. In fact, (3.8a) immediately indicates that for $z \le \xi_1$,

$$f^{(2)}(z) = 2b_2 + 6b_3 z = 0, \quad f^{(3)}(z-) = 6b_3 = 0,$$

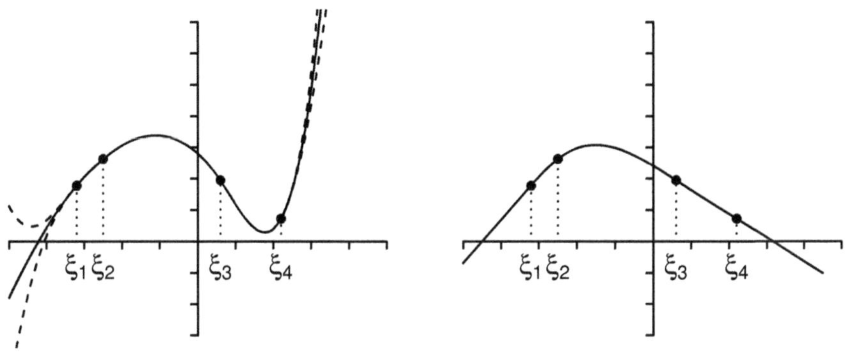

FIGURE 3.6
A cubic spline with 4 knots (left) and a natural cubic spline with the same knots (right), passing the same four points at the knots. Each dashed line on the left indicates a possible cubic spline which deviates from the solid line only outside the boundary knots.

and hence $b_2 = b_3 = 0$. Then (3.8b) implies that for $z \geq \xi_k$,

$$f^{(2)}(z) = 2b_2 + 6b_3 z + \sum_{j=1}^{k} 6c_j(z - \xi_j) = 0, \quad f^{(3)}(z+) = 6b_3 + \sum_{j=1}^{k} 6c_j = 0,$$

and hence $\sum_{j=1}^{k} c_j = 0$ and $\sum_{j=1}^{k} c_j \xi_j = 0$. By incorporating these constraints in (3.7), a natural cubic spline $f(z)$ can be represented as a linear combination of the following k basis functions (Hastie et al., 2009, Section 5.2.1):

$$\phi_{\mathrm{N},1}(z) = 1, \quad \phi_{\mathrm{N},2}(z) = z, \quad \phi_{\mathrm{N},j+2}(z) = d_j(z) - d_{k-1}(z), \quad j = 1, \ldots, k-2, \tag{3.9}$$

where

$$d_j(z) = \frac{\{(z - \xi_j)_+\}^3 - \{(z - \xi_k)_+\}^3}{\xi_k - \xi_j}.$$

For $j = 1, \ldots, k-2$, the basis function $\phi_{\mathrm{N},j+2}(z)$ is zero over $z \leq \xi_1$ and linear (with zero second and third derivatives) over $z > \xi_k$. With k fixed knots, the set of natural cubic splines is k dimensional. A (usually unique) natural cubic spline can be found to interpolate arbitrary function values at the k knots. In particular, for any cubic spline with k knots, there exists a natural cubic spline with the same knots and associated values. See Figure 3.6.

Next we introduce cubic smoothing splines. Assume the data points $Z_1, \ldots, Z_n$ fall in an interval $[a, b]$. For fixed $\lambda > 0$, a cubic smoothing spline is defined as

$$\hat{f}_\lambda = \text{argmin}_{f\in\mathcal{F}} \left\{ \frac{1}{n}\sum_{i=1}^{n}(Y_i - f(Z_i))^2 + \lambda \int_a^b (f''(z))^2 \,\mathrm{d}z \right\}, \tag{3.10}$$

where $\mathcal{F}$ is the set of all twice continuously differentiable functions $f : [a,b] \to \mathbb{R}$. Similarly to penalized linear regression in Section 1.8, the objective in (3.10) combines two terms: the first is a measure of prediction errors from using f, and the second is a penalty term on the curvatures of f. However, the optimization in (3.10) is over an infinite-dimensional function space $\mathcal{F}$. A major result underlying the appeal of smoothing splines is that a solution $\hat{f}_\lambda$ in (3.10) can be represented as a natural cubic spline with knots at the data points $Z_1, \ldots, Z_n$. In Section 3.4, we will discuss main ideas for proving this result, which also informs a different route for implementing smoothing splines from described below.

For the remainder of this section, we describe a direct approach for computing $\hat{f}_\lambda$ and examine basic properties of smoothing splines. From the aforementioned result, an important simplification is that the optimization in (3.10) can be reduced to a finite-dimensional program, searching over all natural cubic splines with knots at $(Z_1, \ldots, Z_n)$. For simplicity, assume that $Z_1, \ldots, Z_n$ are distinct and hence there are n knots. Otherwise, the set of distinct points in $(Z_1, \ldots, Z_n)$ are used as knots and the discussion can be modified accordingly. As noted earlier, such a natural cubic spline can be parameterized as

$$f(z;\theta) = \sum_{j=1}^{n} \theta_j \phi_{\mathrm{N},j}(z) = \phi_{\mathrm{N}}^{\mathrm{T}}(z)\theta,$$

with coefficients $\theta = (\theta_1, \ldots, \theta_n)^{\mathrm{T}}$ and basis functions $\phi_{\mathrm{N}} = (\phi_{\mathrm{N},1}, \ldots, \phi_{\mathrm{N},n})^{\mathrm{T}}$, obtained from (3.9) by setting $k = n$ and $(\xi_1, \ldots, \xi_k) = (Z_1, \ldots, Z_n)$. For fixed $\lambda > 0$, The smoothing spline $\hat{f}_\lambda$ can be determined as $\hat{f}_\lambda = f(\cdot;\hat{\theta}_\lambda)$ with

$$\hat{\theta}_\lambda = \text{argmin}_{\theta\in\mathbb{R}^n} \left\{ \frac{1}{n}\sum_{i=1}^{n}(Y_i - f(Z_i;\theta))^2 + \lambda \int_a^b (f''(z;\theta))^2 \,\mathrm{d}z \right\}. \tag{3.11}$$

The penalty term can be explicitly calculated as

$$\int_a^b (f''(z;\theta))^2 \,\mathrm{d}z = \int_a^b \left(\sum_{j=1}^{n} \theta_j \phi''_{\mathrm{N},j}(z) \right)^2 \mathrm{d}z$$
$$= \sum_{i,j=1}^{n} \theta_i \theta_j \int_a^b (\phi''_{\mathrm{N},j}(z))^2 \,\mathrm{d}z = \theta^{\mathrm{T}}\Omega\theta,$$

where Ω is an $n \times n$ matrix with (i,j)th element $\Omega_{ij} = \int_a^b \phi''_{\mathrm{N},i}(z)\phi''_{\mathrm{N},j}(z)\,\mathrm{d}z$. The preceding calculation also indicates that Ω is nonnegative definite. In the matrix notation, denote

$$\mathbf{X}_{\mathrm{N}} = \begin{pmatrix} \phi_{\mathrm{N},1}(Z_1) & \dots & \phi_{\mathrm{N},n}(Z_1) \\ \vdots & \vdots & \vdots \\ \phi_{\mathrm{N},1}(Z_n) & \dots & \phi_{\mathrm{N},n}(Z_n) \end{pmatrix}, \quad \mathbf{Y} = \begin{pmatrix} Y_1 \\ \vdots \\ Y_n \end{pmatrix}.$$

Assuming that $\mathbf{X}_{\mathrm{N}}^{\mathrm{T}}\mathbf{X}_{\mathrm{N}} + n\lambda\Omega$ is nonsingular, $\hat{\theta}_\lambda$ can be obtained as

$$\begin{aligned} \hat{\theta}_\lambda &= \operatorname{argmin}_{\theta \in \mathbb{R}^n} \left\{ \frac{1}{n} \|\mathbf{Y} - \mathbf{X}_{\mathrm{N}}\theta\|_2^2 + \lambda\theta^{\mathrm{T}}\Omega\theta \right\} \\ &= (\mathbf{X}_{\mathrm{N}}^{\mathrm{T}}\mathbf{X}_{\mathrm{N}} + n\lambda\Omega)^{-1}(\mathbf{X}_{\mathrm{N}}^{\mathrm{T}}\mathbf{Y}). \end{aligned} \tag{3.12}$$

Therefore, $\hat{\theta}_\lambda$ is a generalized ridge estimator in the linear regression with response vector $\mathbf{Y}$, basis matrix $\mathbf{X}_{\mathrm{N}}$, and ridge weight matrix Ω. The form of (3.12) is similar to the ridge estimators (3.4) and (3.6) in the basis-driven approach,

As the fitted function, the smoothing spline is $\hat{f}_\lambda(z) = \phi_{\mathrm{N}}(z)^{\mathrm{T}}\hat{\theta}_\lambda$. The fitted values on the training data, $\hat{\mathbf{f}}_\lambda = (\hat{f}_\lambda(Z_1), \dots, \hat{f}_\lambda(Z_n))^{\mathrm{T}}$, can be written as

$$\hat{\mathbf{f}}_\lambda = \mathbf{X}_{\mathrm{N}}\hat{\theta}_\lambda = \mathbf{S}_\lambda\mathbf{Y},$$

where $\mathbf{S}_\lambda$ is an $n \times n$ matrix, called a smoother matrix:

$$\mathbf{S}_\lambda = \mathbf{X}_{\mathrm{N}}(\mathbf{X}_{\mathrm{N}}^{\mathrm{T}}\mathbf{X}_{\mathrm{N}} + n\lambda\Omega)^{-1}\mathbf{X}_{\mathrm{N}}^{\mathrm{T}}.$$

Two remarks are in order. First, $\mathbf{S}_\lambda$ depends only on the training covariates $(Z_1, \dots, Z_n)$, but not the training responses $(Y_1, \dots, Y_n)$. In other words, $\mathbf{S}_\lambda$ can be computed without requiring access to $(Y_1, \dots, Y_n)$. Second, the fitted vector $\hat{\mathbf{f}}_\lambda$ is linear in the response vector $\mathbf{Y}$; in fact, $\hat{\mathbf{f}}_\lambda$ is defined by left-multiplying $\mathbf{Y}$ with the smoother matrix $\mathbf{S}_\lambda$ (depending on $\mathbf{X}_{\mathrm{N}}$ but not $\mathbf{Y}$). Hence the smoothing spline is called a linear smoother.

The basis properties of the smoother matrix are as follows.

Proposition 3.1 (Smoother matrix for smoothing spline)**.** *For any $\lambda > 0$, the smoother matrix $\mathbf{S}_\lambda$ satisfies the following properties.*

(i) $\mathbf{S}_\lambda$ is symmetric and nonnegative definite;

(ii) $\mathbf{S}_\lambda^2 \leq \mathbf{S}_\lambda$, i.e., $\mathbf{S}_\lambda - \mathbf{S}_\lambda^2$ is nonnegative definite.

(iii) $\mathrm{tr}(\mathbf{S}_\lambda) \leq n$, and $\mathrm{tr}(\mathbf{S}_\lambda)$ is non-increasing in λ, where $\mathrm{tr}(\cdot)$ denotes the trace of a square matrix.

Commonly, $\mathrm{tr}(\mathbf{S}_\lambda)$ is called the degrees of freedom of $\mathbf{S}_\lambda$ and the associated smoothing spline.

Proof. Property (i) is straightforward. To show (ii), we calculate

$$\mathbf{S}_\lambda^2 = \mathbf{X}_\mathrm{N}(\mathbf{X}_\mathrm{N}^\mathrm{T}\mathbf{X}_\mathrm{N} + n\lambda\Omega)^{-1}\mathbf{X}_\mathrm{N}^\mathrm{T} \cdot \mathbf{X}_\mathrm{N}(\mathbf{X}_\mathrm{N}^\mathrm{T}\mathbf{X}_\mathrm{N} + n\lambda\Omega)^{-1}\mathbf{X}_\mathrm{N}^\mathrm{T}.$$

Because Ω is nonnegative definite, we have $\mathbf{X}_\mathrm{N}^\mathrm{T}\mathbf{X}_\mathrm{N} \leq \mathbf{X}_\mathrm{N}^\mathrm{T}\mathbf{X}_\mathrm{N} + n\lambda\Omega$. Left- and right-multiplying both sides with A and A^T shows that $\mathbf{S}_\lambda^2 \leq \mathbf{S}_\lambda$, where $A = \mathbf{X}_\mathrm{N}(\mathbf{X}_\mathrm{N}^\mathrm{T}\mathbf{X}_\mathrm{N} + n\lambda\Omega)^{-1}$. For property (iii), we show $\mathrm{tr}(\mathbf{S}_\lambda) \leq n$ as follows:

$$\begin{aligned}\mathrm{tr}(\mathbf{S}_\lambda) &= \mathrm{tr}(\mathbf{X}_\mathrm{N}^\mathrm{T}\mathbf{X}_\mathrm{N}(\mathbf{X}_\mathrm{N}^\mathrm{T}\mathbf{X}_\mathrm{N} + n\lambda\Omega)^{-1}) \\ &= \mathrm{tr}((\mathbf{X}_\mathrm{N}^\mathrm{T}\mathbf{X}_\mathrm{N} + n\lambda\Omega)^{-1/2}\mathbf{X}_\mathrm{N}^\mathrm{T}\mathbf{X}_\mathrm{N}(\mathbf{X}_\mathrm{N}^\mathrm{T}\mathbf{X}_\mathrm{N} + n\lambda\Omega)^{-1/2}) \\ &\leq \mathrm{tr}(I_{n\times n}) = n.\end{aligned}$$

The first two equalities are obtained using $\mathrm{tr}(A_1A_2) = \mathrm{tr}(A_2A_1)$, and the third equality holds because $(\mathbf{X}_\mathrm{N}^\mathrm{T}\mathbf{X}_\mathrm{N}+n\lambda\Omega)^{-1/2}\mathbf{X}_\mathrm{N}^\mathrm{T}\mathbf{X}_\mathrm{N}(\mathbf{X}_\mathrm{N}^\mathrm{T}\mathbf{X}_\mathrm{N}+n\lambda\Omega)^{-1/2} \leq I_{n\times n}$ because $\mathbf{X}_\mathrm{N}^\mathrm{T}\mathbf{X}_\mathrm{N} \leq \mathbf{X}_\mathrm{N}^\mathrm{T}\mathbf{X}_\mathrm{N} + n\lambda\Omega$ and $\mathrm{tr}(A_1) \leq \mathrm{tr}(A_2)$ for two matrices satisfying $A_1 \leq A_2$. By similar reasoning, it can be shown that $\mathrm{tr}(\mathbf{X}_\lambda)$ is non-increasing in λ: $\mathrm{tr}(\mathbf{S}_\lambda) \leq \mathrm{tr}(\mathbf{S}_{\tilde{\lambda}})$ if $\lambda \geq \tilde{\lambda}$. □

The regression and penalized splines in Section 3.1 are also linear smoothers. For example, the fitted vector from the ridge-penalized spline is $\mathbf{f}_\lambda = \mathbf{X}_\mathrm{T}\hat{\beta}_\lambda = \mathbf{S}_\lambda\mathbf{Y}$, where the smoother matrix is

$$\mathbf{S}_\lambda = \mathbf{X}_\mathrm{T}(\mathbf{X}_\mathrm{T}^\mathrm{T}\mathbf{X}_\mathrm{T} + n\lambda D)^{-1}\mathbf{X}_\mathrm{T}^\mathrm{T}, \tag{3.13}$$

where $\mathbf{X}_\mathrm{T}$ is the basis matrix obtained from the truncated power basis functions ϕ_T. The fitted vector and smoother matrix remain the same if the B-spline basis matrix $\mathbf{X}_\mathrm{B}$ is used. Then properties (i)-(ii) in Proposition 3.1 are valid and property (iii) is valid except for the modification $\mathrm{tr}(\mathbf{S}_\lambda) \leq \min(p, n)$, where $p = 1 + k + q$ is the dimension of ϕ_T or the number of columns in $\mathbf{X}_\mathrm{T}$, and it is allowed that $n \geq p$ or $n < p$. [Exercise: Prove this statement.] Similarly, $\mathrm{tr}(\mathbf{S}_\lambda)$ is called the degrees of freedom of $\mathbf{S}_\lambda$ and the associated penalized spline.

3.4 Reproducing kernel Hilbert spaces

We study nonparametric regression based on reproducing kernel Hilbert spaces (RKHS), referred to as RKHS regression. Our motivation is multi-fold. First, RKHS regression provides a useful approach for nonparametric regression, including cubic smoothing splines as a specific example. The approach can be applied to not only univariate smoothing (with a scalar covariate), but also multivariate smoothing (with multiple covariates). Second, we use RKHSs as a mathematical framework to discuss main ideas for proving the finite-dimensional representation of smoothing splines defined by (3.10), as presented in Section 3.3. Third, the process of proving the representation result also suggests an alternative route for implementing smoothing splines, which is more compact than the approach in Section 3.3.

3.4.1 Hilbert spaces

We introduce basic concepts about Hilbert spaces, which can be seen as function spaces with similar properties as Euclidean spaces (which are in fact finite-dimensional Hilbert spaces). A function space $\mathcal{F}$ is a set where each element is a real function, e.g., $f : [a, b] \to \mathbb{R}$. The concepts of norms and inner products in Euclidean spaces can be generalized as follows.

Definition 3.1. A norm $\|\cdot\|$ on $\mathcal{F}$ is a function, $\mathcal{F} \to [0, \infty)$, satisfying the following conditions.

(i) Linearity: $\|\alpha f\| = |\alpha| \|f\|$ for any $f \in \mathcal{F}$ and $\alpha \in \mathbb{R}$;

(ii) Positive definiteness: $\|f\| \geq 0$, and $\|f\| = 0$ if and only if $f \equiv 0$.

(iii) Triangle inequality: $\|f + g\| \leq \|f\| + \|g\|$ for any $f, g \in \mathcal{F}$.

Definition 3.2. An inner product $\langle \cdot, \cdot \rangle$ on $\mathcal{F}$ is a function, $\mathcal{F} \times \mathcal{F} \to \mathbb{R}$, satisfying the following conditions.

(i) Symmetry: $\langle f, g \rangle = \langle g, f \rangle$ for any $f, g \in \mathcal{F}$;

(ii) Linearity: $\langle \alpha_1 f_1 + \alpha_2 f_2, g \rangle = \alpha_1 \langle f_1, g \rangle + \alpha_2 \langle f_2, g \rangle$ for any $f_1, f_2, g \in \mathcal{F}$ and $\alpha_1, \alpha_2 \in \mathbb{R}$;

(iii) Positive definiteness: $\langle f, f \rangle \geq 0$, and $\langle f, f \rangle = 0$ if and only if $f \equiv 0$.

Example 3.1. The Euclidean space $\mathbb{R}^d$ can be treated as the space of all functions on the discrete domain $\mathcal{X} = \{1, \ldots, d\}$, where each function $f : \mathcal{X} \to \mathbb{R}$ is identified as a vector $f = (f(1), \ldots, f(d))^{\mathrm{T}}$. For any $p \geq 1$, $\|f\|_p = \{\sum_{j=1}^d |f(j)|^p\}^{1/p}$ can be shown to be a norm, satisfying Definition 3.1. An inner product is $\langle f, g \rangle = \sum_{j=1}^d f(j) g(j)$, satisfying Definition 3.2. Among the L_p norms, it is the L_2 norm $\|f\|_2 = \{\sum_{j=1}^d f(j)^2\}^{1/2}$ that is associated with the inner product via $\|f\|_2^2 = \langle f, f \rangle$ as in (3.14).

Given an inner product, a norm can be derived as in (3.14) and called the norm associated with the inner product. The Cauchy–Schwartz inequality can also be deduced as a mere consequence of Definition 3.2.

Proposition 3.2 (Cauchy–Schwartz inequality). *Given an inner product $\langle \cdot, \cdot \rangle$ on $\mathcal{F}$, a norm (satisfying Definition 3.1) can be defined as*

$$\|f\| = \sqrt{\langle f, f \rangle}. \tag{3.14}$$

Moreover, the Cauchy–Schwartz inequality holds:

$$|\langle f, g \rangle| \leq \sqrt{\langle f, f \rangle} \sqrt{\langle g, g \rangle},$$

or equivalently $|\langle f, g \rangle| \leq \|f\| \|g\|$ by the definition of $\|\cdot\|$.

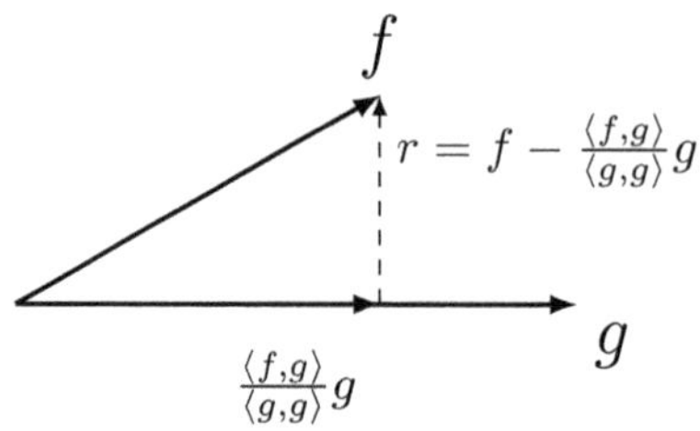

FIGURE 3.7
Illustration of projecting f on g.

Proof. Properties (i) and (ii) in Definition 3.1 can be easily verified. First, we show that the Cauchy–Schwartz inequality holds. The result is trivial if $f \equiv 0$ and $g \equiv 0$. Suppose that $g \not\equiv 0$. Then by Definition 3.2 (iii), $\|g\| \neq 0$ and

$$\langle f - \frac{\langle f, g\rangle}{\langle g, g\rangle} g, f - \frac{\langle f, g\rangle}{\langle g, g\rangle} g\rangle \geq 0. \tag{3.15}$$

By Definition 3.2 (i) and (ii), it can be calculated that

$$\langle f - \frac{\langle f, g\rangle}{\langle g, g\rangle} g, f - \frac{\langle f, g\rangle}{\langle g, g\rangle} g\rangle = \|f\|^2 - \frac{\langle f, g\rangle^2}{\|g\|^2}. \tag{3.16}$$

Combining the two displays (3.15) and (3.16) yields the Cauchy–Schwartz inequality $\langle f, g\rangle^2 \leq \|f\|^2\|g\|^2$. Next, the triangle inequality can be obtained as follows:

$$\begin{aligned}\|f + g\|^2 &= \|f\|^2 + \|g\|^2 + 2\langle f, g\rangle \\ &\leq \|f\|^2 + \|g\|^2 + 2\|f\|\|g\| = (\|f\| + \|g\|)^2,\end{aligned}$$

using Definition 3.2 (i) and (ii) and the Cauchy–Schwartz inequality.
[From the preceding display, for an inner product $\langle \cdot, \cdot\rangle$ and $\| \cdot \|$ defined in (3.14), the triangle inequality and the Cauchy–Schwartz inequality can be deduced from each other (i.e., the two are equivalent).] □

The function $r = f - \frac{\langle f,g\rangle}{\langle g,g\rangle} g$ used in (3.15) and (3.16) is the residual after projecting f on g. See Figure 3.7 for an illustration. In fact, r can be derived as $r = f - \alpha g$ where $\alpha = \frac{\langle f,g\rangle}{\langle g,g\rangle}$ is chosen such that

$$0 = \langle r, g\rangle = \langle f, g\rangle - \alpha\langle g, g\rangle,$$

i.e., r and g are orthogonal to each other. Equation (3.15) can be stated as

$$\|f - \alpha g\|^2 + \|\alpha g\|^2 = \|f\|^2,$$

which corresponds to the Pythagorean theorem in Euclidean geometry.

A formal definition of Hilbert spaces requires completeness as follows. For example, the real line is complete, but the set of all rational numbers is not complete. For most of subsequent discussions, we do not discuss completeness.

Definition 3.3. In a space $\mathcal{F}$ endowed with a norm $\|\cdot\|$, a sequence $\{f_n\}$ is called Cauchy if for every $\epsilon > 0$, there exists an integer $N \geq 1$ such that $\|f_n - f_m\| \leq \epsilon$ for any $n, m \geq N$. The space $\mathcal{F}$ is called complete if every Cauchy sequence in $\mathcal{F}$ has a limit that is also in $\mathcal{F}$.

Definition 3.4. A Hilbert space is a function space which is endowed with an inner product and is complete. The associated norm is defined as (3.14).

Example 3.2. Let $C([a,b]) = \{f : f \text{ is continuous on } [a,b]\}$. An inner product (satisfying Definition 3.2) is $\langle f, g\rangle = \int_a^b fg\,\mathrm{d}x$, defined as a Riemann integral. The associate norm is $\|f\| = (\int_a^b f^2\,\mathrm{d}x)^{1/2}$. However, this space is not complete, hence not a Hilbert space.

Example 3.3. Let $L_2([a,b]) = \{f : f \text{ is measurable on } [a,b] \text{ with } \int_a^b f^2\,\mathrm{d}x < \infty\}$, where the Lebesgue integral is used. Two functions are considered equivalent if they differ only on a set of Lebesgue measure 0 (i.e., they are the same almost everywhere). An inner product is $\langle f, g\rangle = \int_a^b fg\,\mathrm{d}x$, which is finite for $f, g \in L_2([a,b])$ by the Cauchy–Schwartz inequality:

$$\left|\int_a^b fg\,\mathrm{d}x\right|^2 \leq \left(\int_a^b f^2\,\mathrm{d}x\right)\left(\int_a^b g^2\,\mathrm{d}x\right).$$

Definition 3.2 can be shown to be satisfied. In particular, for Definition 3.2 (iii), $\int_a^b f^2\,\mathrm{d}x = 0$ if and only if $f \equiv 0$ almost everywhere. The associated norm is $\|f\| = (\int_a^b f^2\,\mathrm{d}x)^{1/2}$. The space $L_2([a,b])$ is complete, hence a Hilbert space. In fact, $L_2([a,b])$ can be shown to be a completion of $C([a,b])$.

Example 3.4. For a sample space Ω with probability P (where the σ-field is suppressed), let $L_2(\Omega, P) = \{X : X \text{ is a random variable with } E(X^2) < \infty\}$. Two random variables are considered equivalent if they differ with probability 0 (i.e., they are the same almost surely). An inner product is $\langle X, Y\rangle = E(XY)$, which is finite for $X, Y \in L_2(\Omega, P)$ by the Cauchy–Schwartz inequality: $|E(XY)| \leq E^{1/2}(X^2)E^{1/2}(Y^2)$. The associated norm is $\|X\| = E^{1/2}(X^2)$. The space $L_2([a,b])$ is complete, hence a Hilbert space. Examples 3.3 and 3.5 are closely related. The space $L_2([0,1])$ is equivalent to $L_2(\Omega, P)$ for $\Omega = [0,1]$ and P a uniform probability, by treating $f : [0,1] \to \mathbb{R}$ as a random variable.

The space $L_2([a,b])$ or even $C([a,b])$ is too big (or flexible) to model an unknown mean function f^* in nonparametric regression (3.1). For a sample $(Z_1, Y_1), \ldots, (Z_n, Y_n)$ of any size n, there exists $f \in C([a,b])$ (by forming narrow spikes only about the data points and setting f to 0 elsewhere) such that a perfect fit to the sample data is obtained, $f(X_i) = Y_i$ for $i = 1, \ldots, n$, but the norm $\|f\|$, treated as a roughness measure, can be arbitrarily close to 0. Hence minimizing any combination of the prediction errors and the roughness measure would not identify the unknown mean function. To make progress, we need to introduce further constraints on the function space.

Before moving to discuss RKHSs, we present a fundamental result, known as the Riesz representation theorem, for Hilbert spaces. A mapping from a function space to the real line is called a functional.

Definition 3.5. On a function space $\mathcal{F}$, a functional $G : \mathcal{F} \to \mathbb{R}$ is said to be linear if $G(\alpha_1 f_1 + \alpha_2 f_2) = \alpha_1 G(f_1) + \alpha_2 G(f_2)$ for any $f_1, f_2 \in \mathcal{F}$ and $\alpha_1, \alpha_2 \in \mathbb{R}$.

Definition 3.6. On a function space $\mathcal{F}$ with norm $\|\cdot\|$, a functional $G : \mathcal{F} \to \mathbb{R}$ is said to be bounded if there exists a constant $M > 0$ such that $|G(f)| \leq M\|f\|$ for any $f \in \mathcal{F}$.

For a linear functional G, Definition 3.6 can be simplified as follows: G is bounded if there exists a constant $M > 0$ such that $|G(f)| \leq M$ for any $f \in \mathcal{F}$ with $\|f\| = 1$.

Theorem 3.1 (Riesz representation theorem)**.** *For a bounded linear functional G on a Hilbert space $\mathcal{H}$, there exists a unique function $g \in \mathcal{H}$, called the representer of G, such that $G(f) = \langle f, g \rangle$.*

For fixed $g \in \mathcal{H}$, the inner product $\langle \cdot, g \rangle$, treated as a functional on $\mathcal{H}$, is easily shown to be linear by Definition 3.2 (ii) and bounded, $|\langle f, g \rangle| \leq M\|f\|$ with the constant $M = \|g\|$, by the Cauchy–Schwartz inequality. Conversely, the Riesz representation theorem asserts that each bounded linear functional on $\mathcal{H}$ corresponds to the inner product with some fixed g.

The Riesz representation theorem can also be stated for a continuous linear functional. In fact, a linear functional is bounded if and only if it is continuous. See Section 3.9.2.

3.4.2 Reproducing kernel Hilbert spaces

Reproducing kernel Hilbert spaces can be defined by introducing constraints on Hilbert spaces in terms of the evaluation functionals.

Definition 3.7. Let $\mathcal{F}$ be a function space of real functions on a domain $\mathcal{X}$. The evaluation functional F_x at $x \in \mathcal{X}$ is defined as $F_x(f) = f(x)$ for any $f \in \mathcal{F}$.

Definition 3.8. Let $\mathcal{H}$ be a Hilbert space of real functions on a domain $\mathcal{X}$. Then $\mathcal{H}$ is said to be a reproducing kernel Hilbert space (RKHS) if the evaluation functional F_x at each $x \in \mathcal{X}$ is bounded on $\mathcal{H}$, i.e., for each $x \in \mathcal{X}$, there exists a constant $M_x > 0$ (possibly depending on x) such that $|f(x)| = |F_x(f)| \leq M_x\|f\|$ for any $f \in \mathcal{H}$.

Apparently, the evaluation functional F_x at each $x \in \mathcal{X}$ is linear according to Definition 3.5. Hence a RKHS can be equivalently defined as an Hilbert space such that the evaluation functional F_x at each $x \in \mathcal{X}$ is continuous.

Example 3.5. The space $L_2([a,b])$ is not a RKHS. In fact, for any $x_0 \in [a,b]$, the evaluation functional F_{x_0} is not bounded, i.e., there does not exist a constant $M > 0$ such that $|f(x_0)| \le M$ for any $f \in L_2([a,b])$ with $\|f\| = 1$. Intuitively, a function $f \in L_2([a,b])$ can be found such that $\|f\| = 1$ but $f(x_0)$ is arbitrarily large. More formally, an element of $L_2([a,b])$ is an equivalence class of functions, where each two functions differ on a set of Lebesgue measure 0. It can be shown that if the evaluation functionals are defined by choosing representatives from equivalence classes, then the evaluation functionals are not continuous. See `https://math.stackexchange.com/questions/4372270/why-is-l2-not-a-rkhs`.

By the linearity of the evaluation functionals and the Riesz representation theorem (Theorem 3.1), a RKHS $\mathcal{H}$ is a Hilbert space satisfying the following property: for each $x \in \mathcal{X}$, the evaluation functional F_x admits a representer $K_x \in \mathcal{H}$ such that for any $f \in \mathcal{H}$,

$$f(x) = F_x(f) = \langle K_x, f \rangle, \tag{3.17}$$

which is called the reproducing property for a RKHS. The reproducing kernel of $\mathcal{H}$ is a function $K(\cdot,\cdot) : \mathcal{X} \times \mathcal{X} \to \mathbb{R}$ such that $K(x,y) = K_x(y)$ for any $x, y \in \mathcal{X}$.

Example 3.6. To continue from Example 3.1, the Euclidean space $\mathbb{R}^d$, treated as the space of all functions on the discrete domain $\mathcal{X} = \{1,\ldots,d\}$, is a Hilbert space, with the inner product $\langle f, g \rangle = \sum_{j=1}^d f(j)g(j)$. The space is a RKHS, because the reproducing property (3.17) is satisfied: for any $x \in \{1,\ldots,d\}$ and $f : \{1,\ldots,d\} \to \mathbb{R}$,

$$f(x) = \sum_{j=1}^d K_x(j) f(j) = \langle K_x, f \rangle,$$

where $K_x(j) = 1$ if $x = j$ or $= 0$ otherwise, i.e., $K_x(\cdot)$ is identified as the vector with the xth component being 1 and the remaining being 0. The reproducing kernel $K(\cdot,\cdot)$ can be identified as the $d \times d$ identity matrix.

The important role of reproducing kernels is demonstrated by the following result.

Theorem 3.2 (Reproducing kernels). *For a RKHS, the reproducing kernel $K(\cdot,\cdot)$ satisfies the two properties:*

(i) Symmetry: $K(x,y) = K(y,x)$ for $x, y \in \mathcal{X}$;

(ii) Positive semi-definiteness: $\sum_{i=1}^n \sum_{j=1}^n c_i c_j K(x_i, x_j) \ge 0$ for any integer $n \ge 1$, $x_1, \ldots, x_n \in \mathcal{X}$, and $c_1, \ldots, c_n \in \mathbb{R}$.

Conversely, if $K(\cdot,\cdot) : \mathcal{X} \times \mathcal{X} \to \mathbb{R}$ satisfies symmetry and positive semi-definiteness, then there exists a RKHS with K as the reproducing kernel.

Proof. [Necessity in Theorem 3.2] Taking $f = K_y(\cdot)$ in (3.17) gives

$$K(y, x) \stackrel{\text{def}}{=} K_y(x) = \langle K_x, K_y \rangle. \tag{3.18}$$

Exchanging x and y gives $K(x, y) = \langle K_y, K_x \rangle$. Hence $K(x, y) = K(y, x)$ by symmetry of $\langle \cdot, \cdot \rangle$. Next, we calculate

$$\begin{aligned}
\sum_{i=1}^{n}\sum_{j=1}^{n} c_i c_j K(x_i, x_j) &= \sum_{i=1}^{n}\sum_{j=1}^{n} c_i c_j \langle K_{x_i}, K_{x_j} \rangle && \text{[by (3.18)]} \\
&= \langle \sum_{i=1}^{n} c_i K_{x_i}, \sum_{j=1}^{n} c_j K_{x_j} \rangle && \text{[by linearity of } \langle \cdot, \cdot \rangle] \\
&= \| \sum_{i=1}^{n} c_i K_{x_i} \|^2 \ge 0.
\end{aligned}$$

Hence K satisfies the positive semi-definiteness. □

Proof outline. [Sufficiency in Theorem 3.2] The desired RKHS can be defined as the completion of the set of functions in the following form:

$$f = \sum_{i=1}^{n} \alpha_i K_{x_i}(\cdot) \tag{3.19}$$

for some integer $n \ge 1$, $x_1, \ldots, x_n \in \mathcal{X}$, and $\alpha_1, \ldots, \alpha_n \in \mathbb{R}$. The inner product is defined such that for $f = \sum_{i=1}^{n} \alpha_i K_{x_i}(\cdot)$ and $g = \sum_{i=1}^{m} \beta_i K_{y_i}(\cdot)$,

$$\langle f, g \rangle = \sum_{i=1}^{n}\sum_{j=1}^{m} \alpha_i \beta_j K(x_i, y_j). \tag{3.20}$$

The definition (3.20) can also be written as

$$\langle f, g \rangle = \sum_{j=1}^{m} \beta_j \sum_{i=1}^{n} \alpha_i K(x_i, y_j) = \sum_{j=1}^{m} \beta_j f(y_j), \tag{3.21}$$

or

$$\langle f, g \rangle = \sum_{i=1}^{n} \alpha_i \sum_{j=1}^{m} \beta_j K(x_i, y_j) = \sum_{i=1}^{n} \alpha_i g(x_i). \tag{3.22}$$

From (3.21) and (3.22), we note that the definition of $\langle f, g \rangle$ in (3.20) is independent of the particular representations of f and g. Suppose that f and

g can also be represented as $f = \sum_{i=1}^{\tilde{n}} \tilde{\alpha}_i K_{\tilde{x}_i}(\cdot)$ and $g = \sum_{i=1}^{\tilde{m}} \tilde{\beta}_i K_{\tilde{y}_i}(\cdot)$ for $\tilde{\alpha}_1, \ldots, \tilde{\alpha}_{\tilde{n}}, \tilde{\beta}_1, \ldots, \tilde{\beta}_{\tilde{m}} \in \mathbb{R}$ and $\tilde{x}_1, \ldots, \tilde{x}_{\tilde{n}}, \tilde{y}_1, \ldots, \tilde{y}_{\tilde{m}} \in \mathcal{X}$. Then

$$\begin{aligned}
\sum_{i=1}^{\tilde{n}} \sum_{j=1}^{\tilde{m}} \tilde{\alpha}_i \tilde{\beta}_j K(\tilde{x}_i, \tilde{y}_j) &= \sum_{j=1}^{\tilde{m}} \tilde{\beta}_j f(\tilde{y}_j) \\
= \sum_{i=1}^{n} \sum_{j=1}^{\tilde{m}} \alpha_i \tilde{\beta}_j K(x_i, \tilde{y}_j) &= \sum_{i=1}^{n} \alpha_i g(x_i) \\
= \sum_{i=1}^{n} \sum_{j=1}^{m} \alpha_i \beta_j K(x_i, y_j),
\end{aligned}$$

by repeated applications of (3.21) and (3.22).

In the remaining, we consider the set of functions in the form (3.19) over all possible $n \geq 1$, $x_1, \ldots, x_n \in \mathcal{X}$, and $\alpha_1, \ldots, \alpha_n \in \mathbb{R}$, and verify the following properties.

(i) The inner product defined in (3.20) is valid, satisfying Definition 3.2. In fact, the symmetry and positive semi-definiteness of $\langle \cdot, \cdot \rangle$ in (3.20) follow directly from those of K, and the linearity of $\langle \cdot, \cdot \rangle$ follows by algebra. It is left as an exercise to show that for any f in the form (3.19), if $\langle f, f \rangle = 0$, then $f = 0$. See Gu (2013), Lemma 2.4.

(ii) K satisfies the reproducing property. For any f in the form (3.19) and any $x \in \mathcal{X}$,

$$\begin{aligned}
\langle K_x, f \rangle &= \langle K_x, \sum_{i=1}^{n} \alpha_i K_{x_i} \rangle && \text{[by definition of } f] \\
&= \sum_{i=1}^{n} \alpha_i K(x_i, x) && \text{[by definition (3.20)]} \\
&= f(x),
\end{aligned}$$

where in the second equality, (3.20) is applied by writing $K_x = \sum_{i=1}^{m} \beta_i K_{y_i}$ with $m = 1$, $y_1 = x$, and $\beta_1 = 1$.

□

We give two examples to illustrate the construction of RKHSs from kernels. For simplicity, univariate functions are considered on the domain $[a, b]$. The examples can be readily extended to multivariate functions.

Example 3.7. Consider $K(x, y) = 1 + xy$ for $x, y \in [a, b]$. We verify the two properties:

(i) Symmetry: $K(x, y) = 1 + xy = 1 + yx = K(y, x)$;

(ii) Positive semi-definiteness: for any $x_1, \ldots, x_n \in [a, b]$ and $c_1, \ldots, c_n \in \mathbb{R}$,

$$\sum_{i=1}^{n}\sum_{j=1}^{n} c_i c_j K(x_i, x_j) = \sum_{i=1}^{n}\sum_{j=1}^{n} c_i c_j (1 + x_i x_j)$$

$$= \sum_{i=1}^{n}\sum_{j=1}^{n} c_i c_j + \sum_{i=1}^{n}\sum_{j=1}^{n} c_i c_j x_i x_j = \left(\sum_{i=1}^{n} c_i\right)^2 + \left(\sum_{j=1}^{n} c_i x_i\right)^2 \geq 0.$$

Then a RKHS with K as the reproducing kernel is

$$\mathcal{H} = \{f : f(x) = \alpha_0 + \alpha_1 x \text{ for } x \in [a, b] \text{ and } \alpha_0, \alpha_1 \in \mathbb{R}\},$$

with the inner product defined as

$$\langle f, g \rangle = \alpha_0 \beta_0 + \alpha_1 \beta_1,$$

where $f = \alpha_0 + \alpha_1 x$ and $g = \beta_0 + \beta_1 x$. It suffices to verify two items. First, the inner product is valid, satisfying Definition 3.2. This is left as an exercise. Second, the reproducing property (3.17) is satisfied: for any $x \in [a, b]$ and $f(y) = \alpha_0 + \alpha_1 y$ (with y denoting the argument),

$$\begin{aligned} \langle K_x, f \rangle &= \langle 1 + xy, \alpha_0 + \alpha_1 y \rangle \\ &= \alpha_0 + \alpha_1 x = f(x). \end{aligned}$$

This example shows that the space of all linear functions on $[a, b]$ is a RKHS.

Example 3.8. Consider the Gaussian kernel

$$K_\sigma(x, y) = \exp\left(-\frac{(x-y)^2}{2\sigma^2}\right), \quad x, y \in \mathbb{R},$$

where $\sigma > 0$ is fixed. This kernel satisfies the two properties:

(i) Symmetry: $K_\sigma(x, y) = K_\sigma(y, x)$;

(ii) Positive semi-definiteness: it is desired to show that for any $x_1, \ldots, x_n \in \mathbb{R}$ and $c_1, \ldots, c_n \in \mathbb{R}$,

$$\sum_{i=1}^{n}\sum_{j=1}^{n} c_i c_j K_\sigma(x_i, x_j) = \sum_{i=1}^{n}\sum_{j=1}^{n} c_i c_j \exp\left(-\frac{(x_i - x_j)^2}{2\sigma^2}\right) \geq 0.$$

We use the series expansion $\exp(y) = \sum_{k=0}^{\infty} \frac{y^k}{k!}$ to calculate

$$\sum_{i=1}^{n}\sum_{j=1}^{n} c_i c_j \exp\left(-\frac{(x_i - x_j)^2}{2\sigma^2}\right)$$

$$= \sum_{i=1}^{n}\sum_{j=1}^{n} c_i \exp(-\frac{x_i^2}{2\sigma^2}) c_j \exp(-\frac{x_j^2}{2\sigma^2}) \exp(\frac{x_i x_j}{\sigma^2})$$

$$= \sum_{i=1}^{n}\sum_{j=1}^{n} c_i \exp(-\frac{x_i^2}{2\sigma^2}) c_j \exp(-\frac{x_j^2}{2\sigma^2}) \sum_{k=0}^{\infty} \frac{(\frac{x_i x_j}{\sigma^2})^k}{k!}$$

$$= \sum_{k=0}^{\infty}\sum_{i=1}^{n}\sum_{j=1}^{n} c_i \exp(-\frac{x_i^2}{2\sigma^2}) c_j \exp(-\frac{x_j^2}{2\sigma^2}) \frac{(\frac{x_i x_j}{\sigma^2})^k}{k!}$$

$$= \sum_{k=0}^{\infty} \frac{1}{\sigma^{2k} k!} \left(\sum_{i=1}^{n} c_i \exp(-\frac{x_i^2}{2\sigma^2}) x_i^k \right)^2 \geq 0.$$

The interchange of the order of summations in the third equality is justified by Fubini's theorem because

$$\sum_{i=1}^{n}\sum_{j=1}^{n}\sum_{k=0}^{\infty} |c_i| \exp(-\frac{x_i^2}{2\sigma^2}) |c_j| \exp(-\frac{x_j^2}{2\sigma^2}) \frac{(\frac{|x_i x_j|}{\sigma^2})^k}{k!}$$

$$= \sum_{i=1}^{n}\sum_{j=1}^{n} |c_i| \exp(-\frac{x_i^2}{2\sigma^2}) |c_j| \exp(-\frac{x_j^2}{2\sigma^2}) \exp(\frac{|x_i x_j|}{\sigma^2})$$

$$= \sum_{i=1}^{n}\sum_{j=1}^{n} |c_i c_j| \exp\left(-\frac{(|x_i| - |x_j|)^2}{2\sigma^2} \right) < \infty.$$

[In essence, this shows that for a doubly-indexed sequence $(a_{lk} : l = 1, \ldots, L, k = 1, \ldots, \infty)$ with finite L, the order of summations can be interchanged, $\sum_{l=1}^{L}\sum_{k=1}^{\infty} a_{lk} = \sum_{k=1}^{\infty}\sum_{l=1}^{L} a_{lk}$, provided that $\sum_{k=1}^{\infty} |a_{lk}| < \infty$ for each l.]

By Theorem 3.2, there exists a RKHS with K_σ as the reproducing kernel. Formally, this RKHS is the completion of functions in the form

$$\sum_{i=1}^{n} c_i \exp\left(-\frac{(x - x_i)^2}{2\sigma^2} \right)$$

for some integer $n \geq 1$, $x_1, \ldots, x_n \in [a, b]$, and $c_1, \ldots, c_n \in \mathbb{R}$.

Finally, we discuss the function space associated with cubic smoothing splines. A function $f : [a, b] \to \mathbb{R}$ is absolute continuous if it is differentiable almost everywhere, with a derivative f', and the fundamental theorem of calculus holds: $f(x) = f(a) + \int_a^x f'(u) \mathrm{d}u$ for $x \in [a, b]$.

Example 3.9. Consider the space

$$\begin{aligned} C^{(2)}([0,1]) &= \{f : f' \text{ is absolute continuous and } f'' \in L_2([0,1])\} \\ &= \left\{ f : f' \text{ is absolute continuous and } \int_0^1 (f''(x))^2 \,\mathrm{d}x < \infty \right\}, \end{aligned}$$

where $L_2([0,1])$ is defined in Example 3.3. The space $C^{(2)}([0,1])$ can be obtained as a completion of the more easily defined set

$$\{f : f \text{ is twice continuously differentiable on } [0,1]\}.$$

Note that for $f = x^{3/2}$, $f' = (3/2)x^{1/2}$ is absolute continuous on $[0,1]$ but not differentiable at 0 (hence f is not twice differentiable at 0), and $f'' = (3/4)x^{-1/2} \notin L_2([0,1])$. Then the space $C^{(2)}([0,1])$ is a RKHS with the inner product defined as

$$\langle f, g\rangle = f(0)g(0) + f'(0)g'(0) + \int_0^1 f''(x)g''(x)\,\mathrm{d}x.$$

[Exercise: Show that the inner product is valid, satisfying Definition 3.2.] Moreover, the reproducing kernel is

$$K(x,y) = 1 + xy + \int_0^1 (x-u)_+(y-u)_+\,\mathrm{d}u,$$

where $(x-u)_+ = 0$ if $x \le u$ or $x - u$ if $x > u$. We show that the reproducing property (3.17) is satisfied:

$$f(x) = K_x(0)f(0) + K'_x(0)f'(0) + \int_0^1 K''_x(u)f''(u)\,\mathrm{d}u. \tag{3.23}$$

First, we calculate for $x \le y$,

$$\begin{aligned}\int_0^1 (x-u)_+(y-u)_+\,\mathrm{d}u &= \int_0^x (x-u)(y-u)\,\mathrm{d}u\\ = \int_0^x (xy - uy - ux + u^2)\,\mathrm{d}u &= x^2\left(\frac{y}{2} - \frac{x}{6}\right).\end{aligned}$$

Hence by symmetry,

$$\int_0^1 (x-u)_+(y-u)_+\,\mathrm{d}u = \begin{cases} x^2(\frac{y}{2} - \frac{x}{6}), & x \le y,\\ y^2(\frac{x}{2} - \frac{y}{6}), & x > y,\end{cases} \tag{3.24}$$

and

$$K_x(y) = K(x,y) = \begin{cases} 1 + xy + x^2(\frac{y}{2} - \frac{x}{6}), & x \le y,\\ 1 + xy + y^2(\frac{x}{2} - \frac{y}{6}), & x > y.\end{cases}$$

See Figure 3.8 for plots of kernel function (3.24). Next, taking derivatives gives

$$K'_x(y) = \begin{cases} x + \frac{x^2}{2}, & x \le y,\\ x + xy - \frac{y^2}{2}, & x > y,\end{cases}$$

$$K''_x(y) = \begin{cases} 0, & x \le y,\\ x - y, & x > y.\end{cases}$$

Note that $K'_x(y)$ and $K''_x(y)$ are continuous in y, including at $y = x$. Substituting $K_x(0) = 1$ and $K'_x(0) = x$ in (3.23) gives

$$f(x) = f(0) + xf'(0) + \int_0^1 (x-u)_+ f''(u)\,\mathrm{d}u$$

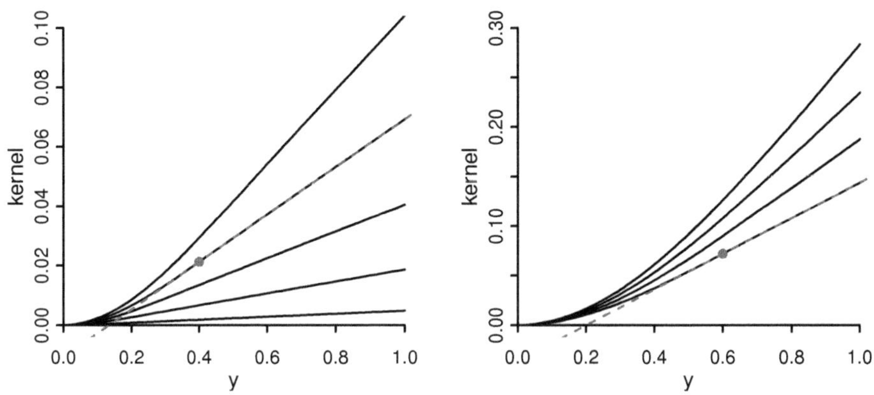

FIGURE 3.8
Plots of kernel function (3.24) over y, denoted as $K_1(x, y)$, for $x = 0.1, 0.2, 0.3, 0.4, 0.5$ (left) and $x = 0.6, 0.7, 0.8, 0.9$ (right), corresponding to curves from bottom to top. For $x = 0.4, 0.6$, straight lines are drawn to match the linear segment over $y \in [x, 1]$.

$$= f(0) + xf'(0) + \int_0^x (x - u) f''(u) \, \mathrm{d}u,$$

which is the second-order Taylor expansion with the remainder in an integral form. Incidentally, the first-order Taylor expansion reads

$$f(x) = f(0) + \int_0^x f'(u) \, \mathrm{d}u,$$

which is the fundamental theorem of (Lebesgue integral) calculus.

3.4.3 RKHS regression

For nonparametric regression, the training data are $\{(X_i, Y_i) : i = 1, \dots, n\}$, where X_i is a covariate vector (not necessarily a scalar) and Y_i a response variable. Consider a *linear* smoothing model

$$Y_i = f^*(X_i) + \varepsilon_i, \quad i = 1, \dots, n,$$

where f^* is an unknown function, assumed to be in a RKHS $\mathcal{H}$, and ε_i is a mean-zero noise variable. For fixed $\lambda > 0$, a fitted function (or RKHS smoother) is defined as

$$\hat{f}_\lambda = \operatorname{argmin}_{f \in \mathcal{H}} \left\{ \frac{1}{n} \sum_{i=1}^n (Y_i - f(X_i))^2 + \lambda \|f\|^2 \right\}, \tag{3.25}$$

where $\|f\|$ is the norm of f in the RKHS. As expected, the objective in (3.25) combines the sample average of squared prediction errors from using f and a

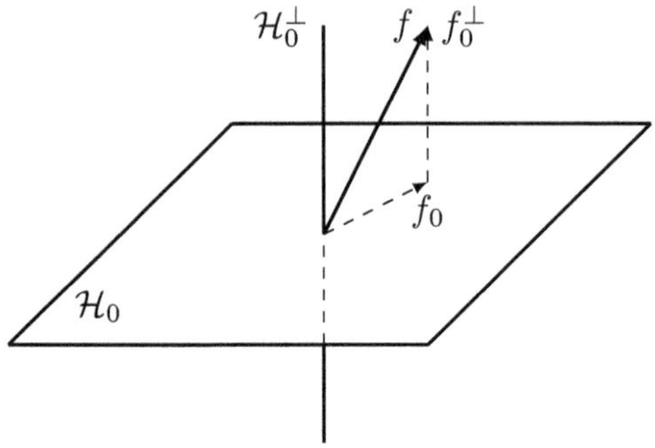

FIGURE 3.9
Illustration of the decomposition $f = f_0 + f_0^\perp$.

penalty term which is the squared RKHS norm of f. In general, the RKHS $\mathcal{H}$ used is an infinite-dimensional function space as in Examples 3.8 and 3.9. Nevertheless, a solution $\hat{f}_\lambda$ in (3.25) can be determined in a finite-dimensional representation as follows.

Theorem 3.3 (Representation theorem for RKHS regression). *In the preceding setting, a solution $\hat{f}_\lambda$ in (3.25) can be represented as*

$$\hat{f}_\lambda(x) = \sum_{i=1}^{n} c_i K(X_i, x),$$

for some constant $c_1, \ldots, c_n \in \mathbb{R}$, independent of x but dependent on (X_i, Y_i), $i = 1, \ldots, n$, where $K(\cdot, \cdot)$ is the reproducing kernel of $\mathcal{H}$.

Proof. Denote $L(f) = \frac{1}{n}\sum_{i=1}^{n}(Y_i - f(X_i))^2$. Define a linear subspace $\mathcal{H}_0$ of $\mathcal{H}$ as

$$\mathcal{H}_0 = \left\{ f \in \mathcal{H} : f = \sum_{i=1}^{n} \alpha_i K(X_i, x) \text{ for some } \alpha_1, \ldots, \alpha_n \in \mathbb{R} \right\}.$$

It is desired to show that there exists $g \in \mathcal{H}_0$ satisfying $L(g) + \lambda\|g\|^2 \leq L(f) + \lambda\|f\|^2$ for any $f \in \mathcal{H}$. Let $\mathcal{H}_0^\perp$ be the ortho-complement of $\mathcal{H}_0$,

$$\mathcal{H}_0^\perp = \{ f \in \mathcal{H} : \langle f, g\rangle = 0 \text{ for all } g \in \mathcal{H}_0 \}.$$

Then $\mathcal{H} = \mathcal{H}_0 \oplus \mathcal{H}_0^\perp$, i.e., for each $f \in \mathcal{H}$, there exist $f_0 \in \mathcal{H}_0$ and $f_0^\perp \in \mathcal{H}_0^\perp$ such that $f = f_0 + f_0^\perp$. [This property can be directly stated as follows: for each $f \in \mathcal{H}$, there exists $f_0 \in \mathcal{H}_0$ such that, with $f_0^\perp = f - f_0$, we have $\langle f_0^\perp, g\rangle = 0$ for all $g \in \mathcal{H}_0$. A proof of this property is omitted.] See Figure 3.9 for an illustration. It suffices to show that for any $f \in \mathcal{H}$ with $f = f_0 + f_0^\perp$,

$$L(f) + \lambda\|f\|^2 \geq L(f_0) + \lambda\|f_0\|^2, \tag{3.26}$$

where the equality holds if and only if $f_0^\perp = 0$. [Exercise: Why does it suffice?] First, we show that

$$\begin{aligned}\|f\|^2 &= \|f_0 + f_0^\perp\|^2 = \|f_0\|^2 + \|f_0^\perp\|^2 + 2\langle f_0, f_0^\perp\rangle \\ &= \|f_0\|^2 + \|f_0^\perp\|^2 \geq \|f_0\|^2,\end{aligned}$$

where the equality holds if and only if $\|f_0^\perp\| = 0$ or equivalently $f_0 = 0$ by the positive definiteness of $\|\cdot\|$. Second, we show that for $i = 1, \ldots, n$,

$$\begin{aligned}f(X_i) &= \langle K_{X_i}, f\rangle = \langle K_{X_i}, f_0 + f_0^\perp\rangle \\ &= \langle K_{X_i}, f_0\rangle + \langle K_{X_i}, f_0^\perp\rangle \\ &= \langle K_{X_i}, f_0\rangle = f_0(X_i),\end{aligned}$$

where $\langle K_{X_i}, f_0^\perp\rangle = 0$ because $K_{X_i} \in \mathcal{H}_0$. Hence $L(f) = L(f_0)$. Combining the preceding results yields the desired inequality (3.26). □

Given the finite-dimensional representation, the RKHS smoother $\hat{f}_\lambda$ can be computed in a closed form. In fact, by Theorem 3.3, the optimization in (3.25) can be restricted to be over a finite-dimensional set of functions in the form

$$f(x; c) = \sum_{i=1}^{n} c_i K(X_i, x),$$

where $c = (c_1, \ldots, c_n)^{\mathrm{T}} \in \mathbb{R}^n$. For fixed $\lambda > 0$, the RKHS smoother $\hat{f}_\lambda$ can be determined as $\hat{f}_\lambda = f(\cdot; \hat{c}_\lambda)$ with

$$\hat{c}_\lambda = \mathrm{argmin}_{c \in \mathbb{R}^n} \left\{ \frac{1}{n} \sum_{i=1}^{n} (Y_i - f(X_i; c))^2 + \lambda \|f(\cdot; c)\|^2 \right\}.$$

In the matrix notation, denote

$$\mathbf{K} = \begin{pmatrix} K(X_1, X_1) & \ldots & K(X_1, X_n) \\ \vdots & \vdots & \vdots \\ K(X_n, X_1) & \ldots & K(X_n, X_n) \end{pmatrix}, \quad \mathbf{Y} = \begin{pmatrix} Y_1 \\ \vdots \\ Y_n \end{pmatrix}.$$

The squared norm of $f(\cdot; c)$ can be calculated as

$$\begin{aligned}\|f(\cdot; c)\|^2 &= \left\| \sum_{i=1}^{n} c_i K_{X_i}(\cdot) \right\|^2 = \sum_{i=1}^{n} \sum_{j=1}^{n} c_i c_j \langle K_{X_i}, K_{X_j}\rangle \\ &= \sum_{i=1}^{n} \sum_{j=1}^{n} c_i c_j K(X_i, X_j) = c^{\mathrm{T}} \mathbf{K} c,\end{aligned} \tag{3.27}$$

using the property $K(x,y) = \langle K_x, K_y \rangle$ as shown in (3.18). Then $\hat{c}_\lambda$ can be obtained as a generalized ridge estimator in the linear regression with response vector $\mathbf{Y}$, basis matric $\mathbf{K}$, and ridge weight matrix $\mathbf{K}$:

$$\hat{c}_\lambda = \text{argmin}_{c \in \mathbb{R}^n} \left\{ \frac{1}{n} \|\mathbf{Y} - \mathbf{K}c\|_2^2 + \lambda c^{\mathrm{T}} \mathbf{K} c \right\}, \tag{3.28}$$

which, by using the symmetry of $\mathbf{K}$ and assuming that $\mathbf{K}$ is nonsingular, can be expressed as

$$\hat{c}_\lambda = (\mathbf{K}^{\mathrm{T}}\mathbf{K} + n\lambda \mathbf{K})^{-1}(\mathbf{K}^{\mathrm{T}}\mathbf{Y}) \tag{3.29a}$$

$$= (\mathbf{K} + n\lambda I)^{-1}\mathbf{Y}. \tag{3.29b}$$

When $\mathbf{K}$ is singular (for example, some points in $X_1, \ldots, X_n$ are duplicated), the expression (3.29a) for $\hat{c}_\lambda$ is not applicable, but the expression (3.29b) remains a valid (non-unique) solution to (3.28). [Exercise: Prove this statement.] See also Exercise 3.5 for related results when using a representation of $\hat{f}_\lambda$ based on distinct points in $(X_1, \ldots, X_n)$.

The vector of fitted values is $\hat{\mathbf{f}}_\lambda = \mathbf{K}\hat{c}_\lambda = \mathbf{K}(\mathbf{K} + n\lambda I)^{-1}\mathbf{Y}$, indicating that for fixed $\lambda > 0$, the RKHS smoother is a linear smoother. The smoother matrix (which is symmetric) can be expressed as

$$\begin{aligned} \mathbf{S}_\lambda &= \mathbf{K}(\mathbf{K} + n\lambda I)^{-1} \\ &= I - n\lambda(\mathbf{K} + n\lambda I)^{-1}. \end{aligned} \tag{3.30}$$

Left-multiplying both sides of (3.29b) by $\mathbf{K} + n\lambda I$ yields

$$\mathbf{Y} = \mathbf{K}\hat{c}_\lambda + n\lambda\hat{c}_\lambda = \hat{\mathbf{f}}_\lambda + n\lambda\hat{c}_\lambda, \tag{3.31}$$

which shows that the residual vector is $\mathbf{Y} - \hat{\mathbf{f}}_\lambda = n\lambda\hat{c}_\lambda$, hence proportional to $\hat{c}_\lambda$. Alternatively, the relationship (3.31) can also be seen from the calculation: $\hat{\mathbf{f}}_\lambda = \mathbf{S}_\lambda\mathbf{Y} = \mathbf{Y} - n\lambda\hat{c}_\lambda$ using the expression (3.30) for $\mathbf{S}_\lambda$.

For analytical and numerical purposes, it is useful to consider an eigenvalue decomposition of $\mathbf{K}$:

$$\mathbf{K} = Q\text{diag}(\rho_1, \ldots, \rho_n)Q^{\mathrm{T}},$$

where Q is an $n \times n$ orthogonal matrix and $\rho_1, \ldots, \rho_n$ are the eigenvalues (which are nonnegative). The columns of Q are the eigenvectors of $\mathbf{K}$. Then it can be directly shown that

$$\begin{aligned} \hat{c}_\lambda &= Q\text{diag}(\frac{1}{\rho_1 + n\lambda}, \ldots, \frac{1}{\rho_n + n\lambda})Q^{\mathrm{T}}\mathbf{Y}, \\ \mathbf{S}_\lambda &= Q\text{diag}(\frac{\rho_1}{\rho_1 + n\lambda}, \ldots, \frac{\rho_n}{\rho_n + n\lambda})Q^{\mathrm{T}}. \end{aligned}$$

The degrees of freedom is $\text{tr}(\mathbf{S}_\lambda) = \sum_{j=1}^n \frac{\rho_j}{\rho_j + n\lambda}$, which is decreasing in λ. When fitting the RKHS regression with multiple choices of λ, using these expressions is convenient because repeated matrix inversion is avoided.

Example 3.10. Ridge linear regression can be recast as an example of RKHS regression. Consider the space of linear functions of the covariate vector:

$$\mathcal{H} = \{f : f(x) = \alpha^{\mathrm{T}} x \text{ for } x \in \mathbb{R}^p \text{ and } \alpha \in \mathbb{R}^p\}.$$

Similarly as in Example 3.7, $\mathcal{H}$ is a RKHS, where the inner product is defined as $\langle f, g\rangle = \alpha^{\mathrm{T}}\gamma$ for $f = \alpha^{\mathrm{T}} x$ and $g = \gamma^{\mathrm{T}} x$, and the reproducing kernel is defined as $K(x, z) = x^{\mathrm{T}} z$. The norm of $f = \alpha^{\mathrm{T}} x$ is $\|f\|^2 = \alpha^{\mathrm{T}}\alpha = \|\alpha\|_2^2$. The linear smoothing model reduces to linear regression

$$Y_i = f^*(X_i) + \varepsilon_i = \beta^{*\mathrm{T}} X_i + \varepsilon_i, \quad i = 1, \ldots, n$$

where $f^*(x) = \beta^{*\mathrm{T}} x$, and ε_i is a mean-zero noise variable. For fixed $\lambda > 0$, the RKHS smoother is defined as $\hat{f}_\lambda(x) = \hat{\beta}_\lambda^{\mathrm{T}} x$, where

$$\hat{\beta}_\lambda = \operatorname{argmin}_{\beta \in \mathbb{R}^p} \left\{ \frac{1}{n} \sum_{i=1}^n (Y_i - \beta^{\mathrm{T}} X_i)^2 + \lambda \|\beta\|_2^2 \right\},$$

that is, $\hat{\beta}_\lambda$ is the ridge estimator $\hat{\beta}_{\lambda, L_2}$ as defined in Section 1.8.2. From Theorem 3.3, $\hat{f}_\lambda(x)$ can be represented for some $\hat{c}_\lambda = (\hat{c}_{\lambda,1}, \ldots, \hat{c}_{\lambda,n})^{\mathrm{T}} \in \mathbb{R}^n$ as

$$\hat{f}_\lambda(x) = \sum_{i=1}^n \hat{c}_{\lambda,i} K(X_i, x) = \sum_{i=1}^n \hat{c}_{\lambda,i} X_i^{\mathrm{T}} x = \hat{c}_\lambda^{\mathrm{T}} \mathbf{X} x,$$

where $\mathbf{X}$ is the covariate matrix $(X_1, \ldots, X_n)^{\mathrm{T}}$. By formula (3.29b),

$$\hat{c}_\lambda = (\mathbf{K} + n\lambda I)^{-1} \mathbf{Y} = (\mathbf{X}\mathbf{X}^{\mathrm{T}} + n\lambda I)^{-1} \mathbf{Y},$$

where $\mathbf{K}$ is the kernel matrix $\{K(X_i, X_j)\}_{i,j=1,\ldots,n} = \mathbf{X}\mathbf{X}^{\mathrm{T}}$. Equating the two expressions, $\hat{f}_\lambda(x) = \hat{\beta}_\lambda^{\mathrm{T}} x$ and $\hat{f}_\lambda(x) = \hat{c}_\lambda^{\mathrm{T}} \mathbf{X} x$ gives

$$\hat{\beta}_\lambda = \mathbf{X}^{\mathrm{T}} \hat{c}_\lambda = \mathbf{X}^{\mathrm{T}} (\mathbf{X}\mathbf{X}^{\mathrm{T}} + n\lambda I)^{-1} \mathbf{Y},$$

which is the second expression of $\hat{\beta}_{\lambda, L_2}$ as stated in (1.58). The previous derivation of formula (3.29b) assumes that $\mathbf{K} = \mathbf{X}\mathbf{X}^{\mathrm{T}}$ is nonsingular (or equivalently, $\mathbf{X}$ is of rank n $(\le p)$). But from the discussion of (1.58), the preceding expression of $\hat{\beta}_\lambda$ is in general valid for $\lambda > 0$, including when $\mathbf{X}$ is of rank $p\,(< n)$ and $\mathbf{K}$ is singular. In this case, $\hat{c}_\lambda$ is not unique but $\hat{\beta}_\lambda$ is unique.

3.4.4 Smoothing splines

We return to cubic smoothing splines with scalar covariates, and give a proof of the natural-cubic-spline representation for cubic smoothing splines as described in Section 3.3. Without loss of generality, assume that the covariate data $(X_1, \ldots, X_n)$ are contained in $[0, 1]$. Otherwise, smoothing splines can be computed by rescaling the covariate data and the tuning parameter λ. See

Exercise 3.9 for details. For fixed $\lambda > 0$, a cubic smoothing spline is defined as

$$\hat{f}_\lambda = \operatorname{argmin}_{f \in \mathcal{H}} \left\{ \frac{1}{n} \sum_{i=1}^{n} (Y_i - f(X_i))^2 + \lambda \int_0^1 (f''(x))^2 \, \mathrm{d}x \right\}, \tag{3.32}$$

where $\mathcal{H} = C^{(2)}([0,1])$ as discussed in Example 3.9. This slightly differs from the earlier definition in (3.10) with $[a,b] = [0,1]$, where the optimization is over the set of all twice continuously differentiable functions, which is not a Hilbert space due to lack of completeness (see Example 3.2). The difference does not affect the representation of a solution $\hat{f}_\lambda$ as a smoothing spline, which is twice continuously differentiable. However, optimization over $C^{(2)}([0,1])$ as a RKHS is needed for a proper proof of the representation.

The smoothing spline in (3.32) is similar to the RKHS smoother with $\mathcal{H} = C^{(2)}([0,1])$ in (3.25), but the penalty term differs from the squared RKHS norm

$$\|f\|^2 = f(0)^2 + (f'(0))^2 + \int_0^1 (f''(x))^2 \, \mathrm{d}x.$$

This difference affects the representation of $\hat{f}_\lambda$, which is stated in the following result.

Theorem 3.4 (Representation theorem for smoothing spline). *In the preceding setting, a solution $\hat{f}_\lambda$ in (3.32) can be represented as a natural cubic spline with knots at $X_1, \ldots, X_n$.*

Proof. First, the reproducing kernel for $C^{(2)}([0,1])$ can be decomposed as

$$K(x,y) = \underbrace{1 + xy}_{K_0(x,y)} + \underbrace{\int_0^1 (x-u)_+ (y-u)_+ \, \mathrm{d}u}_{K_1(x,y)},$$

where by the calculation of (3.24) in Example 3.9,

$$K_1(x,y) = \begin{cases} x^2(\frac{y}{2} - \frac{x}{6}), & x \le y, \\ y^2(\frac{x}{2} - \frac{y}{6}), & x > y. \end{cases} \tag{3.33}$$

By Example 3.7, $K_0(x,y)$ is a reproducing kernel for the RKHS

$$\mathcal{H}_0 = \{f : f = \alpha_0 + \alpha_1 x \text{ for some } \alpha_0, \alpha_1 \in \mathbb{R}\}.$$

Moreover, it can be shown that $K_1(x,y)$ is a reproducing kernel for the RKHS

$$\mathcal{H}_1 = \left\{ f : f(0) = f'(0) = 0 \text{ and } f \in C^{(2)}([0,1]) \right\},$$

with the inner-product $\langle f, g\rangle_{\mathcal{H}_1} = \int_0^1 f''(x)g''(x)\,\mathrm{d}x$ for $f, g \in \mathcal{H}_1$. The associated norm is $\|f\|^2_{\mathcal{H}_1} = \int_0^1 (f''(x))^2\,\mathrm{d}x$ for $f \in \mathcal{H}_1$. The inner product and norm on $\mathcal{H}_1$ agree with those on $\mathcal{H}$. For $f \in \mathcal{H}_0$ and $g \in \mathcal{H}_1$,

$$\langle f, g\rangle = f(0)g(0) + f'(0)g'(0) + \int_0^1 f''(x)g''(x)\,\mathrm{d}x = 0 \tag{3.34}$$

because $f''(x) = 0$ and $g(0) = g'(0) = 1$. Then $\mathcal{H}$ admits an orthogonal decomposition $\mathcal{H} = \mathcal{H}_0 \oplus \mathcal{H}_1$.

Next, we show that $\hat{f}_\lambda$ can be represented as

$$\hat{f}_\lambda = b_0 + b_1 x + \sum_{i=1}^n c_i K_1(X_i, x) \tag{3.35}$$

for some constants $b_0, b_1, c_1, \ldots, c_n \in \mathbb{R}$. Then, by the expression of K_1 in (3.33), $\hat{f}_\lambda$ is a cubic spline with knots at $X_1, \ldots, X_n$. The proof is similar to that of Theorem 3.3. Consider a linear subspace of $\mathcal{H}_1$:

$$\mathcal{H}_{10} = \left\{ f \in \mathcal{H}_1 : f = \sum_{i=1}^n \alpha_i K_1(X_i, x) \text{ for some } \alpha_1, \ldots, \alpha_n \in \mathbb{R} \right\}$$

and the ortho-complement of $\mathcal{H}_{10}$ within $\mathcal{H}_1$:

$$\mathcal{H}_{10}^\perp = \{ f \in \mathcal{H}_1 : \langle f, g\rangle_{\mathcal{H}_1} = 0 \text{ for all } g \in \mathcal{H}_{10} \}.$$

Then $\mathcal{H} = \mathcal{H}_0 \oplus \mathcal{H}_{10} \oplus \mathcal{H}_{10}^\perp$, i.e., $f \in \mathcal{H}$ has an orthogonal decomposition

$$f = \underbrace{f_0}_{\in \mathcal{H}_0} + \underbrace{f_{10}}_{\in \mathcal{H}_{10}} + \underbrace{f_{10}^\perp}_{\in \mathcal{H}_{10}^\perp}.$$

Denote $L(f) = \frac{1}{n}\sum_{i=1}^n (Y_i - f(X_i))^2$ and $J(f) = \int_0^1 (f''(x))^2\,\mathrm{d}x$. It suffices to show that for any $f \in \mathcal{H}$ decomposed as above,

$$L(f) + \lambda J(f) \geq L(f_0 + f_{10}) + \lambda J(f_0 + f_{10}),$$

where the equality holds if and only if $f_{10}^\perp = 0$. We deal with the L and J terms separately as follows.

- We show that $J(f) \geq J(f_0 + f_{10})$:

$$\begin{aligned}
J(f) &= J(f_0 + f_{10} + f_{10}^\perp) = J(f_{10} + f_{10}^\perp) \\
&\qquad [\text{because } f_0'' \equiv 0 \text{ for a linear function } f_0] \\
&= \|f_{10} + f_{10}^\perp\|^2_{\mathcal{H}_1} = \|f_{10}\|^2_{\mathcal{H}_1} + \|f_{10}^\perp\|^2_{\mathcal{H}_1} \qquad [\text{because } \langle f_{10}, f_{10}^\perp\rangle_{\mathcal{H}_1} = 0] \\
&\geq \|f_{10}\|^2_{\mathcal{H}_1} \\
&= J(f_{10}) = J(f_0 + f_{10}), \qquad [\text{because } f_0'' \equiv 0]
\end{aligned}$$

  where the equality holds if and only if $\|f_{10}^\perp\|_{\mathcal{H}_1} = 0$ or equivalently $f_{10}^\perp = 0$.

- We show that $L(f) = L(f_0 + f_{10})$:

$$\begin{aligned} f(X_i) &= \langle K_{X_i}, f_0 + f_{10} + f_{10}^{\perp}\rangle \qquad \text{[by the reproducing property]} \\ &= \langle K_{X_i}, f_0 + f_{10}\rangle + \langle K_{X_i}, f_{10}^{\perp}\rangle \\ &= \langle K_{X_i}, f_0 + f_{10}\rangle \qquad \text{[because } \langle K_{X_i}, f_{10}^{\perp}\rangle = 0\text{]} \\ &= f_0(X_i) + f_{10}(X_i), \qquad \text{[by the reproducing property]} \end{aligned}$$

for $i = 1, \ldots, n$. To verify $\langle K_{X_i}, f_{10}^{\perp}\rangle = 0$ above, we note that $K_{X_i} = K_{0,X_i} + K_{1,X_i}$, $K_{0,X_i} \in \mathcal{H}_0$ (i.e., $K_{0,X_i}(x)$ is linear in x), and $K_{1,X_i} \in \mathcal{H}_1$ (i.e., $K_{1,X_i}(0) = K'_{1,X_i}(0) = 0$). Then $\langle K_{X_i}, f_{10}^{\perp}\rangle = \langle K_{0,X_i}, f_{10}^{\perp}\rangle + \langle K_{1,X_i}, f_{10}^{\perp}\rangle = 0$, because $\langle K_{0,X_i}, f_{10}^{\perp}\rangle = 0$ by property (3.34) and $\langle K_{1,X_i}, f_{10}^{\perp}\rangle = 0$ by the definition of $f_{10}^{\perp}$ and $K_{1,X_i} \in \mathcal{H}_{10}$.
[More directly, $f(X_i) = f_0(X_i) + f_{10}(X_i)$ follows from $f_{10}^{\perp}(X_i) = \langle K_{1,X_i}, f_{10}^{\perp}\rangle = 0$ using the reproducing kernel K_1 on $\mathcal{H}_1$. This does not use the orthogonality (3.34) between $\mathcal{H}_0$ and $\mathcal{H}_1$. See Exercise 3.14.]

Finally, we show that $\hat{f}_\lambda$ is a natural cubic spline with knots at $X_1, \ldots, X_n$. Without loss of generality, suppose that $X_1 \le X_2 \le \cdots \le X_n$. By the representation (3.35) and the expression (3.33), $\hat{f}_\lambda$ is a cubic spline with knots at $X_1, \ldots, X_n$. It remains to show that $\hat{f}_\lambda$ is linear on $[0, X_1]$ and $[X_n, 1]$. In fact, by the expression of K_1 in (3.33), for $x \in [X_n, 1]$ (hence $x \ge X_i$),

$$K_1(X_i, x) = X_i^2(\frac{x}{2} - \frac{X_i}{6})$$

is linear in x for $i = 1, \ldots, n$. Then by the expression (3.35), $\hat{f}_\lambda$ is linear in x for $x \in [X_n, 1]$. Unfortunately, this reasoning does not work for proving that $\hat{f}_\lambda$ is linear in x for $x \in [0, X_1]$. For $x \in [0, X_1]$ (hence $x \le X_i$),

$$K_1(X_i, x) = x^2(\frac{X_i}{2} - \frac{x}{6})$$

is nonlinear in x. Nevertheless, there are at least two ways to show that $\hat{f}_\lambda$ is linear in x for $x \in [0, X_1]$. The first way is related to the computation of $b_0, b_1, c_1, \ldots, c_n$ for $\hat{f}_\lambda$ in (3.35), which will be discussed after the proof. The second way is based on symmetry. The space $\mathcal{H} = C^{(2)}([0, 1])$ can also be decomposed as $\mathcal{H} = \mathcal{H}_0 \oplus \tilde{\mathcal{H}}_1$, where $\tilde{\mathcal{H}}_1$ is the RKHS

$$\tilde{\mathcal{H}}_1 = \left\{f : f(1) = f'(1) = 0 \text{ and } f \in C^{(2)}([0, 1])\right\},$$

with the reproducing kernel

$$\tilde{K}_1(x, y) = \int_0^1 (1 - x - u)_+(1 - y - u)_+ \, du.$$

[In other words, the same space $C^{(2)}([0, 1])$ can be treated as different RKHSs with different inner products and reproducing kernels.] Then $\hat{f}_\lambda$ can be represented as (3.35) with K_1 replaced by $\tilde{K}_1$, and $\tilde{K}(X_1, x)$ can be shown to be linear in x for $x \in [0, X_1]$. Combining these two properties gives that $\hat{f}_\lambda$ is linear in x for $x \in [0, X_1]$. Details are left as Exercise 3.4. □

The proof of Theorem 3.4 suggests an alternative route for computing $\hat{f}_\lambda$, compared with the approach using natural-cubic-spline basis functions in Section 3.3. The optimization in (3.32) can be restricted to be over the finite-dimensional set of functions in the form

$$f(x;\theta) = b_0 + b_1 x + \sum_{i=1}^{n} c_i K_1(X_i, x).$$

where $\theta = (b^{\mathrm{T}}, c^{\mathrm{T}})^{\mathrm{T}}$, $b = (b_0, b_1)^{\mathrm{T}}$ and $c = (c_1, \ldots, c_n)^{\mathrm{T}}$. For fixed $\lambda > 0$, the smoothing spline $\hat{f}_\lambda$ can be obtained as $\hat{f}_\lambda = f(\cdot;\hat{\theta}_\lambda)$ with

$$\hat{\theta}_\lambda = \mathrm{argmin}_{\theta \in \mathbb{R}^{n+2}} \left\{ \frac{1}{n} \sum_{i=1}^{n} (Y_i - f(X_i;\theta))^2 + \lambda \int_0^1 (f''(x;\theta))^2 \,\mathrm{d}x \right\}.$$

In the matrix notation, denote

$$\mathbf{K}_0 = \begin{pmatrix} 1 & X_1 \\ \vdots & \vdots \\ 1 & X_n \end{pmatrix}, \quad \mathbf{K}_1 = \begin{pmatrix} K_1(X_1, X_1) & \ldots & K_1(X_1, X_n) \\ \vdots & \vdots & \vdots \\ K_1(X_n, X_1) & \ldots & K_1(X_n, X_n) \end{pmatrix}, \quad \mathbf{Y} = \begin{pmatrix} Y_1 \\ \vdots \\ Y_n \end{pmatrix}.$$

Similarly to (3.27), the penalty can be calculated as

$$\begin{aligned} \int_0^1 (f''(x;\theta))^2 \,\mathrm{d}x &= \int_0^1 (\sum_{i=1}^{n} c_i K_1''(X_i, x))^2 \,\mathrm{d}x \\ &= \left\| \sum_{i=1}^{n} c_i K_1(X_i, \cdot) \right\|_{\mathcal{H}_1}^2 = \sum_{i=1}^{n} \sum_{j=1}^{n} c_i c_j \langle K_1(X_i, \cdot), K_1(X_j, \cdot) \rangle_{\mathcal{H}_1} \\ &= \sum_{i=1}^{n} \sum_{j=1}^{n} c_i c_j K_1(X_i, X_j) = c^{\mathrm{T}} \mathbf{K}_1 c. \end{aligned}$$

Then $\hat{\theta}_\lambda$ can be obtained as a generalized ridge estimator

$$\begin{aligned} \hat{\theta}_\lambda &= \mathrm{argmin}_{\theta = (b^{\mathrm{T}}, c^{\mathrm{T}})^{\mathrm{T}}} \left\{ \frac{1}{n} \|\mathbf{Y} - \mathbf{K}_0 b - \mathbf{K}_1 c\|_2^2 + \lambda c^{\mathrm{T}} \mathbf{K}_1 c \right\} \\ &= \mathrm{argmin}_\theta \left\{ \frac{1}{n} \|\mathbf{Y} - \mathbf{K}\theta\|_2^2 + \lambda \theta^{\mathrm{T}} \Omega \theta \right\}, \end{aligned} \tag{3.36}$$

where $\mathbf{K} = (\mathbf{K}_0, \mathbf{K}_1)$ and $\Omega = \begin{pmatrix} 0 & 0 \\ 0 & \mathbf{K}_1 \end{pmatrix}$. Assuming that $\mathbf{K}_0$ is of rank 2 and $\mathbf{K}_1$ is nonsingular, $\hat{\theta}_\lambda$ can be determined as

$$\hat{\theta}_\lambda = (\mathbf{K}^{\mathrm{T}} \mathbf{K} + n\lambda\Omega)^{-1} (\mathbf{K}^{\mathrm{T}} \mathbf{Y}), \tag{3.37}$$

and further calculated as (Exercise 3.6)

$$\hat{\theta}_\lambda = \begin{pmatrix} \hat{b}_\lambda \\ \hat{c}_\lambda \end{pmatrix} = \begin{pmatrix} (\mathbf{K}_0^T\mathbf{K}_{1,\lambda}^{-1}\mathbf{K}_0)^{-1}\mathbf{K}_0^T\mathbf{K}_{1,\lambda}^{-1}\mathbf{Y} \\ (\mathbf{K}_{1,\lambda}^{-1} - \mathbf{K}_{1,\lambda}^{-1}\mathbf{K}_0(\mathbf{K}_0^T\mathbf{K}_{1,\lambda}^{-1}\mathbf{K}_0)^{-1}\mathbf{K}_0^T\mathbf{K}_{1,\lambda}^{-1})\mathbf{Y} \end{pmatrix}, \tag{3.38}$$

where $\mathbf{K}_{1,\lambda} = \mathbf{K}_1 + n\lambda I$. When $\mathbf{K}_1$ is singular (for example, some points in $X_1, \ldots, X_n$ are duplicated), the expression (3.37) for $\hat{\theta}_\lambda$ is not applicable, but the expression (3.38) remains a valid (non-unique) solution to (3.36). The vector of fitted values is

$$\begin{aligned}\hat{\mathbf{f}}_\lambda &= \mathbf{K}_0\hat{b}_\lambda + \mathbf{K}_1\hat{c}_\lambda \\ &= \Big\{\mathbf{K}_0(\mathbf{K}_0^T\mathbf{K}_{1,\lambda}^{-1}\mathbf{K}_0)^{-1}\mathbf{K}_0^T\mathbf{K}_{1,\lambda}^{-1} \\ &\quad +\mathbf{K}_1(\mathbf{K}_{1,\lambda}^{-1} - \mathbf{K}_{1,\lambda}^{-1}\mathbf{K}_0(\mathbf{K}_0^T\mathbf{K}_{1,\lambda}^{-1}\mathbf{K}_0)^{-1}\mathbf{K}_0^T\mathbf{K}_{1,\lambda}^{-1})\Big\}\mathbf{Y},\end{aligned}$$

confirming that the smoothing spline is a linear smoother as discussed in Section 3.3. From the expressions of $(\hat{b}_\lambda, \hat{c}_\lambda)$ in (3.38), it also holds that

$$\begin{aligned}\hat{c}_\lambda &= \mathbf{K}_{1,\lambda}^{-1}(\mathbf{Y} - \mathbf{K}_0\hat{b}_\lambda), \\ \mathbf{Y} &= \mathbf{K}_0\hat{b}_\lambda + \mathbf{K}_{1,\lambda}\hat{c}_\lambda = \hat{\mathbf{f}}_\lambda + n\lambda\hat{c}_\lambda,\end{aligned}$$

which echo the relationships (3.29b) and (3.31) in RKHS regression. From the second equation above, the residual vector is $\mathbf{Y} - \hat{\mathbf{f}}_\lambda = n\lambda\hat{c}_\lambda$, hence proportional to $\hat{c}_\lambda$. Moreover, the vector of fitted values $\hat{\mathbf{f}}_\lambda$ can be expressed as

$$\hat{\mathbf{f}}_\lambda = \mathbf{Y} - n\lambda\hat{c}_\lambda, \tag{3.39}$$

and hence the smoother matrix is a symmetric matrix as follows:

$$\mathbf{S}_\lambda = I - n\lambda(\mathbf{K}_{1,\lambda}^{-1} - \mathbf{K}_{1,\lambda}^{-1}\mathbf{K}_0(\mathbf{K}_0^T\mathbf{K}_{1,\lambda}^{-1}\mathbf{K}_0)^{-1}\mathbf{K}_0^T\mathbf{K}_{1,\lambda}^{-1}),$$

which echoes the expression (3.30) for $\mathbf{S}_\lambda$ in RKHS regression.

The expressions of $(\hat{b}_\lambda, \hat{c}_\lambda)$ in (3.38) can be used to give an alternative proof of the fact that $\hat{f}_\lambda$ is linear in x for $x \in [0, X_1]$, where the data points are ordered such that $X_1 \le X_2 \le \cdots \le X_n$. In fact, we show that for $x \in [0, X_1]$,

$$\sum_{i=1}^n \hat{c}_{\lambda,i}K_1(X_i, x) = 0,$$

and hence $\hat{f}_\lambda(x) = \hat{b}_{\lambda,0} + \hat{b}_{\lambda,1}x$. By the expression of K_1 in (3.33), for $x \in [0, X_1]$ (hence $x \le X_i$),

$$\begin{aligned}\sum_{i=1}^n \hat{c}_{\lambda,i}K_1(X_i, x) &= \sum_{i=1}^n \hat{c}_{\lambda,i}x^2(\frac{X_i}{2} - \frac{x}{6}) \\ &= (\sum_{i=1}^n \hat{c}_{\lambda,i}X_i)\frac{x^2}{2} - (\sum_{i=1}^n \hat{c}_{\lambda,i})\frac{x^3}{6}.\end{aligned}$$

By the expression of $\hat{c}_\lambda$ in (3.38), it can be calculated that

$$\begin{pmatrix} \sum_{i=1}^n \hat{c}_{\lambda,i} \\ \sum_{i=1}^n \hat{c}_{\lambda,i} X_i \end{pmatrix} = \mathbf{K}_0^{\mathrm{T}} \hat{c}_\lambda = 0.$$

Combining the preceding two displays gives the desired result.

For numerical implementation, the expressions of $(\hat{b}_\lambda, \hat{c}_\lambda)$ in (3.38) can be simplified using a QR decomposition of $\mathbf{K}_0$. See Wahba (1990), Section 1.3, or Gu (2013), Section 3.1.

Lemma 3.1 (QR decomposition for smoothing spline). *Suppose that $\mathbf{K}_0$ is of rank 2. Consider a QR decomposition of $\mathbf{K}_0$:*

$$\begin{aligned} \mathbf{K}_0 = QR &= \begin{pmatrix} Q_1 & Q_2 \end{pmatrix} \begin{pmatrix} R_1 \\ 0 \end{pmatrix} \\ &= Q_1 R_1, \end{aligned}$$

where Q is an $n \times n$ orthogonal matrix, R is an $n \times 2$ matrix, Q_1 is an $n \times 2$ matrix, and R_1 is a 2×2 (invertible) matrix. Then $(\hat{b}_\lambda, \hat{c}_\lambda)$ in (3.38) can be equivalently expressed as

$$\begin{aligned} \hat{b}_\lambda &= R_1^{-1} (Q_1^{\mathrm{T}} \mathbf{K}_{1,\lambda}^{-1} Q_1)^{-1} Q_1^{\mathrm{T}} \mathbf{K}_{1,\lambda}^{-1} \mathbf{Y}, \\ \hat{c}_\lambda &= Q_2 (Q_2^{\mathrm{T}} \mathbf{K}_{1,\lambda} Q_2)^{-1} Q_2^{\mathrm{T}} \mathbf{Y}, \end{aligned}$$

where $\mathbf{K}_{1,\lambda} = \mathbf{K}_1 + n\lambda I$. From the expression (3.39) *for $\hat{\mathbf{f}}_\lambda$, the associated smoother matrix can be obtained as $\mathbf{S}_\lambda = I - n\lambda Q_2 (Q_2^{\mathrm{T}} \mathbf{K}_{1,\lambda} Q_2)^{-1} Q_2^{\mathrm{T}}$.*

Proof. Substituting $\mathbf{K}_0 = Q_1 R_1$ into (3.38) gives

$$\begin{aligned} \hat{b}_\lambda &= (\mathbf{K}_0^{\mathrm{T}} \mathbf{K}_{1,\lambda}^{-1} \mathbf{K}_0)^{-1} \mathbf{K}_0^{\mathrm{T}} \mathbf{K}_{1,\lambda}^{-1} \mathbf{Y} \\ &= (R_1^{\mathrm{T}} Q_1^{\mathrm{T}} \mathbf{K}_{1,\lambda}^{-1} Q_1 R_1)^{-1} R_1^{\mathrm{T}} Q_1^{\mathrm{T}} \mathbf{K}_{1,\lambda}^{-1} \mathbf{Y} \\ &= R_1^{-1} (Q_1^{\mathrm{T}} \mathbf{K}_{1,\lambda}^{-1} Q_1)^{-1} Q_1^{\mathrm{T}} \mathbf{K}_{1,\lambda}^{-1} \mathbf{Y}, \\ \hat{c}_\lambda &= (\mathbf{K}_{1,\lambda}^{-1} - \mathbf{K}_{1,\lambda}^{-1} \mathbf{K}_0 (\mathbf{K}_0^{\mathrm{T}} \mathbf{K}_{1,\lambda}^{-1} \mathbf{K}_0)^{-1} \mathbf{K}_0^{\mathrm{T}} \mathbf{K}_{1,\lambda}^{-1}) \mathbf{Y} \\ &= (\mathbf{K}_{1,\lambda}^{-1} - \mathbf{K}_{1,\lambda}^{-1} Q_1 (Q_1^{\mathrm{T}} \mathbf{K}_{1,\lambda}^{-1} Q_1)^{-1} Q_1^{\mathrm{T}} \mathbf{K}_{1,\lambda}^{-1}) \mathbf{Y} \\ &= Q_2 (Q_2^{\mathrm{T}} \mathbf{K}_{1,\lambda} Q_2)^{-1} Q_2^{\mathrm{T}} \mathbf{Y}. \end{aligned}$$

The last step in simplifying $\hat{c}_\lambda$ uses the fact that

$$\mathbf{K}_{1,\lambda}^{-1} - \mathbf{K}_{1,\lambda}^{-1} Q_1 (Q_1^{\mathrm{T}} \mathbf{K}_{1,\lambda}^{-1} Q_1)^{-1} Q_1^{\mathrm{T}} \mathbf{K}_{1,\lambda}^{-1} = Q_2 (Q_2^{\mathrm{T}} \mathbf{K}_{1,\lambda} Q_2)^{-1} Q_2^{\mathrm{T}},$$

due to the following result (Exercise 3.7). Let A be a $n \times n$ invertible matrix, B a $n \times k$ matrix of rank k, and C a $n \times (n-k)$ matrix of rank $(n-k)$ satisfying $B^{\mathrm{T}} C = 0$, where $1 \le k \le n-1$. Then

$$A^{-1} - A^{-1} B (B^{\mathrm{T}} A^{-1} B)^{-1} B^{\mathrm{T}} A^{-1} = C (C^{\mathrm{T}} A C)^{-1} C^{\mathrm{T}}. \tag{3.40}$$

□

It is helpful to note the following special case of the identity (3.40). Take

$$B = \begin{pmatrix} I_{k\times k} \\ 0_{(n-k)\times k} \end{pmatrix}, \quad C = \begin{pmatrix} 0_{k\times(n-k)} \\ I_{(n-k)\times(n-k)} \end{pmatrix}.$$

Partition A and A^{-1} are partitioned accordingly:

$$A = \begin{pmatrix} A_{11} & A_{12} \\ A_{21} & A_{22} \end{pmatrix}, \quad A^{-1} = \begin{pmatrix} A^{11} & A^{12} \\ A^{21} & A^{22} \end{pmatrix}.$$

Then (3.40) reduces to

$$\begin{pmatrix} 0 & 0 \\ 0 & A^{22} - A^{21}(A^{11})^{-1}A^{12} \end{pmatrix} = \begin{pmatrix} 0 & 0 \\ 0 & A_{22}^{-1} \end{pmatrix}.$$

or equivalently

$$A^{22} - A^{21}(A^{11})^{-1}A^{12} = A_{22}^{-1}.$$

This can also be verified from the inverse formula (1.32) for a partitioned matrix.

3.5 Numerical demonstration: RKHS regression

We demonstrate implementation of RKHS regression, including smoothing splines and Gaussian kernel regression. As in Section 3.2, we use the ozone data and study univariate smoothing (or curve fitting) of `ozone` against the Dagget pressure gradient (`dpg`).

3.5.1 Smoothing splines

First we use the built-in function `smooth.spline()` from the R package `splines` to fit cubic smoothing splines of `ozone` against `dpg` with several specified degrees of freedom, which are plotted in Figure 3.10. The following R code produces the $(1, 2)$ plot with `df = 8` (except for graphical customization) in Figure 3.10. The curves are drawn by connecting the fitted values at the covariate data points `dpg`.

```
> out2 <- smooth.spline(y=ozone, x=dpg, df=8, all.knots=T)
> plot(dpg, ozone, xlim=c(-80,120), ylim=c(-5,40))
> lines(out2, col="blue")
```

We write the following functions, `kernel.smsp()` and `sm.spline()`, to implement cubic smoothing splines using the kernel basis functions,

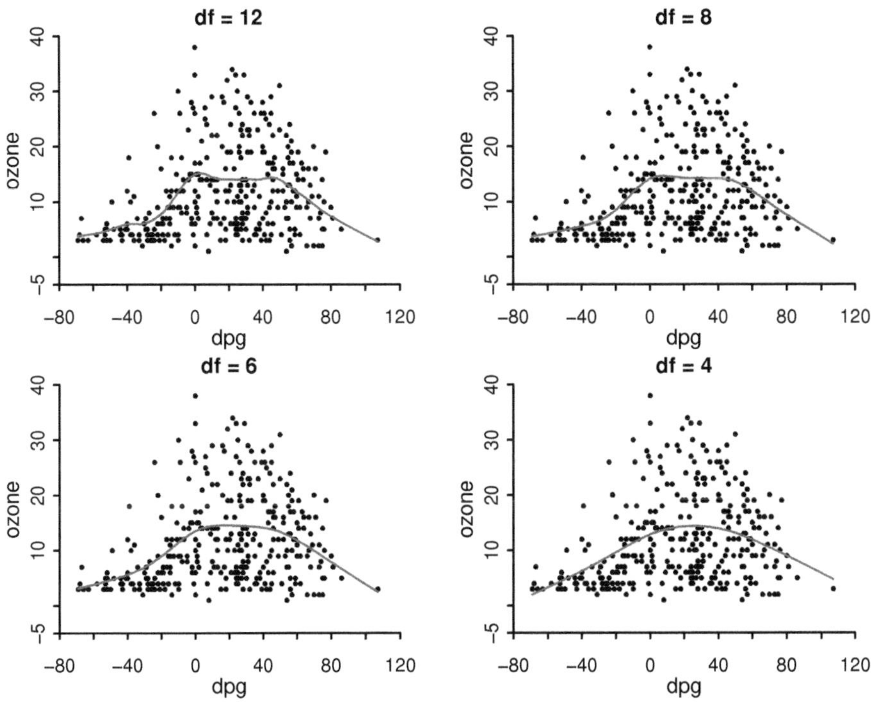

FIGURE 3.10
Plots of cubic smoothing splines for `ozone` against `dpg` in the ozone dataset, with several choices of degrees of freedom, using the built-in function `smooth.spline()`.

$1, x, K_1(X_i; x)$, $i = 1, \ldots, n$, as indicated by the representation (3.35). The function `kernel.smsp()` evaluates the kernel $K_1(x, w)$ for any points x and w supplied as two vectors `x` and `w` and returns a kernel basis matrix (including the linear basis). The function `sm.spline()` computes the coefficient estimates $(\hat{b}_\lambda, \hat{c}_\lambda)$ and the smoother matrix $\mathbf{S}_\lambda$, returned as `(cL,cK)` and `S`, using the formulas based on a QR decomposition in Lemma 3.1. The function `sm.spline()` also returns the fitted values at the data points (`fit`) and the associated degrees of freedom (`df`). See the accompanying R code file for implementation using the original formula (3.38) for $(\hat{b}_\lambda, \hat{c}_\lambda)$.

For both functions `kernel.smsp()` and `sm.spline()`, the kernel basis functions and coefficient estimates are computed after rescaling the covariate data to [0, 1] (as assumed in Section 3.4.4), provided that the data points are contained between the two points specified by `ends`. To be consistent with `smooth.spline()`, for `ends = c(0,1)`, the tuning parameter `lam` corresponds to $n\lambda$ in the notation of Section 3.4.4. See Exercise 3.9 for details about the effects of recaling the covariate and choosing the end points.

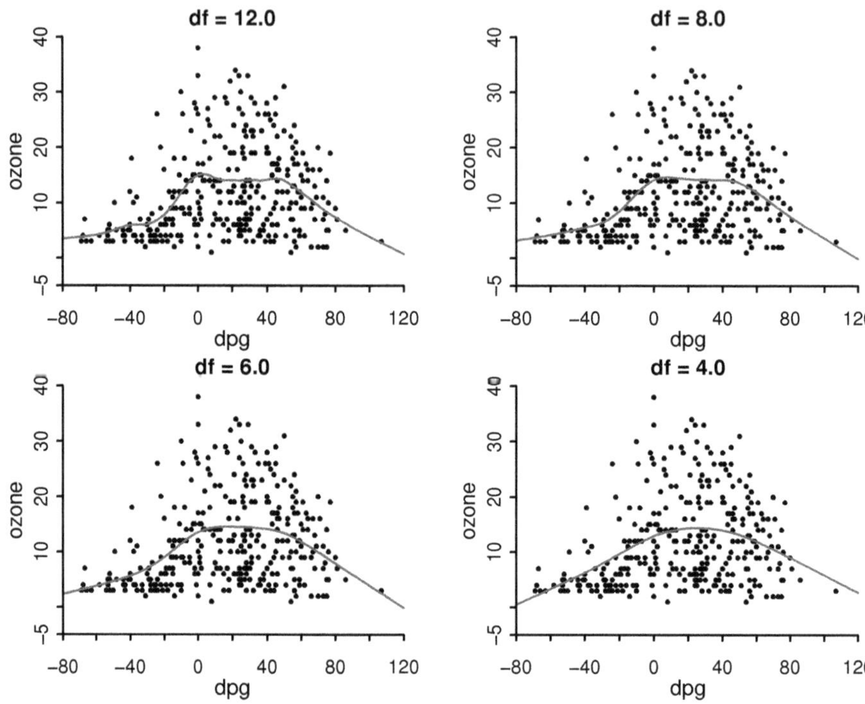

FIGURE 3.11
Plots of cubic smoothing splines for `ozone` against `dpg` in the ozone dataset, using our own function `sm.spline()`, with several tuning parameters leading to approximately the same degrees of freedom as in Figure 3.10.

```
kernel.smsp <- function(x, w, ends=c(0,1)) {
  n <- length(x)
  k <- length(w)
  x <- (x-ends[1])/(ends[2]-ends[1])    # rescaling
  w <- (w-ends[1])/(ends[2]-ends[1])

  out <- matrix(0, n, k)
  for (i in 1:n) {
    for (j in 1:k) {
      if (x[i]<w[j])  out[i,j] <- x[i]^2 *(w[j]/2-x[i]/6)
      else            out[i,j] <- w[j]^2 *(x[i]/2-w[j]/6)
    }
  }
  cbind(x,out)   # include linear basis
}
```

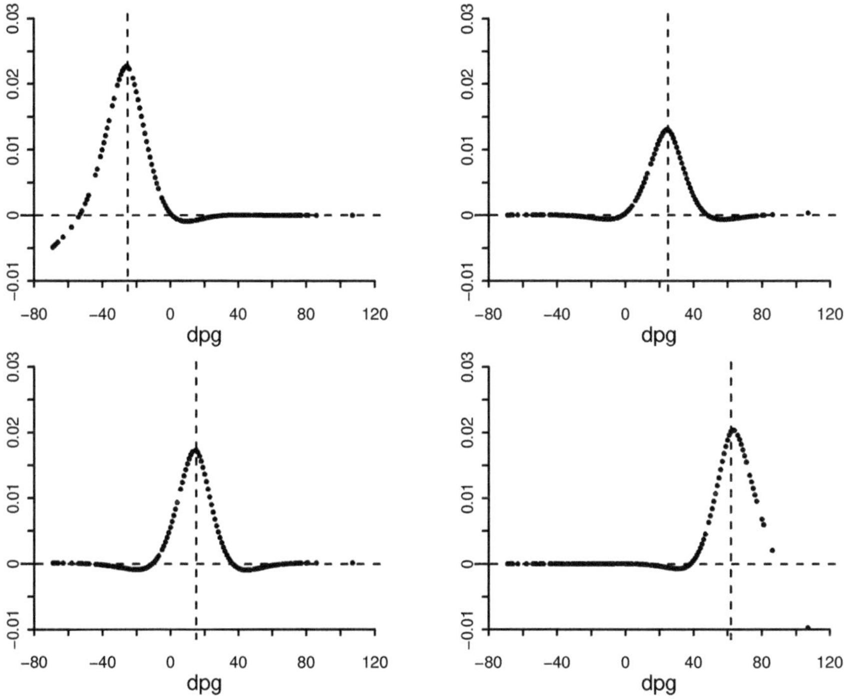

FIGURE 3.12
Plots of rows (or columns) of the smoother matrix $\mathbf{S}_\lambda$ at four data points in `dpg`, -25 and 25 (first row) and 15 and 62 (second row), for the cubic smoothing spline with `df = 7.9` in Figure 3.11. A vertical line is placed at the selected data point in each plot.

```
sm.spline <- function(y, x, lam, ends) {
  x <- (x-ends[1])/(ends[2]-ends[1])    # rescaling
  L <- cbind(1,x)
  n <- length(x)
  K <- kernel.smsp(x, x, ends=c(0,1))[,-1]
  K2 <- K+lam*diag(n)

  out <- qr(L)
  Q <- qr.Q(out, complete=TRUE)
  Q1 <- Q[, 1:dim(L)[2]]
  Q2 <- Q[, (1+dim(L)[2]):n]
  R1 <- qr.R(out, complete=FALSE)
```

```
  cL <- solve(R1, solve(t(Q1)%*%solve(K2,Q1), t(Q1)%*%solve(K2,y)))
  cK <- Q2%*%solve(t(Q2)%*%K2%*%Q2,t(Q2)%*%y)
  fit <- L %*%cL + K %*%cK
  S <- diag(n) - lam *Q2%*%solve(t(Q2)%*%K2%*%Q2,t(Q2))
  df <- sum(diag(S))

  list(cL=cL, cK=cK, fit=fit, S=S, df=df)
}
```

Figure 3.11 shows the cubic smoothing splines obtained using our own function sm.spline(), where the tuning parameters are hand picked to give approximately the same degrees of freedom as in Figure 3.10 obtained using the built-in function smooth.spline(). The curves in Figure 3.11 are, however, drawn by connecting the fitted function values at a set of grid points, not the covariate data points dpg. For example, the following R code produces the (1,2) plot with df = 7.9 (except for graphical customization) in Figure 3.11. It is left as Exercise 3.11 to expand the function sm.spline() for fitting smoothing splines with a pre-specified degree of freedom (instead of a pre-specified tuning parameter λ).

```
> ends <- c(-70, 110)
> x <- seq(-80,120,2)
> bX <- kernel.smsp(x, dpg, ends=ends)
>
> out2 <- sm.spline(ozone, dpg, lam=1.55e-3, ends)
> out2$df
[1] 8.000886
> ffit.smsp2 <- cbind(1,bX) %*% c(out2$cL, out2$cK)
> plot(dpg, ozone, xlim=c(-80,120), ylim=c(-5,40))
> lines(x, ffit.smsp2, col="blue")
```

Figure 3.12 shows the rows (or equivalently columns) of the smoother matrix $\mathbf{S}_\lambda$ at four data points in dpg for the cubic smoothing spline in the (1,2) plot of Figure 3.11. The following R code illustrates how the (2,1) plot in Figure 3.12 can be produced. As indicated by these plots, the ith row of the smoother matrix, plotted against the data points, tends to be bell-shaped, peaking at the data point X_i and then decaying for data points further away from X_i. A caveat is that the elements of the smoother matrix may be negative, especially when the associated data points are near the boundary.

```
> dpg[4]
[1] 15
> plot( dpg, out2$S[4,], xlim=c(-80,120), ylim=c(-0.01,0.03), ylab="")
> abline(v=dpg[4], h=0, lty=2)
```

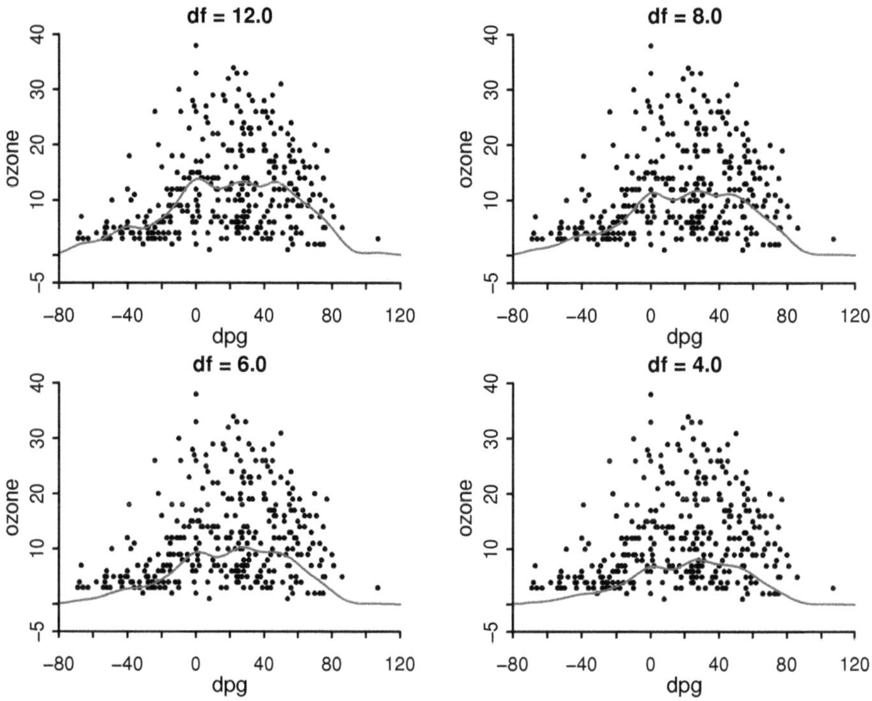

FIGURE 3.13
Plots of fitted functions from Gaussian kernel regression for `ozone` against `dpg` in the ozone dataset, using `kernel.gauss()`, with `sigma=10` and several tuning parameters leading to approximately the same degrees of freedom as in Figures 3.10 and 3.11.

3.5.2 Gaussian kernel regression

We write the following functions, `kernel.gauss()` and `rkhs.gauss()`, to implement RKHS regression using the Gaussian kernel in Example 3.8. The function `kernel.gauss()` evaluates the kernel $K_\sigma(x, w)$, with σ specified by the argument `sigma`, for any points x and w supplied as two vectors `x` and `w` and returns a kernel basis matrix. The function `rkhs.gauss()` computes the coefficient vector $\hat{c}_\lambda$ and the smoother matrix $\mathbf{S}_\lambda$, returned as `coef` and `S`, using the formulas based on an eigenvalue decomposition of the kernel matrix in Section 3.4.3. Similarly as in `sm.spline()`, the tuning parameter `lam` corresponds to $n\lambda$. The function `rkhs.gauss()` also returns the fitted values at the data points (`fit`) and the associated degrees of freedom (`df`).

```
kernel.gauss <- function(x, w, sigma) {
  n <- length(x)
  k <- length(w)
```

```
  out <- matrix(0, n, k)
  for (i in 1:n) {
    for (j in 1:k) {
      out[i,j] <- exp( -(x[i]-w[j])^2/sigma^2 )
    }
  }
  out
}

rkhs.gauss <- function(y, x, lam, sigma) {
  n <- length(x)
  K <- kernel.gauss(x, x, sigma)

  eigenK <- eigen(K)
  Q <- eigenK$vec
  L <- eigenK$val

  coef <- Q %*% ((t(Q) /(L+lam)) %*%y)
  fit <- K%*%coef
  S <- Q %*% (t(Q) *(L/(L+lam)))
  df <- sum(diag(S))

  list(coef=coef, fit=fit, S=S, df=df)
}
```

Figure 3.13 shows the fitted functions from Gaussian kernel regression using the function `rkhs.gauss()`, where `sigma=10` and the tuning parameters are hand picked to give approximately the same degrees of freedom as the smoothing splines in Figures 3.10 and 3.11. For example, the following R code produces the (1,2) plot with `df = 8.0` (except for graphical customization) in Figure 3.13.

```
> x <- seq(-80,120,2)
> bX <- kernel.gauss(x, dpg, sigma=10)
>
> out2 <- rkhs.gauss(ozone, dpg, lam=15.8, sigma=10)
> out2$df
[1] 7.999901
> ffit.gauss2 <- bX %*% out2$coef
> plot(dpg, ozone, xlim=c(-80,120), ylim=c(-5,40))
> lines(x, ffit.gauss2, col="blue")
```

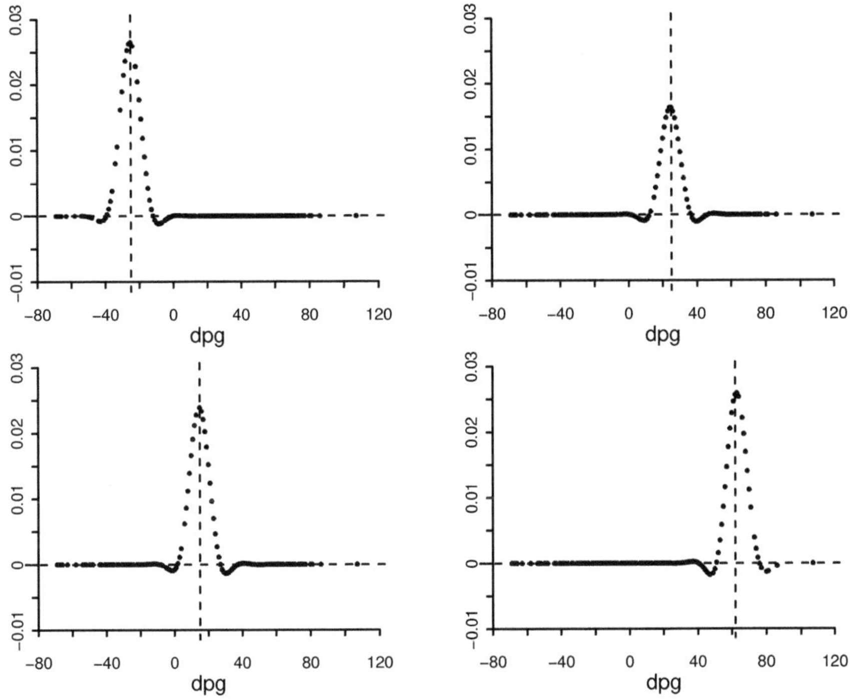

FIGURE 3.14
Plots of rows (or columns) of the smoother matrix $\mathbf{S}_\lambda$ at four data points in `dpg`, −25 and 25 (first row) and 15 and 62 (second row), for Gaussian kernel regression with `df = 8.0` in Figure 3.13. A vertical line is placed at the selected data point in each plot.

Figure 3.14 shows the rows (or equivalently columns) of the smoother matrix $\mathbf{S}_\lambda$ at four data points in `dpg` for Gaussian kernel regression in the (1,2) plot of Figure 3.13. The rows of the smoother matrix, plotted against the data points, exhibit a similar bell shape as in Figure 3.12. The spread of the bell shape is affected by the choice of `sigma`.

3.6 Lasso penalized and total variation splines

3.6.1 Lasso penalized splines

We return to the smoothing model (3.3) based on polynomial splines with degree q and k knots, $(\xi_1, \ldots, \xi_k)$, as follows:

$$Y_i = \phi_{\mathrm{T}}(Z_i)^{\mathrm{T}} \beta^* + \varepsilon_i, \quad i = 1, \ldots, n,$$

where $\beta^* = (b_0^*, \ldots, b_q^*, c_1^*, \ldots, c_k^*)^{\mathrm{T}}$ is a vector of $p = 1 + q + k$ unknown coefficients and ϕ_{T} is a vector of truncated power basis functions,

$$\phi_{\mathrm{T}}(z) = (1, z, \ldots, z^q, \{(z - \xi_1)_+\}^q, \ldots, \{(z - \xi_k)_+\}^q)^{\mathrm{T}}.$$

As an alternative to the ridge estimator (3.4), it is interesting to use the (generalized) Lasso estimator

$$\hat{\beta}_\lambda = \mathrm{argmin}_{\beta \in \mathbb{R}^p} \left\{ \frac{1}{2n} \sum_{i=1}^{n} (Y_i - \phi_{\mathrm{T}}(Z_i)^{\mathrm{T}} \beta)^2 + \lambda \sum_{j=1}^{k} |c_j| \right\}, \tag{3.41}$$

where $\lambda > 0$ is a tuning parameter. The fitted function, $\hat{f}_\lambda = \phi_{\mathrm{T}}(z)\hat{\beta}_\lambda$, is called a Lasso penalized spline. The Lasso penalty is placed only on the coefficients $(c_1, \ldots, c_k)$ associated with the truncated basis functions, $\{(z - \xi_j)_+\}^q$ for $j = 1, \ldots, k$. A nonzero coefficient c_j indicates a change of the qth derivatives across the knot ξ_j. Hence the Lasso penalty is used to prevent too frequent changes of the qth derivatives.

Similarly as the standard Lasso estimator in Section 1.8, the Lasso estimator $\hat{\beta}_\lambda$ in (3.41) has a sparsity property that depending on the tuning parameter λ, some penalized components of $\hat{\beta}_\lambda$, i.e., some of $(\hat{c}_{\lambda,1}, \ldots, \hat{c}_{\lambda,k})$, may be exactly zero, so that the corresponding knots are effectively dropped. In this way, Lasso penalized splines enable automatic selection of knots, as controlled by λ. There is no closed-form expression for $\hat{\beta}_\lambda$, but the numerical algorithms for Lasso penalized linear regression in Section 1.8 can be readily extended to allow only penalization on $(c_1, \ldots, c_k)$.

3.6.2 Total variation splines

When the maximum possible set of knots are used from the data points, Lasso penalized splines can also be formulated in a function-driven approach, similar to smoothing splines, but with the penalty on a function related to the total variation. The function-driven formulation facilitates the study of convergence properties for the fitted function within the associated class of functions. See Mammen and van de Geer (1997) for the relevant theory.

For a function $f : [a, b] \to \mathbb{R}$, the total variation (TV) is defined as

$$\mathrm{TV}(f) = \sup \left\{ \sum_{i=1}^{l-1} |f(z_{i+1}) - f(z_i)| : z_1 < \cdots < z_l \in [a, b],\ l \geq 2 \right\}.$$

If f is absolute continuous on $[a, b]$, then $\mathrm{TV}(f) = \int_a^b |f'(z)|\, \mathrm{d}z$. For fixed $\lambda > 0$, a total variation spline of order $m\,(\geq 1)$ is defined as

$$\hat{f}_\lambda = \mathrm{argmin}_{f \in \mathcal{F}} \left\{ \frac{1}{2n} \sum_{i=1}^{n} (Y_i - f(Z_i))^2 + \lambda \mathrm{TV}(f^{(m-1)}) \right\}, \tag{3.42}$$

where $\mathcal{F} = \mathcal{V}^{(m)}([a,b])$ is the set of all functions $f : [a,b] \to \mathbb{R}$ such that, if $m \geq 2$, the $(m-2)$th derivative $f^{(m-2)}$ is absolute continuous and $\text{TV}(f^{(m-1)}) < \infty$. Of special interest is the case of $m = 1$ or 2.

- $\mathcal{V}^{(1)}([a,b])$ is the set of all functions $f : [a,b] \to \mathbb{R}$ such that $\text{TV}(f) < \infty$.
- $\mathcal{V}^{(2)}([a,b])$ is the set of all functions $f : [a,b] \to \mathbb{R}$ such that f is absolute continuous and $\text{TV}(f^{(1)}) < \infty$.

Similarly to smoothing splines, a solution $\hat{f}_\lambda$ in (3.42) in the case of $m = 1$ or 2 can be obtained in a finite-dimensional representation as a polynomial spline of degree 0 or 1 respectively with knots from the data points $Z_1, \ldots, Z_n$. For $m \geq 3$, a solution $\hat{f}_\lambda$ can be defined as a polynomial spline of degree $m-1$ but the knots may be outside the data points (Mammen & van de Geer, 1997). A spline of degree 0 with k knots $\xi_1 < \cdots < \xi_k$ is defined as a piecewise constant function in the form

$$f(z) = b_0 + \sum_{j=1}^{k} c_j \{(z - \xi_j)_+\}^0,$$

where $(a_+)^0 = 0$ if $a < 0$ or 1 if $a \geq 0$. The basis functions, $1, \{(z - \xi_1)_+\}^0, \ldots, \{(z - \xi_k)_+\}^0$ are called truncated constant basis functions.

Theorem 3.5 (Representation theorem for TV spline). *Suppose that $Z_1 \leq \cdots \leq Z_n$. A solution $\hat{f}_\lambda$ in (3.42) can be represented as a polynomial spline of degree 0 with $n-1$ knots at $Z_2, \ldots, Z_n$ if $m = 1$ or as a polynomial spline of degree 1 with $n-2$ knots at $Z_2, \ldots, Z_{n-1}$ if $m = 2$.*

[Such representations are not unique. For example, in the case of $m = 1$, for a solution $\hat{f}_\lambda$ represented as a piecewise constant function, a piecewise linear function can be defined such that the two functions agree on the data points $Z_1, \ldots, Z_n$ and admit the same TV. Then the piecewise linear function is also a solution.]

Proof. Denote by $R(f)$ the penalized objective in (3.42). Consider the case $m = 1$. For any $f \in \mathcal{V}^{(1)}([a,b])$, define $\tilde{f}$ as a piecewise constant function (Figure 3.15):

$$\tilde{f}(z) = \begin{cases} f(Z_1), & a \leq z < Z_2, \\ f(Z_i), & Z_i \leq z < Z_{i+1}, i = 2, \ldots, n-1, \\ f(Z_n), & Z_n \leq z \leq b. \end{cases}$$

Then $\tilde{f}(Z_i) = f(Z_i)$ for $i = 1, \ldots, n$, but $\text{TV}(\tilde{f}) = \sum_{i=1}^{n-1} |f(Z_{i+1}) - f(Z_i)| \leq \text{TV}(f)$ and hence $R(\tilde{f}) \leq R(f)$, which implies the desired result for $m = 1$.

Consider the case $m = 2$. For any $f \in \mathcal{V}^{(2)}([a,b])$, define $\tilde{f}$ as a piecewise linear function (Figure 3.15):

- $\tilde{f}(Z_i) = f(Z_i)$, $i = 1, \ldots, n$, and

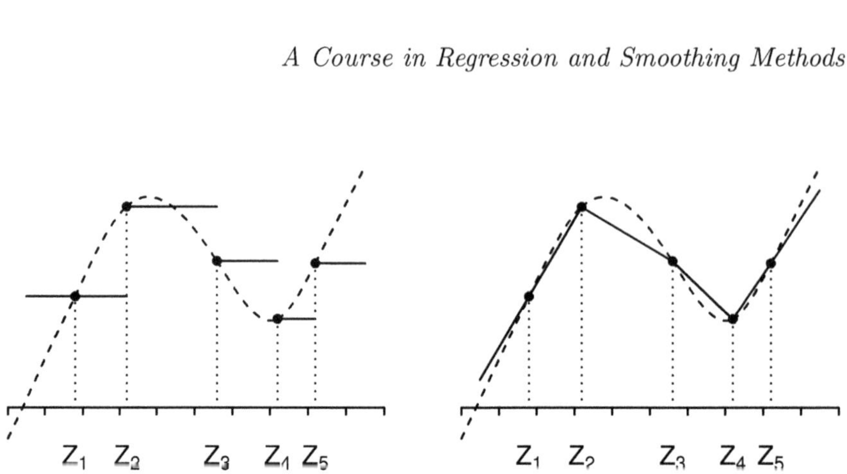

FIGURE 3.15
Construction of a piecewise constant function (left) and a piecewise linear function (right) for a function f (dashed), as in the proof of Theorem 3.5.

- $\tilde{f}(z)$ is linear with respective slopes in the intervals $[a, Z_2]$, $[Z_i, Z_{i+1}]$, $i = 2, \ldots, n-2$, and $[Z_{n-1}, b]$.

Then $\mathrm{TV}(\tilde{f}^{(1)}) = \sum_{i=1}^{n-1} |g_{i+1} - g_i|$, where g_i is the slope of $\tilde{f}$ in the interval $[Z_i, Z_{i+1}]$:

$$g_i = \frac{\tilde{f}(Z_{i+1}) - \tilde{f}(Z_i)}{Z_{i+1} - Z_i} = \frac{f(Z_{i+1}) - f(Z_i)}{Z_{i+1} - Z_i}.$$

By the mean value theorem, for $i = 1, \ldots, n-1$, there exists $z_i \in [Z_i, Z_{i+1}]$ such that $f^{(1)}(z_i) = g_i$, i.e.,

$$f^{(1)}(z_i) = \frac{f(Z_{i+1}) - f(Z_i)}{Z_{i+1} - Z_i}.$$

Then $\mathrm{TV}(\tilde{f}^{(1)}) = \sum_{i=1}^{n-1} |g_{i+1} - g_i| = \sum_{i=1}^{n-1} |f^{(1)}(z_{i+1}) - f^{(1)}(z_i)| \leq \mathrm{TV}(f^{(1)})$. Therefore $\tilde{f}(Z_i) = f(Z_i)$ for $i = 1, \ldots, n$, but $R(\tilde{f}) \leq R(f)$, which implies the desired result for $m = 2$. □

From Theorem 3.5, the total variation spline $\hat{f}_\lambda$ in the case of $m = 1$ or 2 can be computed as a Lasso penalized spline described earlier. For $m = 1$, a piecewise constant function with $n-1$ knots at $Z_2, \ldots, Z_n$ can be parameterized as

$$\phi_{\mathrm{T}}(z)^{\mathrm{T}}\beta = b_0 + \sum_{j=1}^{n-1} c_j\{(z - Z_{j+1})_+\}^0,$$

where $\phi_{\mathrm{T}}(z) = (1, \{(z - Z_2)_+\}^0, \ldots, \{(z - Z_n)_+\}^0)^{\mathrm{T}}$ and $\beta = (b_0, c_1, \ldots, c_{n-1})^{\mathrm{T}}$. For $f = \phi_{\mathrm{T}}(\cdot)^{\mathrm{T}}\beta$, $\mathrm{TV}(f)$ reduces to the Lasso penalty

$\sum_{j=1}^{n-1} |c_j|$. Then $\hat{f}_\lambda$ in (3.42) can be obtained as $\hat{f}_\lambda = \phi_{\mathrm{T}}(z)^{\mathrm{T}}\hat{\beta}_\lambda$, where $\hat{\beta}_\lambda$ is a (generalized) Lasso estimator,

$$\hat{\beta}_\lambda = \mathrm{argmin}_{\beta \in \mathbb{R}^n} \left\{ \frac{1}{2n} \sum_{i=1}^{n} (Y_i - \phi_{\mathrm{T}}(Z_i)^{\mathrm{T}}\beta)^2 + \lambda \sum_{j=1}^{n-1} |c_j| \right\}.$$

For $m = 2$, a linear spline with $n - 2$ knots at $Z_2, \ldots, Z_{n-1}$ can be parameterized as

$$\phi_{\mathrm{T}}(z)^{\mathrm{T}}\beta = b_0 + b_1 z + \sum_{j=1}^{n-2} c_j (z - Z_{j+1})_+,$$

where $\phi_{\mathrm{T}}(z) = (1, z, (z - Z_2)_+, \ldots, (z - Z_{n-1})_+)^{\mathrm{T}}$ and $\beta = (b_0, b_1, c_1, \ldots, c_{n-2})^{\mathrm{T}}$. For $f = \phi_{\mathrm{T}}(\cdot)^{\mathrm{T}}\beta$, $\mathrm{TV}(f^{(1)})$ reduces to the Lasso penalty $\sum_{j=1}^{n-2} |c_j|$. Then $\hat{f}_\lambda$ in (3.42) can be obtained as $\hat{f}_\lambda = \phi_{\mathrm{T}}(z)^{\mathrm{T}}\hat{\beta}_\lambda$, where

$$\hat{\beta}_\lambda = \mathrm{argmin}_{\beta \in \mathbb{R}^n} \left\{ \frac{1}{2n} \sum_{i=1}^{n} (Y_i - \phi_{\mathrm{T}}(Z_i)^{\mathrm{T}}\beta)^2 + \lambda \sum_{j=1}^{n-2} |c_j| \right\}.$$

In each case, the fitted function $\hat{f}_\lambda$ is a Lasso penalized spline, with the number of basis functions p in $\phi_{\mathrm{T}}(\cdot)$ equal to the number of data points n.

To accommodate the case $m \geq 3$, a practical approach for the minimization in (3.42) is to restrict f to be a spline of degree $m-1$ with knots contained in the set of data points. Such a restricted solution $\hat{f}_\lambda$ is called a restricted total variation spline or still a total variation spline of order m. By the discussion in Section 3.1, a spline of degree $q = m - 1$ with knots from the data points can be parameterized as

$$\phi_{\mathrm{T}}(z)^{\mathrm{T}}\beta = b_0 + b_1 z + \cdots + b_q z^q + \sum_{j=1}^{k} c_j \{(z - \xi_j)_+\}^q,$$

where $k = n - 1 - q = n - m$, $\phi_{\mathrm{T}}(z) = (1, z, \cdots, z^q, \{(z - \xi_1)_+\}^q, \ldots, \{(z - \xi_k)_+\}^q)^{\mathrm{T}}$ and $\beta = (b_0, b_1, \ldots, b_q, c_1, \ldots, c_k)^{\mathrm{T}}$, both of dimension n, and, with $Z_1 \leq \cdots \leq Z_n$ the ordered data points,

$$(\xi_1, \ldots, \xi_k) = \begin{cases} (Z_{1+\frac{q+1}{2}}, \cdots, Z_{n-\frac{q+1}{2}}) & \text{if } q = m-1 \text{ is odd}, \\ (Z_{2+\frac{q}{2}}, \cdots, Z_{n-\frac{q}{2}}) & \text{if } q = m-1 \text{ is even}. \end{cases} \tag{3.43}$$

For $f = \phi_{\mathrm{T}}(\cdot)^{\mathrm{T}}\beta$, $\mathrm{TV}(f^{(m-1)})$ reduces to the Lasso penalty $\sum_{j=1}^{q} |c_j|$. Then $\hat{f}_\lambda$ in the restricted version of (3.42) can be obtained as $\hat{f}_\lambda = \phi_{\mathrm{T}}(z)^{\mathrm{T}}\hat{\beta}_\lambda$, where $\phi_{\mathrm{T}}(z)$ is the truncated power basis vector defined above and $\hat{\beta}_\lambda$ is the (generalized) Lasso estimator in (3.41). Hence, with knot restriction, $\hat{f}_\lambda$ is a Lasso penalized spline for general m.

3.7 Numerical demonstration: Total variation splines

We demonstrate implementation of total variation splines and more broadly Lasso penalized splines. As in Sections 3.2 and 3.5, we use the ozone data and study univariate smoothing of `ozone` against the Dagget pressure gradient (`dpg`).

We write the following functions, `tv.spline0()` and `tv.spline()`, to implement Lasso penalized splines using truncated power basis functions and the Lasso estimator of the coefficient vector in (3.41). The two functions are structured in a similar manner to `pen.spline0()` and `pen.spline()` for implementing ridge penalized splines in Section 3.2. The function `tv.spline0()` requires, as an input argument, the basis matrix `X` at the data points. The function `tv.spline()` combines `tr.pow()` and `tv.spline0()` and serves as a higher-level interface, with the covariate data points `x` and the covariate knots, `knots`, as input arguments. As `tr.pow()` computes truncated power basis functions for `degree` at least 1, our code excludes the case of fitting piecewise constant splines ($m = 1$). But this case can be handled by extending `tr.pow()` to cover truncated constant basis functions.

In `tv.spline0()`, to fit Lasso linear regression, we apply the function `my.lasso()` implementing the coordinate descent from Section 1.9.2 or `glm.regu()` implementing the active-set descent from the R package `RCAL`. The argument `lam` for `my.lasso()` can be either a scalar as used in Section 1.9.2 or a vector of the column dimension of basis matrix `X`. The coordinate descent algorithm tends to produce less exact estimates (even with `niter=1e+4` and `eps=1e-6`) than active-set descent, which is set to the default method.

```
tv.spline0 <- function(y, X, degree=1, lam,
                       init=NULL, niter=1e+4, eps=1e-6, method="ASD") {
  k <- dim(X)[2]
  lam2 <- rep(lam, times=k)
  lam2[1:degree] <- 0
  if (method=="CD") {
    out <- my.lasso(y, X, lam=lam2, intercept=T, init, niter, eps)
    coef <- out$bet
  } else {  #method=="ASD"
    out <- glm.regu(y, X, loss="gaus", rhos=lam2)  #from package RCAL
    coef <- c(out$inter, out$bet)
  }
  fit <- as.vector(cbind(1,X)%*% coef)
  list(coef=coef, fit=fit)
}
```

```
tv.spline <- function(y, x, degree=1, knots=NULL, ends, lam,
                      init=NULL, niter=1e+4, eps=1e-6, method="ASD") {
  if (is.null(knots))
    knots <- tv.knots(x, degree)
  X <- tr.pow(x, degree, knots, ends)
  out <- tv.spline0(y, X, degree, lam, init, niter, eps, method)
  list(coef=out$coef, fit=out$fit)
}
```

Total variation splines are implemented by calling `tv.spline()` with the argument `knots=NULL`. In this case, the following function computes the knots as defined in (3.43).

```
tv.knots <- function(x, degree=1) {
  n <- length(x)
  ox <- sort(x)
  if (degree%%2 ==1) {
    knots <- ox[(1+(degree+1)/2):(n-(degree+1)/2)]
  } else {
    knots <- ox[(2+degree/2):(n-degree/2)]
  }
  knots
}
```

Figure 3.16 shows the total variation linear splines ($m = 2$ or $q = 1$ in Section 3.6.2) obtained using `tv.spline()`. The tuning parameters are hand picked such that the numbers of nonzero penalized coefficients (i.e., the numbers of selected knots) plus the number of non-penalized coefficients 2 (i.e., constant and linear coefficients) match the degrees of freedom of cubic smoothing splines in Figure 3.10. The following R code produces the (1,2) plot with `nonzero = 6` (except for graphical customization) in Figure 3.16.

```
> knots <- tv.knots(dpg, degree=1)
> ends <- c(-70, 110)
> x <- seq(-80,120,2)
> bX <- tr.pow(x, degree=1, knots)
>
> out2 <- tv.spline(ozone, dpg, ends=ends, lam=1.7e-3)
> plot(dpg, ozone, xlim=c(-80,120), ylim=c(-5,40))
> ffit.tvsp2 <- cbind(1,bX) %*% out2$coef
> lines(x, ffit.tvsp2, col="blue")
```

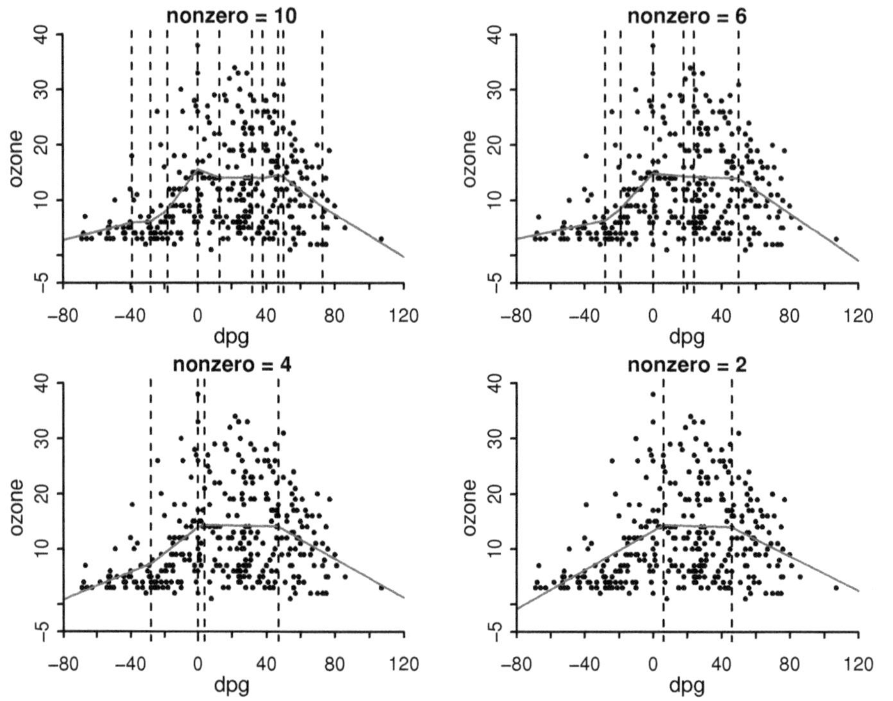

FIGURE 3.16
Plots of total variation linear splines for `ozone` against `dpg` in the ozone dataset, using `tv.spline()`, with several tuning parameters leading to the numbers of nonzero penalized coefficients (absolute values > `1e-3`) equal to the degrees of freedom minus 2 in Figure 3.10. Vertical lines are placed at the data points as knots associated with nonzero coefficients.

```
> abline(v=knots[abs(out2$coef[-(1:2)]) > 1e-3 ], lty=2)
>
> cbind(knots, out2$coef[-(1:2)])[abs(out2$coef[-(1:2)]) > 1e-3, ]
     knots
[1,]   -28  31.148001
[2,]   -19  17.174687
[3,]     0 -65.211663
[4,]    18   1.249683
[5,]    24   2.189637
[6,]    50 -36.543543
```

3.8 Generalized nonparametric regression

The discussion in the preceding sections is based on the linear smoothing model (3.1), which is suitable to Gaussian-like response variables conditionally on the covariates. To handle other types of response variables (for example, binary responses), it is natural to consider a *generalized* smoothing (or generalized nonparametric regression) model with two components. The first component is a mean restriction

$$E(Y_i|X_i) = g^{-}\{f^*(X_i)\}, \quad i = 1, \ldots, n, \tag{3.44}$$

where $f^*(\cdot)$ is an unknown function, and $g(\cdot)$ is a *known* monotonic "link" function. The second component is an exponential-family assumption (2.2) for the conditional distribution of Y_i given X_i in the form

$$f(y;\theta,\phi) = \exp\left\{\frac{y\theta - a(\theta)}{\phi} + c(y,\phi)\right\},$$

where θ and $\phi > 0$ are two scalar parameters and $a(\cdot)$ and $c(\cdot)$ are *known*, differentiable functions. Combining the two components implies that the parameter $\theta = \theta(X)$ is related to the mean $\mu(X) = E(Y|X)$ and the predictor $f^*(Z)$ as in (2.6):

$$\theta(X) = (a')^{-}\{\mu(X)\} = (a')^{-}\{g^{-}(f^*(X))\}.$$

For simplicity, $\phi = \phi(X)$ is assumed to be a constant ϕ^*.

The methods from the preceding sections can be readily extended to the generalized smoothing model (3.44), by the replacing the mean-squared-error loss with the average negative log-likelihood function after dropping the constant factor $\frac{1}{\phi^*}$:

$$K_n(f) = \frac{-1}{n}\sum_{i=1}^{n}\{Y_i\theta(X_i;f) - a(\theta(X_i;f))\},$$

where $\theta(X_i;f) = (a')^{-}\{g^{-}(f(X_i))\}$. For logistic regression, the likelihood loss is

$$K_n(f) = \frac{-1}{n}\sum_{i=1}^{n}\left\{Y_i f(X_i) - \log(1 + \mathrm{e}^{f(X_i)})\right\}.$$

In the following, we describe the finite-dimensional representations for cubic smoothing splines and total variation splines with scalar covariates X.

For fixed $\lambda > 0$, a cubic smoothing spline on $[a,b]$ is defined as

$$\hat{f}_\lambda = \mathrm{argmin}_{f\in\mathcal{H}}\left\{K_n(f) + \frac{\lambda}{2}\int_a^b (f''(x))^2\,\mathrm{d}x\right\},$$

where $\mathcal{H} = C^{(2)}([a,b])$ as discussed in Example 3.9. Then a solution $\hat{f}_\lambda$ can be represented as a natural cubic spline with knots at $X_1, \ldots, X_n$. The result can be proved in a similar manner as Theorem 3.4, because the loss function $K_n(f)$ depends on f only through its values at the data points $X_1, \ldots, X_n$.

For fixed $\lambda > 0$, a total variation spline of order $m\,(\geq 1)$ is defined as

$$\hat{f}_\lambda = \text{argmin}_{f \in \mathcal{F}} \left\{ K_n(f) + \lambda \text{TV}(f^{(m-1)}) \right\},$$

where $\mathcal{F} = \mathcal{V}^{(m)}([a,b])$ as used in (3.42). Then a solution $\hat{f}_\lambda$ in the case of $m = 1$ or 2 can be represented as a polynomial spline of degree 0 or 1 respectively with knots from the data points $X_1, \ldots, X_n$ as stated in Theorem 3.5. The result can also be proved in a similar manner as Theorem 3.5. For $m \geq 3$, a solution $\hat{f}_\lambda$ can be defined as a polynomial spline of degree $m-1$ but the knots may be outside the data points.

Given the finite-dimensional representation, the cubic smoothing spline or total variation spline can be computed by using suitable basis functions and then proceeding similarly as in ridge or Lasso penalized regression discussed in Section 2.5.

3.9 Appendices

3.9.1 On assessment of prediction errors

We discuss assessment of prediction errors for a smoothing method, including penalized linear regression as a special case. Suppose that a training dataset consists of observations (X_i, Y_i), $i = 1, \ldots, n$, which are independent and identically distributed from a probability distribution. Consider a linear smoothing model:

$$Y_i = f^*(X_i) + \varepsilon_i, \quad i = 1, \ldots, n,$$

where $f^*(x) = E(Y_i|X_i = x)$, assumed to be from some parametric or nonparametric function class $\mathcal{F}$, Y_i is a response variable, X_i is a covariate vector, and ε_i is a noise variable such that $E(\varepsilon_i|X_i) = 0$. For simplicity, assume homoscedastic noises: $E(\varepsilon_i^2|X_i) = \sigma^2$ for a constant σ^2. Let $\hat{f}(\cdot)$ be an estimator of $f^*(\cdot)$ based on the training data. The predictive performance of $\hat{f}(\cdot)$ can be measured in two distinct manners. Throughout this section, the function class $\mathcal{F}$ used to derive $\hat{f}$ may be misspecified, i.e., the assumption that $f^* \in \mathcal{F}$ may be incorrect.

The out-of-sample conditional prediction mean-squared-error (MSE) of $\hat{f}$ is defined as

$$\text{Err}_{\text{out}}(\hat{f}|\mathbf{X}) = E\{(Y_0 - \hat{f}(X_0))^2 | X_1, \ldots, X_n\},$$

where (X_0, Y_0) is a new observation, independent of and identically distributed as the training data, and the expectation is taken over both (X_0, Y_0) and $(Y_1, \ldots, X_n)$, conditionally on $\mathbf{X} = (X_1, \ldots, X_n)^{\mathrm{T}}$. The in-sample conditional prediction MSE of $\hat{f}$ is defined as

$$\mathrm{Err}_{\mathrm{in}}(\hat{f}|\mathbf{X}) = E\left\{ \frac{1}{n} \sum_{i=1}^{n} (\tilde{Y}_i - \hat{f}(X_i))^2 \Big| X_1, \ldots, X_n \right\},$$

where $\tilde{Y}_i$ is a new response, independent of and identically distributed as Y_i, given X_i for $i = 1, \ldots, n$, and the expectation is taken over both $\tilde{Y}_i$ and Y_i, $i = 1, \ldots, n$, conditionally on $\mathbf{X}$. The conditional prediction MSEs can be further averaged over $(X_1, \ldots, X_n)$, leading to $\mathrm{Err}_{\mathrm{out}}(\hat{f}) = E\{(Y_0 - \hat{f}(X_0))^2\}$ and $\mathrm{Err}_{\mathrm{in}}(\hat{f}) = E\{\frac{1}{n}\sum_{i=1}^{n}(\tilde{Y}_i - \hat{f}(X_i))^2\}$.

The prediction MSEs can be expressed in terms of the estimation MSEs of $\hat{f}$ against the mean function f^*, as in the following proposition. The proof is left as an exercise.

Proposition 3.3 (Prediction and estimation MSEs). *The conditional prediction MSEs can be equivalently expressed as follows:*

$$\mathrm{Err}_{out}(\hat{f}|\mathbf{X}) = \sigma^2 + E\{(\hat{f}(X_0) - f^*(X_0))^2 | X_1, \ldots, X_n\},$$
$$\mathrm{Err}_{in}(\hat{f}|\mathbf{X}) = \sigma^2 + E\left\{ \frac{1}{n} \sum_{i=1}^{n} (\hat{f}(X_i) - f^*(X_i))^2 \Big| X_1, \ldots, X_n \right\}.$$

Similar expressions hold for the unconditional prediction MSEs.

Similarly as described in Sections 1.9.3 and 2.6.2, the training error of $\hat{f}$ is defined as

$$\mathrm{err}(\hat{f}) = \frac{1}{n} \sum_{i=1}^{n} (Y_i - \hat{f}(X_i))^2.$$

The test error of $\hat{f}$ is defined as

$$\widehat{\mathrm{Err}}(\hat{f}) = \frac{1}{N-n} \sum_{i=n+1}^{N} (Y_i - \hat{f}(X_i))^2,$$

where (X_i, Y_i), $i = n+1, \ldots, N$, called a test dataset, are independent and identically distributed observations as the training dataset. It can be directly shown that $\widehat{\mathrm{Err}}(\hat{f})$ is an unbiased estimator of the out-of-sample prediction error $\mathrm{Err}_{\mathrm{out}}(\hat{f})$: $E\{\widehat{\mathrm{Err}}(\hat{f})\} = \mathrm{Err}_{\mathrm{out}}(\hat{f})$. A similar idea would also apply to the in-sample prediction error $\mathrm{Err}_{\mathrm{in}}(\hat{f})$ provided new observations could be obtained with the fixed covariates $X_1, \ldots, X_n$.

There is an interesting relationship between the training error $\mathrm{err}(\hat{f})$ and the in-sample prediction error $\mathrm{Err}_{\mathrm{in}}(\hat{f}|\mathbf{X})$ for a linear smoother $\hat{f}$, where the fitted values on the observed covariates $\hat{\mathbf{f}} = (\hat{f}(X_1), \ldots, \hat{f}(X_n))^{\mathrm{T}}$ can be represented as $\hat{\mathbf{f}} = \mathbf{SY}$, where $\mathbf{Y} = (Y_1, \ldots, Y_n)^{\mathrm{T}}$ and $\mathbf{S}$ is a smoother matrix, depending on $\mathbf{X} = (X_1^{\mathrm{T}}, \ldots, X_n)^{\mathrm{T}}$ but not on $\mathbf{Y}$.

Proposition 3.4 (Unbiased estimation of in-sample prediction error). *For a linear smoother $\hat{f}$ satisfying $\hat{\mathbf{f}} = \mathbf{SY}$, we have*

$$\mathrm{Err}_{in}(\hat{f}|\mathbf{X}) = E\left\{\mathrm{err}(\hat{f}) + \frac{2\sigma^2}{n}\mathrm{tr}(\mathbf{S})\Big|X_1, \ldots, X_n\right\}. \tag{3.45}$$

In other words, $\mathrm{err}(\hat{f}) + \frac{2\sigma^2}{n}\mathrm{tr}(\mathbf{S})$ *is an unbiased estimator of* $\mathrm{Err}_{in}(\hat{f}|\mathbf{X})$ *for known σ^2.*

Proof. Denote $\mathbf{f}^* = (f^*(X_1), \ldots, f^*(X_n))^{\mathrm{T}}$. Below, all expectations are taken conditionally on $(X_1, \ldots, X_n)$. First, we show

$$E\left\{\frac{1}{n}\sum_{i=1}^{n}(\hat{f}(X_i) - f^*(X_i))^2\right\} = \frac{1}{n}\|(\mathbf{S} - I)\mathbf{f}^*\|_2^2 + \frac{\sigma^2}{n}\mathrm{tr}(\mathbf{SS}^{\mathrm{T}}). \tag{3.46}$$

In fact, the left-hand side can be decomposed by the bias and variance:

$$E\left\{\frac{1}{n}\sum_{i=1}^{n}(\hat{f}(X_i) - f^*(X_i))^2\right\} = \frac{1}{n}\sum_{i=1}^{n} E\Big\{(\hat{f}(X_i) - f^*(X_i))^2\Big\}$$
$$= \frac{1}{n}\sum_{i=1}^{n}\Big\{(E(\hat{f}(X_i)) - f^*(X_i))^2 + \mathrm{var}(\hat{f}(X_i))\Big\}.$$

The two terms involved can be calculated as follows:

$$\sum_{i=1}^{n}(E(\hat{f}(X_i)) - f^*(X_i))^2 = \|E(\hat{\mathbf{f}}) - \mathbf{f}^*\|_2^2$$
$$= \|E(\mathbf{SY}) - \mathbf{f}^*\|_2^2 \qquad [\text{because } \hat{\mathbf{f}} = \mathbf{SY}]$$
$$= \|\mathbf{S}E(\mathbf{Y}) - \mathbf{f}^*\|_2^2 \qquad [\text{because } \mathbf{S} \text{ does not depend on } \mathbf{Y}]$$
$$= \|\mathbf{Sf}^* - \mathbf{f}^*\|_2^2 \qquad [\text{because } E(\mathbf{Y}) = \mathbf{f}^*],$$

and

$$\sum_{i=1}^{n}\mathrm{var}(\hat{f}(X_i)) = \mathrm{tr}(\mathrm{var}(\hat{\mathbf{f}}))$$
$$= \mathrm{tr}(\mathrm{var}(\mathbf{SY})) \qquad [\text{because } \hat{\mathbf{f}} = \mathbf{SY}]$$
$$= \mathrm{tr}(\mathbf{S}\mathrm{var}(\mathbf{Y})\mathbf{S}^{\mathrm{T}}) \qquad [\text{because } \mathbf{S} \text{ does not depend on } \mathbf{Y}]$$
$$= \sigma^2\mathrm{tr}(\mathbf{SS}^{\mathrm{T}}) \qquad [\text{because } \mathrm{var}(\mathbf{Y}) = \sigma^2 I].$$

Combining the preceding three displays yields (3.46).

Second, we show that

$$E\{\text{err}(\hat{f})\} = \frac{1}{n}\|(I-\mathbf{S})\mathbf{f}\|_2^2 + \frac{\sigma^2}{n}(n + \text{tr}(\mathbf{S}\mathbf{S}^\text{T}) - 2\text{tr}(\mathbf{S})). \quad (3.47)$$

In fact, $nE\{\text{err}(\hat{f})\}$ can be calculated as follows:

$$\begin{aligned} nE\{\text{err}(\hat{f})\} &= \sum_{i=1}^n E\{(Y_i - \hat{f}(X_i))^2\} = E\{\|\mathbf{Y} - \hat{\mathbf{f}}\|_2^2\} \\ &= E\{\|\mathbf{Y} - \mathbf{S}\mathbf{Y}\|_2^2\} = E\{\mathbf{Y}^\text{T}(I-\mathbf{S})^\text{T}(I-\mathbf{S})\mathbf{Y}\} \\ &= \mathbf{f}^{*\text{T}}(I-\mathbf{S})^\text{T}(I-\mathbf{S})\mathbf{f}^* + \text{tr}((I-\mathbf{S})^\text{T}(I-\mathbf{S})\text{var}(\mathbf{Y})) \\ &= \|(I-\mathbf{S})\mathbf{f}^*\|_2^2 + \sigma^2(n + \text{tr}(\mathbf{S}^\text{T}\mathbf{S}) - 2\text{tr}(\mathbf{S})). \end{aligned}$$

Justification of each step is left as an exercise.

Finally, combining (3.46) and (3.47) and noting the expression of $\text{Err}_\text{in}(\hat{f}|\mathbf{X})$ in Proposition 3.3 leads to (3.45) as desired. □

Proposition 3.4 provides a justification for interpreting $\text{tr}(\mathbf{S})$ as the degrees of freedom for a linear smoother, including least-squares linear regression, ridge-penalized linear regression, ridge-penalized splines and smoothing splines with a fixed tuning parameter λ. In particular, for least-squares linear regression with $\mathbf{X}^\text{T}\mathbf{X}$ nonsingular, the smoother matrix is $\mathbf{S} = \mathbf{X}(\mathbf{X}^\text{T}\mathbf{X})^{-1}\mathbf{X}^\text{T}$ and satisfies $\text{tr}(\mathbf{S}) = p$ (Section 1.8.2). In this case, (3.45) shows that $\text{Err}_\text{in}(\hat{f}|\mathbf{X}) = E\{\text{err}(\hat{f}) + 2\sigma^2\frac{p}{n}|\mathbf{X}\}$; this leads to the C_p statistic (Mallows, 1973)

$$C_p = \text{err}(\hat{f}) + 2\hat{\sigma}^2\frac{p}{n},$$

where $\hat{\sigma}^2$ is an estimator of σ^2. The C_p statistic is typically used for selecting among several subset regressions: linear regression with a subset of covariates is preferred if the smallest C_p statistic is achieved, where $\hat{\sigma}^2$ can be the usual estimated noise variance from the full regression, i.e., linear regression with the full set of covariates. Similarly, for a linear smoother $\hat{f}_\lambda$ with smoother matrix $\mathbf{S}_\lambda$ depending on tuning parameter λ, a C_p statistic is

$$C_p(\lambda) = \text{err}(\hat{f}_\lambda) + \frac{2\hat{\sigma}^2}{n}\text{tr}(\mathbf{S}_\lambda),$$

where $\hat{\sigma}^2$ is an estimated noise variance obtained with a fixed λ close to 0.

As described in Sections 1.9.3 and 2.6.2, cross validation (CV) provides a practical approach for assessing prediction errors. As a special case of K-fold cross validation, leave-one-out cross validation proceeds as follows with a tuning parameter λ:

- For $j = 1, \ldots, n$, compute the fitted function $\hat{f}_\lambda(\cdot; I_{(-j)})$ based on the data $\{(X_i, Y_i) : i \neq j, i = 1, \ldots, n\}$, i.e., the original dataset with jth observation removed.

- Compute the leave-one-out CV statistic

$$\mathrm{CV}(\lambda) = \frac{1}{n}\sum_{j=1}^{n}(Y_j - \hat{f}_\lambda(X_j; I_{(-j)}))^2.$$

For a linear smoother defined through penalized regression, the leave-one-out CV statistic can be obtained without computing the fitted functions on the leave-one-out datasets. Examples of such linear smoothers include ridge-penalized splines and smoothing splines.

Proposition 3.5 (Leave-one-out CV for linear smoothers). *For a tuning parameter $\lambda \ge 0$, suppose that a linear smoother can be defined through penalized regression:*

$$\hat{f}_\lambda = \mathrm{argmin}_{f\in\mathcal{F}}\left\{\sum_{i=1}^{n}(Y_i - f(X_i))^2 + \lambda\,\mathrm{pen}(f)\right\},$$

where $\mathrm{pen}(f)$ *is a penalty on f such as the ridge penalty,* $\mathrm{pen}(f) = \|\beta\|_2^2/2$ *for $f = \beta^{\mathrm{T}}x$, in linear regression. Let $\mathbf{S}_\lambda$ be the smoother matrix of $\hat{f}_\lambda$ such that* $(\hat{f}_\lambda(X_1), \ldots, \hat{f}_\lambda(X_n))^{\mathrm{T}} = \mathbf{S}_\lambda(Y_1, \ldots, Y_n)^{\mathrm{T}}$, *where $\mathbf{S}_\lambda$ depends on $(X_1, \ldots, X_n)$ but not on $(Y_1, \ldots, Y_n)$. Then the leave-one-out CV statistic can be expressed as*

$$\mathrm{CV}(\lambda) = \frac{1}{n}\sum_{j=1}^{n}\left(\frac{Y_j - \hat{f}_\lambda(X_j)}{1 - (\mathbf{S}_\lambda)_{jj}}\right)^2. \tag{3.48}$$

[Note: For the definition of $\hat{f}_\lambda$, the penalty term $\lambda\,\mathrm{pen}(f)$ *is added to the sum of squared errors instead of the mean of squared errors, which differs from our previous definitions, for example* (3.25) *and* (3.32)*. This choice is needed for inequality* (3.50) *in the proof.]*

Proof. A key property underlying (3.48) is that for $j = 1, \ldots, n$,

$$\hat{f}_\lambda(X_j; I_{(-j)}) = \frac{\sum_{i\neq j}(\mathbf{S}_\lambda)_{ji}Y_i}{1 - (\mathbf{S}_\lambda)_{jj}}. \tag{3.49}$$

To show property (3.49), for each j, consider a new response vector $(Y_1^{(j)}, \ldots, Y_n^{(j)})$, associated with $(X_1, \ldots, X_n)$, where $Y_i^{(j)} = Y_i$ if $i \neq j$ or $\hat{f}_\lambda(X_j; I_{(-j)})$ if $i = j$. Then $\hat{f}_\lambda(\cdot; I_{(-j)})$ gives the fitted function based on the new data $\{(X_i, Y_i^{(j)}) : i = 1, \ldots, n\}$, i.e., $\hat{f}_\lambda(\cdot; I_{(-j)})$ is a minimizer of $\sum_{i=1}^{n}(Y_i^{(j)} - f(X_i))^2 + \lambda\,\mathrm{pen}(f)$. In fact, for any $f \in \mathcal{F}$,

$$\sum_{i=1}^{n}(Y_i^{(j)} - f(X_i))^2 + \lambda\,\mathrm{pen}(f)$$

$$
\begin{aligned}
&\geq \sum_{i \neq j} (Y_i - f(X_i))^2 + \lambda \,\mathrm{pen}(f) \\
&\qquad\qquad [\text{because } (Y_j^{(j)} - f(X_j))^2 \geq 0 \text{ and } Y_i^{(j)} = Y_i \text{ for } i \neq j] \\
&\geq \sum_{i \neq j} (Y_i - \hat{f}_\lambda(X_i; I_{(-j)}))^2 + \lambda \,\mathrm{pen}(\hat{f}_\lambda(\cdot; I_{(-j)})) \\
&\qquad\qquad [\text{by the definition of } \hat{f}_\lambda(\cdot; I_{(-j)})] \qquad (3.50) \\
&= \sum_{i=1}^{n} (Y_i^{(j)} - \hat{f}_\lambda(X_i; I_{(-j)}))^2 + \lambda \,\mathrm{pen}(\hat{f}_\lambda(\cdot; I_{(-j)})) \\
&\qquad\qquad [\text{because } Y_j^{(j)} = \hat{f}_\lambda(X_j; I_{(-j)})].
\end{aligned}
$$

Then by the definition of $\mathbf{S}_\lambda$,

$$
\begin{aligned}
\hat{f}_\lambda(X_j; I_{(-j)}) &= \sum_{i=1}^{n} (\mathbf{S}_\lambda)_{ji} Y_i^{(j)} \\
&= (\mathbf{S}_\lambda)_{jj} Y_j^{(j)} + \sum_{i \neq j} (\mathbf{S}_\lambda)_{ji} Y_i^{(j)} \\
&= (\mathbf{S}_\lambda)_{jj} \hat{f}_\lambda(X_j; I_{(-j)}) + \sum_{i \neq j} (\mathbf{S}_\lambda)_{ji} Y_i,
\end{aligned}
$$

which yields (3.49). Then (3.48) follows because for $j = 1, \ldots, n$,

$$
\begin{aligned}
Y_j - \hat{f}_\lambda(X_j; I_{(-j)}) &= Y_j - \frac{\sum_{i \neq j} (\mathbf{S}_\lambda)_{ji} Y_i}{1 - (\mathbf{S}_\lambda)_{jj}} \qquad [\text{by (3.49)}] \\
&= \frac{Y_j - \sum_{i=1}^{n} (\mathbf{S}_\lambda)_{ji} Y_i}{1 - (\mathbf{S}_\lambda)_{jj}} \\
&= \frac{Y_j - \hat{f}_\lambda(X_j)}{1 - (\mathbf{S}_\lambda)_{jj}} \qquad [\text{by the definition of } \mathbf{S}_\lambda].
\end{aligned}
$$

□

Replacing $1 - (\mathbf{S}_\lambda)_{jj}$ for $j = 1, \ldots, n$ in (3.48) by the average $1 - \frac{\mathrm{tr}(\mathbf{S}_\lambda)}{n}$ leads to the generalized CV (GCV) statistic

$$
\mathrm{GCV}(\lambda) = \frac{1}{n} \sum_{j=1}^{n} \left(\frac{Y_j - \hat{f}_\lambda(X_j)}{1 - \frac{\mathrm{tr}(\mathbf{S}_\lambda)}{n}} \right)^2 = \frac{\mathrm{err}(\hat{f}_\lambda)}{(1 - \frac{\mathrm{tr}(\mathbf{S}_\lambda)}{n})^2}. \qquad (3.51)
$$

By the approximation $\frac{1}{(1-x)^2} \approx 1 - 2x$ for $x \approx 0$, $\mathrm{GCV}(\lambda)$ can be related to $C_p(\lambda)$ as follows:

$$
\begin{aligned}
\mathrm{GCV}(\lambda) &\approx \mathrm{err}(\hat{f}_\lambda)(1 + 2\frac{\mathrm{tr}(\mathbf{S}_\lambda)}{n}) \\
&= \mathrm{err}(\hat{f}_\lambda) + \frac{2\hat{\sigma}^2}{n} \mathrm{tr}(\mathbf{S}_\lambda),
\end{aligned}
$$

which is $C_p(\lambda)$ with the estimator $\hat{\sigma}^2 = \mathrm{err}(\hat{f}_\lambda)$.

3.9.2 On bounded linear functionals

We provide additional details about bounded linear functionals. On a function space $\mathcal{F}$ with norm $\|\cdot\|$, a functional $G : \mathcal{F} \to \mathbb{R}$ is said to be continuous if for any $f_0 \in \mathcal{F}$ and $\epsilon > 0$, there exists $\delta > 0$, possibly depending on (f_0, ϵ), such that $|G(f) - G(f_0)| \leq \epsilon$ for any $f \in \mathcal{F}$ satisfying $\|f - f_0\| \leq \delta$. In other words, a continuous functional G satisfies that if $f \to f_0$, then $G(f) \to G(f_0)$.

Lemma 3.2 (Bounded linear functionals). *On a normed space $\mathcal{F}$, a linear functional G is bounded if and only if it is continuous.*

Proof. First, suppose that G is bounded, satisfying $|G(f)| \leq M\|f\|$ for any $f \in \mathcal{F}$. Then for any $f_0, f \in \mathcal{F}$, we have by the linearity and boundedness of G,

$$|G(f) - G(f_0)| = |G(f - f_0)| \leq M\|f - f_0\|.$$

For any $\epsilon > 0$, let $\delta = \frac{\epsilon}{M}$. Then the preceding display implies that for any $f \in \mathcal{F}$ satisfying $\|f - f_0\| \leq \delta$, we have $|G(f) - G(F_0)| \leq M\|f - f_0\| \leq M\delta = \epsilon$.

Second, suppose that G is continuous. Note that $G(0) = 0$ by the linearity of G. By the continuity of G at $f_0 = 0$, for $\epsilon = 1$, there exists $\delta > 0$ such that $|G(f)| = |G(f) - G(0)| \leq 1$ for any $f \in \mathcal{F}$ satisfying $\|f\| \leq \delta$. For any $f \neq 0$, let $\tilde{f} = \delta \frac{f}{\|f\|}$ satisfying $\|\tilde{f}\| = \delta$. Then $|G(\tilde{f})| \leq 1$, and hence by the linearity of G,

$$|G(f)| = \frac{\|f\|}{\delta}|G(\tilde{f})| \leq \frac{\|f\|}{\delta}.$$

This indicates that G is bounded by Definition 3.6 with $M = \frac{1}{\delta}$. □

Exercises

Ex 3.1. Show that a natural cubic spline with k knots in (3.7) can be represented as a linear combination of the k basis functions stated in (3.9). [This is Exercise 5.4 in Hastie et al. (2009).]

Ex 3.2. Find explicit expressions of Ω_{ij}, $i, j = 1, \ldots, n$, the elements of Ω used in (3.12). Organize your results in a way that can be used in programming (Exercise 3.10).

Ex 3.3.

(i) Consider the reproducing kernel

$$K(t, x) = (1 + tx)^2.$$

Determine the corresponding RKHS. What is the inner product?

(ii) For an integer $r \geq 1$, consider the reproducing kernel

$$K(t, x) = (1 + tx)^r.$$

Determine the corresponding RKHS. What is the inner product?

Ex 3.4.

(i) Consider the reproducing kernel

$$K(t, x) = \int_0^1 (1 - x - u)_+(1 - t - u)_+ \, du, \quad t, x \in [0, 1].$$

Determine the corresponding RKHS. What is the inner product?

(ii) Use the result in (i) to prove that the cubic spline f obtained from minimizing

$$\frac{1}{n}\sum_{i=1}^{n}(Y_i - f(X_i))^2 + \lambda \int_0^1 \{f''(x)\}^2 \, dx$$

is linear on $[0, X_{(1)}]$, where $X_{(1)} = \min(X_1, \ldots, X_n)$.

Ex 3.5. Consider RKHS regression with training data $(X_1, Y_1), \ldots, (X_n, Y_n)$ in Section 3.4.3. Let $(\xi_1, \ldots, \xi_k)$ be the set of distinct points in $(X_1, \ldots, X_n)$ and use the representation

$$f(x; \mathring{c}) = \sum_{i=1}^{k} \mathring{c}_i K(\xi_i, x).$$

For fixed $\lambda > 0$, the RKHS smoother $\hat{f}_\lambda$ can be determined as $\hat{f}_\lambda = f(\cdot; \hat{\mathring{c}}_\lambda)$ with

$$\hat{\mathring{c}}_\lambda = \text{argmin}_{\mathring{c} \in \mathbb{R}^k} \left\{ \frac{1}{n}\sum_{i=1}^{n}(Y_i - f(X_i; \mathring{c}))^2 + \lambda \|f(\cdot; \mathring{c})\|^2 \right\}.$$

Denote $\mathbf{Y} = (Y_1, \ldots, Y_n)^{\mathrm{T}}$, $\mathbf{K} = (K(X_i, X_j))_{i,j=1,\ldots,n}$, and $\mathring{\mathbf{K}} = (K(\xi_i, \xi_j))_{i,j=1,\ldots,k}$.

(i) Let A be the $n \times k$ matrix with (i, j)th element $1\{X_i = \xi_j\}$, equal to 1 if $X_i = \xi_j$ or 0 otherwise. Show that $\mathbf{K} = A\mathring{\mathbf{K}}A^{\mathrm{T}}$, and $\hat{\mathring{c}}_\lambda$ can be obtained as

$$\hat{\mathring{c}}_\lambda = \text{argmin}_{\mathring{c} \in \mathbb{R}^k} \left\{ \frac{1}{n}\|\mathbf{Y} - A\mathring{\mathbf{K}}\mathring{c}\|_2^2 + \lambda \mathring{c}^{\mathrm{T}}\mathring{\mathbf{K}}\mathring{c} \right\}. \tag{3.52}$$

(ii) Show that a valid (possibly non-unique) solution to (3.52) is

$$\hat{\mathring{c}}_\lambda = (W\mathring{\mathbf{K}} + n\lambda I)^{-1}A^{\mathrm{T}}\mathbf{Y}, \tag{3.53}$$

where $W = A^{\mathrm{T}}A = \text{diag}(n_1, \ldots, n_k)$ and n_j is the number of replicates of ξ_j in $(X_1, \ldots, X_n)$.

(iii) Show that $\hat{c}_\lambda$ in (3.29b) and $\hat{\mathring{c}}_\lambda$ in (3.53) are related such that

$$\hat{\mathring{c}}_\lambda = A^{\mathrm{T}}\hat{c}_\lambda, \quad \hat{c}_\lambda = AW^{-1}\hat{\mathring{c}}_\lambda.$$

As a result, the vector of fitted values $\hat{\mathbf{f}}_\lambda$ remains the same, computed from either $\hat{c}_\lambda$ or $\hat{\mathring{c}}_\lambda$: $\mathbf{K}\hat{c}_\lambda = A\mathring{\mathbf{K}}\hat{\mathring{c}}_\lambda$.

Ex 3.6.

(i) Assume that $\mathbf{K}_0$ is of rank 2 and $\mathbf{K}_1$ is nonsingular. Show that the generalized ridge estimator $\hat{\theta}_\lambda$ in (3.37) can be expressed as (3.38).

(ii) Assume that $\mathbf{K}_0$ is of rank 2, but $\mathbf{K}_1$ may be singular (for example, some points in $X_1, \ldots, X_n$ are duplicated). Show that the expression (3.38) remains a valid (non-unique) solution to the problem (3.36).

Ex 3.7. Let A be a $n \times n$ invertible matrix, B a $n \times k$ matrix of rank k, and C a $n \times (n-k)$ matrix of rank $(n-k)$ satisfying $B^{\mathrm{T}}C = 0$, where $1 \leq k \leq n-1$. Then, as stated in (3.40),

$$A^{-1} - A^{-1}B(B^{\mathrm{T}}A^{-1}B)^{-1}B^{\mathrm{T}}A^{-1} = C(C^{\mathrm{T}}AC)^{-1}C^{\mathrm{T}}.$$

Ex 3.8. For the ridge-penalized spline, split the basis matrix as $\mathbf{X}_{\mathrm{T}} = (\mathbf{X}_{\mathrm{T},0}, \mathbf{X}_{\mathrm{T},1})$, where

$$\mathbf{X}_{\mathrm{T},0} = \begin{pmatrix} 1 & Z_1 & \ldots & Z_1^q \\ \vdots & \vdots & \vdots & \vdots \\ 1 & Z_n & \ldots & Z_n^q \end{pmatrix},$$

$$\mathbf{X}_{\mathrm{T},1} = \begin{pmatrix} \{(Z_1 - \xi_1)_+\}^q & \ldots & \{(Z_1 - \xi_k)_+\}^q \\ \vdots & \vdots & \vdots \\ \{(Z_n - \xi_1)_+\}^q & \ldots & \{(Z_n - \xi_k)_+\}^q \end{pmatrix}.$$

Suppose that $\mathbf{X}_{\mathrm{T},0}$ is of rank $q+1$, and consider a QR decomposition

$$\mathbf{X}_{\mathrm{T},0} = QR = (Q_1 \quad Q_2)\begin{pmatrix} R_1 \\ 0 \end{pmatrix},$$

where Q is an $n \times n$ orthogonal matrix and R is an $n \times (q+1)$ matrix. Show that the smoother matrix $\mathbf{S}_\lambda$ in (3.13) can be expressed as

$$\mathbf{S}_\lambda = I - n\lambda Q_2(Q_2^{\mathrm{T}}\mathbf{K}_{1,\lambda}Q_2)^{-1}Q_2^{\mathrm{T}},$$

where $\mathbf{K}_{1,\lambda} = \mathbf{X}_{\mathrm{T},1}\mathbf{X}_{\mathrm{T},1}^{\mathrm{T}} + n\lambda I$.

[Hint: Consult the proof of Lemma 3.1 as well as Lemma 4.3 later. The estimator $\hat{\beta}_\lambda$ in (3.5) can be determined as $\hat{\beta}_\lambda = (\hat{b}_\lambda^{\mathrm{T}}, \hat{c}_\lambda^{\mathrm{T}})^{\mathrm{T}}$ with

$$\hat{b}_\lambda = (\mathbf{X}_{\mathrm{T},0}^{\mathrm{T}}\mathbf{K}_{1,\lambda}^{-1}\mathbf{X}_{\mathrm{T},0})^{-1}\mathbf{X}_{\mathrm{T},0}^{\mathrm{T}}\mathbf{K}_{1,\lambda}^{-1}\mathbf{Y},$$

$$\begin{aligned}\hat{c}_\lambda &= \mathbf{X}_{\mathrm{T},1}^{\mathrm{T}} \mathbf{K}_{1,\lambda}^{-1} (\mathbf{Y} - \mathbf{X}_{\mathrm{T},0} \hat{b}_\lambda) \\ &= H_{1,\lambda}^{-1} \mathbf{X}_{\mathrm{T},1}^{\mathrm{T}} (\mathbf{Y} - \mathbf{X}_{\mathrm{T},0} \hat{b}_\lambda),\end{aligned}$$

where $H_{1,\lambda} = \mathbf{X}_{\mathrm{T},1}^{\mathrm{T}} \mathbf{X}_{\mathrm{T},1} + n\lambda I$.]

Ex 3.9. Let (Y_i, Z_i), $i = 1, \ldots, n$, be response and scalar covariate data. For $\lambda > 0$, a cubic smoothing spline is defined as in (3.10):

$$\hat{f}_\lambda = \operatorname{argmin}_{f \in C^{(2)}[a,b]} \left\{ \frac{1}{n} \sum_{i=1}^{n} (Y_i - f(Z_i))^2 + \lambda \int_a^b (f''(z))^2 \, \mathrm{d}z \right\},$$

where $[a, b]$ is an interval containing $(Z_1, \ldots, Z_n)$.

(i) Let $X_i = \frac{Z_i - a}{b - a} \in [0, 1]$, $i = 1, \ldots, n$. For $\kappa > 0$, let

$$\hat{g}_\kappa = \operatorname{argmin}_{g \in C^{(2)}[0,1]} \left\{ \frac{1}{n} \sum_{i=1}^{n} (Y_i - g(X_i))^2 + \kappa \int_0^1 (g''(x))^2 \, \mathrm{d}x \right\}.$$

Find κ such that $\hat{f}_\lambda(z)$ can be obtained as $\hat{f}_\lambda(z) = \hat{g}_\kappa(\frac{z-a}{b-a})$.

(ii) Let $[\tilde{a}, \tilde{b}]$ be another interval containing $(Z_1, \ldots, Z_n)$. For $\tilde{\lambda} > 0$, let

$$\check{f}_{\tilde{\lambda}} = \operatorname{argmin}_{f \in C^{(2)}[\tilde{a},\tilde{b}]} \left\{ \frac{1}{n} \sum_{i=1}^{n} (Y_i - f(Z_i))^2 + \tilde{\lambda} \int_{\tilde{a}}^{\tilde{b}} (f''(z))^2 \, \mathrm{d}z \right\}.$$

Find $\tilde{\lambda}$ such that $\hat{f}_\lambda(z) = \check{f}_{\tilde{\lambda}}(z)$ for $z \in [a, b] \cup [\tilde{a}, \tilde{b}]$. This shows that a smoothing spline is invariant to the choice of an interval containing the covariate data, provided that the tuning parameter is properly rescaled.

Ex 3.10. Write your own R program to fit cubic smoothing splines, using the set of basis functions in (3.9), derived from truncated cubic bases.

For the dataset `LAozone.data`, show that you can replicate the following results obtained using the R function `sm.spline()` from Section 3.5.1.

Note: The function `sm.spline()` rescales the covariate data to [0, 1] according to the endpoints specified by `ends`. The following output from `sm.spline()` depends on the choices of `ends` and `lam` used. See Exercise 3.9 for related results.

```
> id <- order(dpg)
> ends <- c(-70, 110)
> out2 <- sm.spline(ozone, dpg, lam=1.6e-3, ends)
> out2$df
[1] 7.944442
```

```
> cbind(dpg[id], out2$fit[id])
       [,1]      [,2]
  [1,]  -69  3.693788
  [2,]  -68  3.745327
  [3,]  -67  3.796796
  [4,]  -66  3.848201
  [5,]  -63  4.004241
......
[201,]   30 14.185795
[202,]   31 14.186708
[203,]   31 14.186708
[204,]   31 14.186708
[205,]   31 14.186708
......
[326,]   77  8.203964
[327,]   80  7.568825
[328,]   81  7.359372
[329,]   86  6.331382
[330,]  107  2.315213
```

Ex 3.11. Expand your R program in Exercise 3.10 to allow fitting with a pre-specified degree of freedom (instead of a pre-specified tuning parameter λ). Present *in details* the formulas you used. You may numerically find λ for a given degree of freedom.

Show that you can replicate the following results obtained using the built-in R function `smooth.spline()`. See Section 3.9.1 for the definition of GCV.

```
> out2 <- smooth.spline(y=ozone, x=dpg, df=8, all.knots=T)
> out2
......
Smoothing Parameter  spar= 0.908881  lambda= 0.001657816 (11 iterations)
Equivalent Degrees of Freedom (Df): 8.001308
Penalized Criterion: 5091.517
GCV: 55.20649
>
> cbind(out2$x, out2$y)
       [,1]      [,2]
  [1,]  -69  3.699389
  [2,]  -68  3.750829
  [3,]  -67  3.802196
  [4,]  -66  3.853494
```

```
  [5,]  -63  4.009240
......
 [81,]   31 14.182825
 [82,]   32 14.182365
 [83,]   33 14.178773
 [84,]   34 14.172438
 [85,]   35 14.165329
......
[124,]   77  8.197236
[125,]   80  7.562373
[126,]   81  7.353058
[127,]   86  6.326147
[128,]  107  2.321575
```

Ex 3.12. Investigate and report on how `bs()` and `ns()` are created in the R package `splines`, that is, describe the underlying algebraic formulas. In addition, describe how you can use the basis functions from `ns()` to fit cubic smoothing splines and replicate the result in Exercise 3.10.

Ex 3.13. Let

$$C^{(1)}[0,1] = \left\{ f : f \text{ is absolute continuous and } \int_0^1 (f'(x))^2 \, \mathrm{d}x < \infty \right\}.$$

The linear smoothing spline is defined as

$$\hat{f}_\lambda = \operatorname{argmin}_{f \in C^{(1)}[0,1]} \left\{ \frac{1}{n} \sum_{i=1}^{n} (Y_i - f(X_i))^2 + \lambda \int (f'(x))^2 \, \mathrm{d}x \right\},$$

where $X_1, \ldots, X_n$ are assumed to fall in $[0, 1]$ and $\lambda > 0$ is a tuning parameter. [Ignore the completeness requirement.]

(i) Show that $C^{(1)}[0, 1]$ is a reproducing kernel Hilbert space with an appropriate inner product and a reproducing kernel.

(ii) Show that $\hat{f}_\lambda$ is a linear spline with knots at $X_1, \ldots, X_n$. What is the behavior of $\hat{f}_\lambda$ outside $[\min_i X_i, \max_i X_i]$?

(iii) Write your R code to compute $\hat{f}_\lambda$, while allowing $X_1, \ldots, X_n$ to fall in an arbitrary interval $[a, b]$, not just $[0, 1]$.

(iv) For the dataset `LAozone.data`, use your R code to fit linear smoothing splines of `ozone` against `dpg` with degrees of freedom (`df`) 4, 6, 8, 12. Draw your four fitted functions over `dpg` range $(-80, 120)$ along with the original data in a single figure. For `df`$=8$, report the fitted values $\hat{f}_\lambda$ evaluated at the 5 smallest and 5 largest `dpg` values.

Ex 3.14. Let (X_i, Y_i), $i = 1, \ldots, n$, be a set of n observations, where X_i is a scalar covariate and Y_i is a scalar response. For fixed $\sigma > 0$ and $\lambda > 0$, consider

$$(\hat{b}_0, \hat{b}_1, \hat{g}) = \text{argmin}_{b_0, b_1 \in \mathbb{R}, g \in \mathcal{H}_\sigma} \left\{ \frac{1}{n} \sum_{i=1}^{n} (Y_i - b_0 - b_1 X_i - g(X_i))^2 + \lambda \|g\|^2 \right\}, \tag{3.54}$$

where $\mathcal{H}_\sigma$ is the RKHS defined by the Gaussian kernel

$$K_\sigma(x, z) = \exp\left(-\frac{(x-z)^2}{2\sigma^2} \right), \quad x, z \in \mathbb{R},$$

and $\|g\|$ is the norm of g in the RKHS. Let $\hat{f}(x) = \hat{b}_0 + \hat{b}_1 x + \hat{g}(x)$.

(i) Show that $\hat{g}$ in (3.54) can be represented as

$$\hat{g}(x) = \sum_{i=1}^{n} c_i K_\sigma(X_i, x),$$

for some constants $c_1, \ldots, c_n \in \mathbb{R}$, independent of x.

(ii) The coefficient vectors $b = (b_0, b_1)^{\mathrm{T}}$ and, associated with $\hat{g}$ from (i), $c = (c_1, \ldots, c_n)^{\mathrm{T}}$, denoted as $\hat{b}$ and $\hat{c}$, can be determined as

$$(\hat{b}, \hat{c}) = \text{argmin}_{b \in \mathbb{R}^2, c \in \mathbb{R}^n} \left\{ \frac{1}{n} \sum_{i=1}^{n} (Y_i - b_0 - b_1 X_i - g(X_i; c))^2 + \lambda \, \text{pen}(c) \right\},$$

where $g(X_i; c) = \sum_{i=1}^{n} c_i K_\sigma(X_i, x)$, and $\text{pen}(c)$ is a penalty term. Give an explicit expression for $\text{pen}(c)$, and derive a closed-form formula for $(\hat{b}, \hat{c})$.

(iii) Show that $\hat{f}$ is a linear smoother, i.e., the vector of fitted values $\hat{\mathbf{f}} = (\hat{f}(X_1), \ldots, \hat{f}(X_n))^{\mathrm{T}}$ can be obtained as $\hat{\mathbf{f}} = \mathbf{SY}$ for an $n \times n$ matrix $\mathbf{S}$, independent of $\mathbf{Y} = (Y_1, \ldots, Y_n)^{\mathrm{T}}$. Give an explicit expression for $\mathbf{S}$.

(iv) Write R code to implement $\hat{f}$ using your formula for $(\hat{b}, \hat{c})$ and $\mathbf{S}$. For your code, include (σ, λ) as tuning parameters, and provide the degree of freedom (`df`), defined as the trace of $\mathbf{S}$, as part of output.

(v) Apply your R code to the ozone dataset (`LAozone.data`), with the covariate `dpg` and response `ozone`. Please use $\sigma = 10$ and manually pick the choices of λ to achieve `df`$= 4, 6, 8, 12$ up to the first decimal. For example, for `df=8`, your actual `df` should fall in $[7.95, 8.05)$. Draw your four fitted functions over `dpg` range $(-80, 120)$ along with the original data in a single figure. For `df`$= 8$, report the fitted values $\hat{f}$ at the 5 smallest and 5 largest `dpg` values.

(vi) Repeat your application in (v), except with $\sigma = 40$ in the Gaussian kernel.

4

Generalized linear mixed regression

In this chapter, we study linear and generalized linear mixed models through the following topics. First, for linear mixed models, we discuss estimation of fixed effects and variance parameters and prediction of random effects, and explain the use of extended likelihood to jointly handle estimation and prediction. Second, we discuss a connection between linear mixed models and ridge-penalized and smoothing splines and then exploit the connection to select tuning parameters in the smoothing methods. Third, for generalized linear mixed models, we provide an overview of estimation and prediction and subsequently focus on Laplace approximation based methods. We provide numerical demonstration in R, using both existing packages and self-contained code for implementation and application of main methods.

Linear and generalized linear mixed models, also called multi-level or hierarchical models, have been developed to analyze grouped data by incorporating random effects. Such topics have been studied in many books, including Searle et al. (1992), Pinheiro and Bates (2000), Gelman and Hill (2006), McCulloch et al. (2008), and Lee et al. (2017) among others. Ruppert et al. (2003), Chapter 3, provided a compact account of linear mixed models with applications to smoothing, which is discussed in Section 4.3 here.

4.1 Linear mixed models

4.1.1 Data and model

Data are obtained in the form

$$(X_{ij}, Y_{ij}), \quad i = 1, \ldots, m, j = 1, \ldots, n_i,$$

DOI: 10.1201/9781003680727-4

i.e., there are m groups indexed by $i = 1, \ldots, m$, and the ith group consists of n_i observations, (X_{ij}, Y_{ij}) for $j = 1, \ldots, n_i$, with X_{ij} a covariate vector and Y_{ij} a response variables. For example, in growth curve data, Y_{ij} represents a growth measurement (e.g., weight) and X_{ij} represents the age at occasion j for an individual subject i. We give a list of examples to illustrate how various regression analyses can be conducted for such grouped data. For simplicity, the covariates X_{ij}'s are assumed to be scalar and fixed.

(i) *Separate analysis.* Fit m separate lines to the m groups:

$$Y_{ij} = \beta_{i0} + \beta_{i1} X_{ij} + \varepsilon_{ij}, \quad i = 1, \ldots, m, j = 1, \ldots, n_i,$$
$$(\varepsilon_{ij} : i = 1, \ldots, m, j = 1, \ldots, n_i) \overset{\text{IID}}{\sim} \mathrm{N}(0, \sigma_\varepsilon^2).$$

In total, there are $2m$ coefficient parameters, (β_{i0}, β_{i1}) for $i = 1, \ldots, m$, and 1 variance parameter σ_ε^2.

(ii) *Parallel analysis.* Fit m parallel lines to the m groups:

$$Y_{ij} = \beta_{i0} + \beta_1 X_{ij} + \varepsilon_{ij}, \quad i = 1, \ldots, m, j = 1, \ldots, n_i,$$
$$(\varepsilon_{ij} : i = 1, \ldots, m, j = 1, \ldots, n_i) \overset{\text{IID}}{\sim} \mathrm{N}(0, \sigma_\varepsilon^2).$$

In total, there are $m + 1$ coefficient parameters, $(\beta_{10}, \ldots, \beta_{m0}, \beta_1)$, and 1 variance parameter σ_ε^2.

(iii) *Separate analysis with random effects.* Assume that

$$\begin{aligned} & Y_{ij} = \beta_{i0} + \beta_{i1} X_{ij} + \varepsilon_{ij}, \quad i = 1, \ldots, m, j = 1, \ldots, n_i, \\ & (\varepsilon_{ij} : i = 1, \ldots, m, j = 1, \ldots, n_i) \overset{\text{IID}}{\sim} \mathrm{N}(0, \sigma_\varepsilon^2), \\ & (\beta_{i0} : i = 1, \ldots, m) \overset{\text{IID}}{\sim} \mathrm{N}(b_0, \sigma_0^2), \\ & (\beta_{i1} : i = 1, \ldots, m) \overset{\text{IID}}{\sim} \mathrm{N}(b_1, \sigma_1^2), \\ & (\beta_{i0} : i = 1, \ldots, m) \text{ and } (\beta_{i1} : i = 1, \ldots, m) \text{ are mutually independent} \\ & \qquad \text{and are independent of } (\varepsilon_{ij} : i = 1, \ldots, m, j = 1, \ldots, n_i). \end{aligned} \tag{4.1}$$

In total, there are 2 coefficient parameters, (b_0, b_1), and 3 variance parameters $(\sigma_0^2, \sigma_1^2, \sigma_\varepsilon^2)$.

(iv) *Parallel analysis with random effects.* Assume that

$$\begin{aligned} & Y_{ij} = \beta_{i0} + \beta_1 X_{ij} + \varepsilon_{ij}, \quad i = 1, \ldots, m, j = 1, \ldots, n_i, \\ & (\varepsilon_{ij} : i = 1, \ldots, m, j = 1, \ldots, n_i) \overset{\text{IID}}{\sim} \mathrm{N}(0, \sigma_\varepsilon^2), \\ & (\beta_{i0} : i = 1, \ldots, m) \overset{\text{IID}}{\sim} \mathrm{N}(b_0, \sigma_0^2), \\ & \qquad \text{independently of } (\varepsilon_{ij} : i = 1, \ldots, m, j = 1, \ldots, n_i). \end{aligned} \tag{4.2}$$

In total, there are 2 coefficient parameters, (b_0, β_1), and 2 variance parameters $(\sigma_0^2, \sigma_\varepsilon^2)$. [Question: What if we also assume $\beta_1 \sim \mathrm{N}(b_1, \sigma_1^2)$, independently of β_{i0}'s and ε_{ij}'s?]

In Example (iii), β_{i0} and β_{i1} are called a random intercept and a random slope respectively. In Example (iv), β_{i0} is a random intercept, while β_1 is a fixed slope. Each random intercept β_{i0} or random slope β_{i1} can be decomposed into a fixed component, b_0 or b_1, and a mean-zero random component. It is helpful to rewrite the preceding models with random coefficients by separating the fixed and (mean-zero) random components.

In Example (iii), let

$$\beta_{i0} = b_0 + u_{i0}, \quad \beta_{i1} = b_1 + u_{i1}, \quad i = 1, \ldots, m,$$

where $(u_{i0} : i = 1, \ldots, m) \overset{\text{IID}}{\sim} \mathrm{N}(0, \sigma_0^2)$ and $(u_{i1} : i = 1, \ldots, m) \overset{\text{IID}}{\sim} \mathrm{N}(0, \sigma_1^2)$. Then (4.1) can be written as

$$Y_{ij} = b_0 + b_1 X_{ij} + u_{i0} + u_{i1} X_{ij} + \varepsilon_{ij}, \quad i = 1, \ldots, m, j = 1, \ldots, n_i, \tag{4.3}$$

and called a linear mixed model, where (b_0, b_1) are called fixed effects and (u_{i0}, u_{i1}) are called random effects. In the matrix notation, this can be written as

$$\begin{pmatrix} Y_{11} \\ \vdots \\ Y_{1n_1} \\ \vdots \\ Y_{m1} \\ \vdots \\ Y_{mn_m} \end{pmatrix} = \begin{pmatrix} 1 & X_{11} \\ \vdots & \vdots \\ 1 & X_{1n_1} \\ \vdots & \vdots \\ 1 & X_{m1} \\ \vdots & \vdots \\ 1 & X_{mn_m} \end{pmatrix} \begin{pmatrix} b_0 \\ b_1 \end{pmatrix}$$

$$+ \begin{pmatrix} 1 & & 0 & X_{11} & & 0 \\ \vdots & & \vdots & \vdots & & \vdots \\ 1 & & 0 & X_{1n_1} & & 0 \\ & \ddots & & & \ddots & \\ 0 & & 1 & 0 & & X_{m1} \\ \vdots & & \vdots & & & \vdots \\ 0 & & 1 & 0 & & X_{mn_m} \end{pmatrix} \begin{pmatrix} u_{10} \\ \vdots \\ u_{m0} \\ u_{11} \\ \vdots \\ u_{m1} \end{pmatrix} + \begin{pmatrix} \varepsilon_{11} \\ \vdots \\ \varepsilon_{1n_1} \\ \vdots \\ \varepsilon_{m1} \\ \vdots \\ \varepsilon_{mn_m} \end{pmatrix},$$

or in a compact way

$$\underbrace{\mathbf{Y}}_{N\times 1} = \underbrace{\mathbf{X}}_{N\times 2} \underbrace{b}_{2\times 1} + \underbrace{\mathbf{Z}}_{N\times(2m)} \underbrace{u}_{(2m)\times 1} + \underbrace{\varepsilon}_{N\times 1},$$

where $N = \sum_{i=1}^m n_i$, and ε and u are independent satisfying

$$\varepsilon \sim \mathrm{N}(0, \sigma_\varepsilon^2 I_N), \quad u \sim \mathrm{N}\left(0, \begin{pmatrix} \sigma_0^2 I_m & 0 \\ 0 & \sigma_1^2 I_m \end{pmatrix}\right).$$

Similarly, in Example (iv), (4.2) can be written as

$$Y_{ij} = b_0 + \beta_1 X_{ij} + u_{i0} + \varepsilon_{ij}, \quad i = 1, \ldots, m, j = 1, \ldots, n_i, \tag{4.4}$$

where (b_0, β_1) are fixed effects and u_{i0} is a random effect. In the matrix notation, this can be written as

$$\begin{pmatrix} Y_{11} \\ \vdots \\ Y_{1n_1} \\ \vdots \\ Y_{m1} \\ \vdots \\ Y_{mn_m} \end{pmatrix} = \begin{pmatrix} 1 & X_{11} \\ \vdots & \vdots \\ 1 & X_{1n_1} \\ \vdots & \vdots \\ 1 & X_{m1} \\ \vdots & \vdots \\ 1 & X_{mn_m} \end{pmatrix} \begin{pmatrix} b_0 \\ \beta_1 \end{pmatrix} + \begin{pmatrix} 1 & & 0 \\ \vdots & & \vdots \\ 1 & & 0 \\ & \ddots & \\ 0 & & 1 \\ \vdots & & \vdots \\ 0 & & 1 \end{pmatrix} \begin{pmatrix} u_{10} \\ \vdots \\ u_{m0} \end{pmatrix} + \begin{pmatrix} \varepsilon_{11} \\ \vdots \\ \varepsilon_{1n_1} \\ \vdots \\ \varepsilon_{m1} \\ \vdots \\ \varepsilon_{mn_m} \end{pmatrix},$$

or in a compact way

$$\underbrace{\mathbf{Y}}_{N\times 1} = \underbrace{\mathbf{X}}_{N\times 2} \underbrace{b}_{2\times 1} + \underbrace{\mathbf{Z}}_{N\times m} \underbrace{u}_{m\times 1} + \underbrace{\boldsymbol{\varepsilon}}_{N\times 1},$$

where $b = (b_0, \beta_1)^{\mathrm{T}}$, and $\boldsymbol{\varepsilon}$ and u are independent satisfying

$$\boldsymbol{\varepsilon} \sim \mathrm{N}(0, \sigma_\varepsilon^2 I_N), \quad u \sim \mathrm{N}(0, \sigma_0^2 I_m).$$

In general, a linear mixed model can be formulated as

$$\underbrace{\mathbf{Y}}_{N\times 1} = \underbrace{\mathbf{X}}_{N\times p} \underbrace{b}_{p\times 1} + \underbrace{\mathbf{Z}}_{N\times m} \underbrace{u}_{m\times 1} + \underbrace{\boldsymbol{\varepsilon}}_{N\times 1}, \tag{4.5}$$

where $\mathbf{Y}$ is a response vector, $\mathbf{X}$ and $\mathbf{Z}$ are covariate matrices (or model matrices), b is a vector of fixed coefficients, u is a vector of random coefficients, and $\boldsymbol{\varepsilon}$ is a noise vector, satisfying

$$\begin{pmatrix} u \\ \boldsymbol{\varepsilon} \end{pmatrix} \sim \mathrm{N}\left(0, \begin{pmatrix} D & 0 \\ 0 & G \end{pmatrix}\right). \tag{4.6}$$

The variance matrices D and G are parameterized by some variance parameters, denoted as τ, for example, $\tau = (\sigma_0^2, \sigma_1^2, \sigma_\varepsilon^2)$ in Example (iii) above. For simplicity, the dependency of (D, G) on τ is suppressed in the notation.

There are at least two advantages from using mixed models, which are also called multi-level or hierarchical models. First, mixed modeling facilitates the prediction of a response at different levels: for a new observation in an existing group (the "group level") or for a new observation in a new group (the "population level"). For example, for mixed model (4.3), let $(\hat{b}_0, \hat{b}_1)$ be some estimators of (b_0, b_1) and $(\hat{u}_{i0}, \hat{u}_{i1})$ be some predictors of (u_{i0}, u_{i1}). The prediction of a response with covariates x is $\hat{b}_0 + \hat{b}_1 x + \hat{u}_{i0} + \hat{u}_{i1} x$ in the ith group for $i = 1, \ldots, m$, whereas the prediction is $\hat{b}_0 + \hat{b}_1 x$ in a new group distinct from the observed m groups. Second, even in any observed group, the prediction from mixed modeling can be seen as shrinkage estimation (or empirical Bayes estimation), pooling information across all the observed groups. See Examples 4.1 and 4.3 later for explicit expressions in simple settings. In fact, mixed model (4.3) can be interpreted as placing independent normal priors on the group-specific coefficients (β_{i0}, β_{i1}), and the prior parameters (b_0, b_1) and (σ_0^2, σ_1^2) can be estimated from the data and used in the posterior prediction involving (β_{i0}, β_{i1}) (hence the name "empirical Bayes"). Potential improvement from empirical Bayes estimation, compared with group-specific analysis (i.e., separate analysis in Example (i) earlier), can be traced to James–Stein shrinkage estimation of normal means (James & Stein, 1961; Efron & Morris, 1973). See Section 4.6.1 for an introduction.

4.1.2 Estimation of fixed effects and variance parameters

For convenience, we consider $(\mathbf{X}, \mathbf{Z})$ being fixed or conditioned on. Then only $\mathbf{Y}$ is random in the data. We use the method of maximum likelihood and estimate b and (D, G) based on the marginal distribution of $\mathbf{Y}$, which is determined from (4.5) and (4.6) by marginalizing over u and ε.

Lemma 4.1 (Marginal distribution of $\mathbf{Y}$). *The (marginal) distribution of* $\mathbf{Y}$ *from (4.5) and (4.6) is* $\mathrm{N}(\mathbf{X}b, V_{D,G})$, *where* $V_{D,G} = \mathbf{Z}D\mathbf{Z}^{\mathrm{T}} + G$.

Proof. From (4.5) and (4.6), it is straightforward to calculate

$$\begin{aligned} E(\mathbf{Y}) &= \mathbf{X}b, \\ \mathrm{var}(\mathbf{Y}) &= \mathrm{var}(\mathbf{Z}u + \varepsilon) = \mathrm{var}(\mathbf{Z}u) + \mathrm{var}(\varepsilon) = \mathbf{Z}D\mathbf{Z}^{\mathrm{T}} + G. \end{aligned}$$

Finally, $\mathbf{Y}$ is multivariate normal because it is a linear transformation of the multivariate norm vector (u, ε). □

Based on the marginal distribution of $\mathbf{Y}$, the log-likelihood function is

$$\begin{aligned} \ell(b, D, G) &= \log f(\mathbf{Y}; b, D, G) \\ &= -\frac{1}{2} \log |2\pi V_{D,G}| - \frac{1}{2} (\mathbf{Y} - \mathbf{X}b)^{\mathrm{T}} V_{D,G}^{-1} (\mathbf{Y} - \mathbf{X}b), \end{aligned} \tag{4.7}$$

where $f(\mathbf{y}; b, D, G)$ is the marginal density of $\mathbf{Y}$ depending on (b, D, G), and $|A|$ or $\det(A)$ denotes the determinant for a matrix A. Given (D, G), the maximum likelihood (ML) estimator of b or, equivalently the generalized least squares estimator, is

$$\begin{aligned} \hat{b}_{D,G} &= \mathrm{argmax}_b \, \ell(b, D, G) \\ &= \mathrm{argmin}_b \, (\mathbf{Y} - \mathbf{X}b)^{\mathrm{T}} V_{D,G}^{-1} (\mathbf{Y} - \mathbf{X}b). \end{aligned}$$

Setting the gradient of the objective function to 0 yields

$$\hat{b}_{D,G} = (\mathbf{X}^{\mathrm{T}} V_{D,G}^{-1} \mathbf{X})^{-1} (\mathbf{X}^{\mathrm{T}} V_{D,G}^{-1} \mathbf{Y}). \tag{4.8}$$

The profile log-likelihood of (D, G) is

$$\begin{aligned}\ell_{\mathrm{p}}(D, G) &= \max_b \ \ell(b, D, G) = \ell(\hat{b}_{D,G}, D, G) \\ &= -\frac{1}{2}\log|2\pi V_{D,G}| - \frac{1}{2}(\mathbf{Y} - \mathbf{X}\hat{b}_{D,G})^{\mathrm{T}} V_{D,G}^{-1}(\mathbf{Y} - \mathbf{X}\hat{b}_{D,G}). \qquad (4.9)\end{aligned}$$

The ML estimator of (D, G) is

$$(\hat{D}, \hat{G}) = \mathrm{argmax}_{D,G}\ \ell_{\mathrm{p}}(D, G),$$

and the ML estimator of b is $\hat{b}_{\hat{D},\hat{G}}$, defined by evaluating $\hat{b}_{D,G}$ with $(D, G) = (\hat{D}, \hat{G})$.

4.1.3 Restricted ML estimation of variance parameters

For estimating (D, G), the profile log-likelihood $\ell_{\mathrm{p}}(D, G)$ is obtained by maximizing the log-likelihood $\ell(b, D, G)$ over b, considered as a nuisance parameter. A potential improvement is the adjusted profile log-likelihood defined as

$$\begin{aligned}\ell_{\mathrm{ap}}(D, G) &= \ell_{\mathrm{p}}(D, G) - \frac{1}{2}\log\left|\frac{\mathbf{X}^{\mathrm{T}} V_{D,G}^{-1}\mathbf{X}}{2\pi}\right| \\ &= \ell(\hat{b}_{D,G}, D, G) - \frac{1}{2}\log\left|\frac{\mathbf{X}^{\mathrm{T}} V_{D,G}^{-1}\mathbf{X}}{2\pi}\right|.\end{aligned} \qquad (4.10)$$

The adjustment term up to an additive constant, $\frac{1}{2}\log|\mathbf{X}^{\mathrm{T}} V_{D,G}^{-1}\mathbf{X}|$, is half of the log determinant of the (constant) negative Hessian of $\ell(b, D, G)$ in b. The restricted ML (REML) estimator of (D, G) is defined by maximizing the adjusted profile log-likelihood:

$$(\tilde{D}, \tilde{G}) = \mathrm{argmax}_{D,G}\ \ell_{\mathrm{ap}}(D, G).$$

The adjusted profile log-likelihood can be derived in several conceptually distinct ways. In the following, we describe the derivation using the conditional likelihood, based on the conditional density of $\mathbf{Y}$ given $\hat{b}_{D,G}$, which turns out to depend on (D, G) but not b. See Exercise 4.3 for another derivation using the integrated likelihood with a flat prior on b.

Specifically, the density of $\mathbf{Y} \sim \mathrm{N}(\mathbf{X}b, V_{D,G})$ is

$$f_{\mathbf{Y}}(\mathbf{y}) = |2\pi V_{D,G}|^{-1/2}\exp\{-\frac{1}{2}(\mathbf{y} - \mathbf{X}b)^{\mathrm{T}} V_{D,G}^{-1}(\mathbf{y} - \mathbf{X}b)\},$$

where the subscript $_{\mathbf{Y}}$ is added for clarity, and the dependency on (b, D, G) is suppressed. The estimator $\hat{b}_{D,G} = (\mathbf{X}^{\mathrm{T}} V_{D,G}^{-1}\mathbf{X})^{-1}(\mathbf{X}^{\mathrm{T}} V_{D,G}^{-1}\mathbf{Y})$ is also multivariate normal, and the mean vector and variance matrix can be directly calculated:

$$\begin{aligned}E(\hat{b}_{D,G}) &= b, \\ \mathrm{var}(\hat{b}_{D,G}) &= (\mathbf{X}^{\mathrm{T}} V_{D,G}^{-1}\mathbf{X})^{-1}.\end{aligned} \qquad (4.11)$$

Then the density of $\hat{b}_{D,G}$ is

$$f_{\hat{b}_{D,G}}(\hat{b}) = \left|\frac{\mathbf{X}^{\mathrm{T}}V_{D,G}^{-1}\mathbf{X}}{2\pi}\right|^{1/2} \exp\{-\frac{1}{2}(\hat{b}-b)^{\mathrm{T}}(\mathbf{X}^{\mathrm{T}}V_{D,G}^{-1}\mathbf{X})(\hat{b}-b)\},$$

where $\hat{b}$ denotes the argument of the density function. Hence the conditional density of $\mathbf{Y}$ given $\hat{b}_{D,G}$ (which is a function of $\mathbf{Y}$) is

$$\begin{aligned} f(\mathbf{y}|\hat{b}) &= \frac{f_{\mathbf{Y}}(\mathbf{y})}{f_{\hat{b}_{D,G}}(\hat{b})} \\ &= |2\pi V_{D,G}|^{-1/2}\left|\frac{\mathbf{X}^{\mathrm{T}}V_{D,G}^{-1}\mathbf{X}}{2\pi}\right|^{-1/2} \\ &\quad \times \exp\{-\frac{1}{2}(\mathbf{y}-\mathbf{X}b)^{\mathrm{T}}V_{D,G}^{-1}(\mathbf{y}-\mathbf{X}b) + \frac{1}{2}(\hat{b}-b)^{\mathrm{T}}(\mathbf{X}^{\mathrm{T}}V_{D,G}^{-1}\mathbf{X})(\hat{b}-b)\}, \end{aligned}$$

provided $\hat{b} = (\mathbf{X}^{\mathrm{T}}V_{D,G}^{-1}\mathbf{X})^{-1}(\mathbf{X}^{\mathrm{T}}V_{D,G}^{-1}\mathbf{y})$. [Otherwise, $f(\mathbf{y}|\hat{b}) = 0$.] By direct calculation, this can be simplified to (Exercise 4.2)

$$f(\mathbf{y}|\hat{b}) = |2\pi V_{D,G}|^{-1/2}\left|\frac{\mathbf{X}^{\mathrm{T}}V_{D,G}^{-1}\mathbf{X}}{2\pi}\right|^{-1/2} \exp\{-\frac{1}{2}(\mathbf{y}-\mathbf{X}\hat{b})^{\mathrm{T}}V_{D,G}^{-1}(\mathbf{y}-\mathbf{X}\hat{b})\}.$$

Evaluating the conditional density at $\mathbf{y} = \mathbf{Y}$ and $\hat{b} = \hat{b}_{D,G}$ and taking the logarithm yields the adjusted profile log-likelihood

$$\begin{aligned} &\log f(\mathbf{Y}|\hat{b}_{D,G}) \\ &= -\frac{1}{2}\log\left|\frac{\mathbf{X}^{\mathrm{T}}V_{D,G}^{-1}\mathbf{X}}{2\pi}\right| - \frac{1}{2}\log|2\pi V_{D,G}| - \frac{1}{2}(\mathbf{Y}-\mathbf{X}\hat{b}_{D,G})^{\mathrm{T}}V_{D,G}^{-1}(\mathbf{Y}-\mathbf{X}\hat{b}_{D,G}) \\ &= \ell_{\mathrm{ap}}(D,G). \end{aligned}$$

The adjusted profile log-likelihood $\ell_{\mathrm{ap}}(D,G)$ is a particular example of the adjusted profile log-likelihood of Cox and Reid (1987) with orthogonal parameters. See Lee et al. (2017), Section 1.9, for a related discussion. In general, let $\ell(\theta,\eta)$ be a log-likelihood with θ the parameter of interest and η the nuisance parameter. The profile log-likelihood of θ is $\ell_{\mathrm{p}}(\theta) = \ell(\theta,\hat{\eta}_\theta)$, where $\hat{\eta}_\theta = \mathrm{argmax}_\eta \ell(\theta,\eta)$. Given that θ and η are orthogonal in the sense that $E(\frac{\partial \ell}{\partial\theta\partial\eta}) = 0$, the adjusted profile log-likelihood of Cox and Reid (1987) is defined as

$$\ell_{\mathrm{ap}}(\theta) = \ell_{\mathrm{p}}(\theta) - \frac{1}{2}\log\det\left(-\frac{\partial \ell}{\partial\eta^2}\right)\Big|_{\eta=\hat{\eta}_\theta}, \tag{4.12}$$

For the log-likelihood $\ell(b,D,G)$, the parameters $\theta = (D,G)$ and $\eta = b$ can be verified to be orthogonal: for any component θ_j of θ,

$$E\left(\frac{\partial \ell}{\partial\theta_j\partial\eta}\right) = E\left\{\mathbf{X}^{\mathrm{T}}\frac{\partial}{\partial\theta_j}(V_{D,G}^{-1})(\mathbf{Y}-\mathbf{X}b)\right\} = 0.$$

Then application of (4.12) leads to $\ell_{\text{ap}}(D,G)$ up to an additive constant.

4.1.4 Prediction of random effects

We use the following result about optimal prediction. For two random vectors T and S, the mean squared error (MSE) of predicting T by a function $g(S)$ of S is defined as $\text{MSE}(g) = E\{(T - g(S))^2\}$. The optimal choice of $g(\cdot)$ in minimizing $\text{MSE}(g)$ is the conditional expectation function, $g_{\text{opt}}(s) = E(T|S = s)$:

$$\text{MSE}(g) \geq \text{MSE}(g_{\text{opt}}) = E\{\text{var}(T|S)\},$$

for any measurable function $g(\cdot)$. [Exercise: Prove this result.] The conditional distribution of u given $\mathbf{Y}$ from (4.5) and (4.6) is determined as follows.

Lemma 4.2 (Conditional distribution of u given $\mathbf{Y}$). *The conditional distribution of u given $\mathbf{Y}$ is multivariate normal, with*

$$E(u|\mathbf{Y}) = D\mathbf{Z}^{\mathrm{T}}V_{D,G}^{-1}(\mathbf{Y} - \mathbf{X}b), \tag{4.13a}$$

$$\text{var}(u|\mathbf{Y}) = D - D\mathbf{Z}^{\mathrm{T}}V_{D,G}^{-1}\mathbf{Z}D, \tag{4.13b}$$

where $V_{D,G} = \mathbf{Z}D\mathbf{Z}^{\mathrm{T}} + G$. Equivalently, the mean and variance can be expressed as

$$E(u|\mathbf{Y}) = \Lambda_{D,G}^{-1}\mathbf{Z}^{\mathrm{T}}G^{-1}(\mathbf{Y} - \mathbf{X}b), \tag{4.14a}$$

$$\text{var}(u|\mathbf{Y}) = \Lambda_{D,G}^{-1}, \tag{4.14b}$$

where $\Lambda_{D,G} = \mathbf{Z}^{\mathrm{T}}G^{-1}\mathbf{Z} + D^{-1}$. [Compared with (4.13a)*, the expression* (4.14a) *is more convenient in the proof of Lemma 4.3.]*

Proof. From (4.5) and (4.6), we express $(\mathbf{Y}, u)$ in terms of the "source" (u, ε):

$$\begin{pmatrix} \mathbf{Y} \\ u \end{pmatrix} = \begin{pmatrix} \mathbf{X}b \\ 0 \end{pmatrix} + \begin{pmatrix} \mathbf{Z} & I \\ I & 0 \end{pmatrix} \begin{pmatrix} u \\ \varepsilon \end{pmatrix}.$$

Then $(\mathbf{Y}^{\mathrm{T}}, u^{\mathrm{T}})^{\mathrm{T}}$ is jointly multivariate normal with

$$E\begin{pmatrix} \mathbf{Y} \\ u \end{pmatrix} = \begin{pmatrix} \mathbf{X}b \\ 0 \end{pmatrix},$$

$$\text{var}\begin{pmatrix} \mathbf{Y} \\ u \end{pmatrix} = \begin{pmatrix} \mathbf{Z} & I \\ I & 0 \end{pmatrix} \begin{pmatrix} D & 0 \\ 0 & G \end{pmatrix} \begin{pmatrix} \mathbf{Z}^{\mathrm{T}} & I \\ I & 0 \end{pmatrix} = \begin{pmatrix} V_{D,G} & \mathbf{Z}G \\ G\mathbf{Z}^{\mathrm{T}} & G \end{pmatrix}.$$

The conditional distribution of u given $\mathbf{Y}$, as stated in Lemma 4.2, follows from the standard properties of multivariate normal distributions.

To show the expressions (4.14a)–(4.14b), we use the identity $D\mathbf{Z}^{\mathrm{T}}V_{D,G}^{-1} = \Lambda_{D,G}^{-1}\mathbf{Z}^{\mathrm{T}}G^{-1}$, i.e.,

$$D\mathbf{Z}^{\mathrm{T}}(\mathbf{Z}D\mathbf{Z}^{\mathrm{T}} + G)^{-1} = (\mathbf{Z}^{\mathrm{T}}G^{-1}\mathbf{Z} + D^{-1})^{-1}\mathbf{Z}^{\mathrm{T}}G^{-1}, \tag{4.15}$$

which holds because $(\mathbf{Z}^{\mathrm{T}}G^{-1}\mathbf{Z}+D^{-1})D\mathbf{Z}^{\mathrm{T}} = \mathbf{Z}^{\mathrm{T}}G^{-1}(\mathbf{Z}D\mathbf{Z}^{\mathrm{T}}+G)$ by direct calculation. Then the second expression of $E(u|\mathbf{Y})$ holds:

$$\begin{aligned} E(u|\mathbf{Y}) &= D\mathbf{Z}^{\mathrm{T}}V_{D,G}^{-1}(\mathbf{Y}-\mathbf{X}b), \\ &= \Lambda_{D,G}^{-1}\mathbf{Z}^{\mathrm{T}}G^{-1}(\mathbf{Y}-\mathbf{X}b). \end{aligned}$$

The second expression of $\mathrm{var}(u|\mathbf{Y})$ also holds:

$$\begin{aligned} \mathrm{var}(u|\mathbf{Y}) &= D - D\mathbf{Z}^{\mathrm{T}}V_{D,G}^{-1}\mathbf{Z}D \\ &= D - \Lambda_{D,G}^{-1}\mathbf{Z}^{\mathrm{T}}G^{-1}\mathbf{Z}D \\ &= D - \Lambda_{D,G}^{-1}(\Lambda_{D,G}-D^{-1})D = \Lambda_{D,G}^{-1}, \end{aligned}$$

where the third equality follows because $\mathbf{Z}^{\mathrm{T}}G^{-1}\mathbf{Z} = \Lambda_{D,G} - D^{-1}$. □

The expressions (4.14a)–(4.14b) can also be shown by examining the conditional density of u given $\mathbf{Y}$, which is proportional in u to the joint density of $(\mathbf{Y},u)$: [Exercise]

$$\begin{aligned} E(u|\mathbf{Y}) &= \mathrm{argmax}_u\ h(b,u,D,G), \\ \mathrm{var}(u|\mathbf{Y}) &= \left\{ \frac{\partial^2}{\partial u \partial u^{\mathrm{T}}} h(b,u,D,G) \right\}^{-1}, \end{aligned}$$

where $h(b,u,D,G)$ is the log joint density of $(\mathbf{Y},u)$ as in (4.17) later. See a related discussion about the factorization of the joint density of $(\mathbf{Y},u)$ after Lemma 4.3.

Given (D,G), the point predictor of u is

$$\begin{aligned} \hat{u}_{D,G} &= D\mathbf{Z}^{\mathrm{T}}V_{D,G}^{-1}(\mathbf{Y}-\mathbf{X}\hat{b}_{D,G}) \\ &= \Lambda_{D,G}^{-1}\mathbf{Z}^{\mathrm{T}}G^{-1}(\mathbf{Y}-\mathbf{X}\hat{b}_{D,G}), \end{aligned} \tag{4.16}$$

where $\hat{b}_{D,G}$ is defined in (4.8). With unknown (D,G), the point predictor of u is defined by substituting the ML estimator $(\hat{D},\hat{G})$ or REML estimator $(\tilde{D},\tilde{G})$ for (D,G) in $\hat{u}_{D,G}$.

Given (D,G), the estimator/preductor $(\hat{b}_{D,G},\hat{u}_{D,G})$ is called the best linear unbiased prediction (BLUP). The estimated version, with (D,G) replaced by the ML or REML estimator, is called the empirical BLUP (EBLUP). See Robinson (1991) for more information.

4.1.5 Extended likelihood

Given (D,G), the point estimator and predictor $(\hat{b}_{D,G},\hat{u}_{D,G})$ are derived above in *two steps*: $\hat{b}_{D,G}$ is derived by maximizing the likelihood of b based on $\mathbf{Y}$, and then $\hat{u}_{D,G}$ is derived by taking the conditional expectation of u given

$\mathbf{Y}$. Remarkably, the pair $(\hat{b}_{D,G}, \hat{u}_{D,G})$ can also be *jointly* characterized as a maximizer of the following objective function

$$
\begin{aligned}
h(b,u,D,G) &= \log f(\mathbf{Y},u;b,D,G) \\
&= -\frac{1}{2}(\mathbf{Y}-\mathbf{X}b-\mathbf{Z}u)^{\mathrm{T}}G^{-1}(\mathbf{Y}-\mathbf{X}b-\mathbf{Z}u) - \frac{1}{2}u^{\mathrm{T}}D^{-1}u \quad (4.17)\\
&\quad -\frac{1}{2}\log|2\pi G| - \frac{1}{2}\log|2\pi D|,
\end{aligned}
$$

where $f(\mathbf{y},u;b,D,G)$ is the joint density of $(\mathbf{Y},u)$, depending on (b,D,G). The last two terms, $-\frac{1}{2}\log|2\pi G| - \frac{1}{2}\log|2\pi D|$, depend only on (D,G) and can be ignored when maximizing over (b,u). The function $h(b,u,D,G)$ is *not* a log-likelihood function of fixed parameters based on observed data (which is converted from the log density of observed data with fixed parameters), and is called an extended log-likelihood function of (b,u,D,G) (Lee et al., 2017).

The use of the objective function (4.17) for estimation/prediction of (b,u) can be traced to Henderson (1953) and Henderson et al. (1959). The system of equations (4.18a)–(4.18b) or, in a matrix form, (4.19) is called Henderson's mixed model equations.

Lemma 4.3 (Joint determination of estimator and predictor). *In the preceding setting,* $(\hat{b}_{D,G}, \hat{u}_{D,G})$ *can be equivalently obtained as*

$$(\hat{b}_{D,G}, \hat{u}_{D,G}) = \mathrm{argmax}_{b,u}\, h(b,u,D,G),$$

where $\hat{b}_{D,G}$ *and* $\hat{u}_{D,G}$ *are defined in (4.8) and (4.16):*

$$
\begin{aligned}
\hat{b}_{D,G} &= (\mathbf{X}^{\mathrm{T}}V_{D,G}^{-1}\mathbf{X})^{-1}(\mathbf{X}^{\mathrm{T}}V_{D,G}^{-1}\mathbf{Y}),\\
\hat{u}_{D,G} &= D\mathbf{Z}^{\mathrm{T}}V_{D,G}^{-1}(\mathbf{Y}-\mathbf{X}\hat{b}_{D,G}).
\end{aligned}
$$

Proof. Taking the derivatives of $h(b,u,D,G)$ and setting to 0 yields

$$
\begin{aligned}
\frac{\partial h}{\partial b} &= \mathbf{X}^{\mathrm{T}}G^{-1}(\mathbf{Y}-\mathbf{X}b-\mathbf{Z}u) = 0, && (4.18a)\\
\frac{\partial h}{\partial u} &= \mathbf{Z}^{\mathrm{T}}G^{-1}(\mathbf{Y}-\mathbf{X}b-\mathbf{Z}u) - D^{-1}u = 0. && (4.18b)
\end{aligned}
$$

From (4.18b), we obtain

$$
\begin{aligned}
\hat{u}(b) &= (\mathbf{Z}^{\mathrm{T}}G^{-1}\mathbf{Z}+D^{-1})^{-1}\mathbf{Z}^{\mathrm{T}}G^{-1}(\mathbf{Y}-\mathbf{X}b)\\
&= \Lambda_{D,G}^{-1}\mathbf{Z}^{\mathrm{T}}G^{-1}(\mathbf{Y}-\mathbf{X}b)\\
&= D\mathbf{Z}^{\mathrm{T}}V_{D,G}^{-1}(\mathbf{Y}-\mathbf{X}b),
\end{aligned}
$$

where the last equality follows from the equivalence of the expressions (4.13a) and (4.14a), which in turn follows from the identity (4.15). Substituting $\hat{u}(b)$ for u in (4.18a) gives

$$0 = \mathbf{X}^{\mathrm{T}}G^{-1}(\mathbf{Y}-\mathbf{X}b) - \mathbf{X}^{\mathrm{T}}G^{-1}\mathbf{Z}\hat{u}(b)$$

$$= \mathbf{X}^{\mathrm{T}} G^{-1}(\mathbf{Y} - \mathbf{X}b) - \mathbf{X}^{\mathrm{T}} G^{-1}\mathbf{Z}D\mathbf{Z}^{\mathrm{T}} V_{D,G}^{-1}(\mathbf{Y} - \mathbf{X}b)$$
$$= \mathbf{X}^{\mathrm{T}} G^{-1}(\mathbf{Y} - \mathbf{X}b) - \mathbf{X}^{\mathrm{T}} G^{-1}(V_{D,G} - G) V_{D,G}^{-1}(\mathbf{Y} - \mathbf{X}b)$$
$$= \mathbf{X}^{\mathrm{T}} V_{D,G}^{-1}(\mathbf{Y} - \mathbf{X}b),$$

where the third equality follows because $\mathbf{Z}D\mathbf{Z}^{\mathrm{T}} = V_{D,G} - G$. Combining the preceding two displays leads to $(\hat{b}_{D,G}, \hat{u}_{D,G})$ as stated. □

The proceeding proof can be seen as solving (4.18a)–(4.18b) by the method of substitution. We comment on an alternative proof (or a proof organized in a different manner). Equations (4.18a)–(4.18b) can be put as a linear equation in $(b^{\mathrm{T}}, u^{\mathrm{T}})^{\mathrm{T}}$ jointly:

$$\begin{pmatrix} \frac{\partial h}{\partial b} \\ \frac{\partial h}{\partial u} \end{pmatrix} = \begin{pmatrix} \mathbf{X}^{\mathrm{T}} G^{-1}\mathbf{Y} \\ \mathbf{Z}^{\mathrm{T}} G^{-1}\mathbf{Y} \end{pmatrix} - \underbrace{\begin{pmatrix} \mathbf{X}^{\mathrm{T}} G^{-1}\mathbf{X} & \mathbf{X}^{\mathrm{T}} G^{-1}\mathbf{Z} \\ \mathbf{Z}^{\mathrm{T}} G^{-1}\mathbf{X} & \mathbf{Z}^{\mathrm{T}} G^{-1}\mathbf{Z} + D^{-1} \end{pmatrix}}_{H_{D,G}} \begin{pmatrix} b \\ u \end{pmatrix} = 0, \quad (4.19)$$

where $H_{D,G}$ is the (constant) Hessian of $h(b, u, D, G)$ in (b, u). A solution for $(b^{\mathrm{T}}, u^{\mathrm{T}})^{\mathrm{T}}$ can be obtained as the right-hand side below, so that it remains to verify

$$\begin{pmatrix} \hat{b}_{D,G} \\ \hat{u}_{D,G} \end{pmatrix} = H_{D,G}^{-1} \begin{pmatrix} \mathbf{X}^{\mathrm{T}} G^{-1}\mathbf{Y} \\ \mathbf{Z}^{\mathrm{T}} G^{-1}\mathbf{Y} \end{pmatrix}. \quad (4.20)$$

In fact, by applying the matrix inverse formulas in Exercise 1.3, it can be shown that

$$V_{D,G}^{-1} = G^{-1} - G^{-1}\mathbf{Z}\Lambda_{D,G}^{-1}\mathbf{Z}^{\mathrm{T}} G^{-1},$$

and

$$H_{D,G}^{-1} \quad (4.21a)$$
$$= \begin{pmatrix} I & 0 \\ -\Lambda_{D,G}^{-1}\mathbf{Z}^{\mathrm{T}} G^{-1}\mathbf{X} & I \end{pmatrix} \begin{pmatrix} (\mathbf{X}^{\mathrm{T}} V_{D,G}^{-1}\mathbf{X})^{-1} & 0 \\ 0 & \Lambda_{D,G}^{-1} \end{pmatrix} \begin{pmatrix} I & -\mathbf{X}^{\mathrm{T}} G^{-1}\mathbf{Z}\Lambda_{D,G}^{-1} \\ 0 & I \end{pmatrix} \quad (4.21b)$$
$$= \begin{pmatrix} (\mathbf{X}^{\mathrm{T}} V_{D,G}^{-1}\mathbf{X})^{-1} & -(\mathbf{X}^{\mathrm{T}} V_{D,G}^{-1}\mathbf{X})^{-1}\mathbf{X}^{\mathrm{T}} G^{-1}\mathbf{Z}\Lambda_{D,G}^{-1} \\ * & \Lambda_{D,G}^{-1} + \Lambda_{D,G}^{-1}\mathbf{Z}^{\mathrm{T}} G^{-1}\mathbf{X}(\mathbf{X}^{\mathrm{T}} V_{D,G}^{-1}\mathbf{X})^{-1}\mathbf{X}^{\mathrm{T}} G^{-1}\mathbf{Z}\Lambda_{D,G}^{-1} \end{pmatrix}, \quad (4.21c)$$

where $V_{D,G} = \mathbf{Z}D\mathbf{Z}^{\mathrm{T}} + G$ and $\Lambda_{D,G} = \mathbf{Z}^{\mathrm{T}} G^{-1}\mathbf{Z} + D^{-1}$. Then (4.20) can be verified by direct matrix manipulations. The details are left as Exercise 4.4.

The joint characterization of $(\hat{b}_{D,G}, \hat{u}_{D,G})$ can be explained as follows, depending on specific properties of linear mixed model (4.5)–(4.6). By the factorization of the joint density of $(\mathbf{Y}, u)$, the function $h(b, u) = h(b, u, D, G)$ with fixed (D, G) can be expressed as

$$h(b, u) = \log\{f(u|\mathbf{Y}; b) f(\mathbf{Y}; b)\},$$

where $f(\mathbf{Y}; b) = f(\mathbf{Y}; b, D, G)$ is the marginal density of $\mathbf{Y}$, $f(u|\mathbf{Y}; b, D, G) = f(u|\mathbf{Y}; b)$ is the conditional density of u given $\mathbf{Y}$, and both are multivariate normal densities as discussed earlier. The dependency on (D, G) is suppressed in the notation. First, given b, maximizing $h(b, u)$ over u amounts to finding the mode of u, denoted as $\hat{u}_b$, from $f(u|\mathbf{Y}; b)$, which is the same as the expectation of u from $f(u|\mathbf{Y}; b)$ due to normality. [The mode of a multivariate normal density is also the expectation from that density.] Second, maximizing $\max_u h(b, u) = h(b, \hat{u}_b)$ over b is equivalent to maximizing $f(\mathbf{Y}; b)$ over b, because $f(\hat{u}_b|\mathbf{Y}; b)$ depends only on $\text{var}(u|\mathbf{Y})$ and hence not on b (see Lemma 4.2). [A multivariate normal density evaluated at the mode depends only on the variance of that density, not the mean.] Combining the two pieces shows that jointly maximizing $h(b, u)$ leads to $(\hat{b}_{D,G}, \hat{u}_{D,G})$, where $\hat{u}_{D,G} = \hat{u}(\hat{b}_{D,G})$.

From preceding discussion, the log-likelihood of (b, D, G) in (4.7) can be equivalently expressed in term of the extended log-likelihood $h(\cdot)$ as

$$\begin{aligned} \ell(b, D, G) &= \log f(\mathbf{Y}; b, D, G) \\ &= h(b, \hat{u}_b, D, G) - \log f(\hat{u}_b|\mathbf{Y}; b, D, G) \\ &= \max_u \ h(b, u, D, G) - \frac{1}{2}\log\left|\frac{\Lambda_{D,G}}{2\pi}\right|, \end{aligned} \tag{4.22}$$

where $\Lambda_{D,G} = \mathbf{Z}^{\mathrm{T}}G^{-1}\mathbf{Z}+D^{-1}$ as in Lemma 4.2. Then the profile log-likelihood of (D, G) in (4.9) is

$$\ell_{\mathrm{p}}(D, G) = \max_{b,u} \ h(b, u, D, G) - \frac{1}{2}\log\left|\frac{\Lambda_{D,G}}{2\pi}\right|. \tag{4.23}$$

The adjusted profile log-likelihood of (D, G) in (4.10) is

$$\begin{aligned} \ell_{\mathrm{ap}}(D, G) &= \max_{b,u} \ h(b, u, D, G) - \frac{1}{2}\log\left|\frac{\Lambda_{D,G}}{2\pi}\right| - \frac{1}{2}\log\left|\frac{\mathbf{X}^{\mathrm{T}}V_{D,G}^{-1}\mathbf{X}}{2\pi}\right| \\ &= \max_{b,u} \ h(b, u, D, G) - \frac{1}{2}\log|H_{D,G}| + \text{constant}, \end{aligned} \tag{4.24}$$

where the second line follows because $|H_{D,G}| = |\Lambda_{D,G}|\cdot|\mathbf{X}^{\mathrm{T}}V_{D,G}^{-1}\mathbf{X}|$ by (4.21b), and hence $\log|H_{D,G}| = \log|\Lambda_{D,G}| + \log|\mathbf{X}^{\mathrm{T}}V_{D,G}^{-1}\mathbf{X}|$. See also Exercise 1.4.

Interestingly, the preceding expressions, (4.22) and (4.24), indicate that $\ell(b, D, G)$ and $\ell_{\mathrm{ap}}(D, G)$ can be obtained by profiling the extended log-likelihood $h(\cdot)$ over u or over (b, u) respectively, in the form of the Cox–Reid adjusted profile log-likelihood (4.12) for a usual log-likelihood, similarly as how $\ell_{\mathrm{ap}}(D, G)$ is derived from $\ell(b, D, G)$. First, $\ell(b, D, G)$ is, up to an additive constant, obtained by maximizing (or profiling) $h(b, u, D, G)$ over u and then subtracting $\frac{1}{2}\log|\Lambda_{D,G}|$, half of the log determinant of the (constant) negative Hessian of $h(b, u, D, G)$ in u. Moreover, $\ell_{\mathrm{ap}}(D, G)$ is, up to an additive constant, obtained by maximizing (or profiling) $h(b, u, D, G)$ over (b, u) and then subtracting $\frac{1}{2}\log|H_{D,G}|$, half of the log determinant of the (constant) negative Hessian of $h(b,u,D,G)$ in (b,u). Therefore, the extended log-likelihood $h(\cdot)$ can be profiled similarly as a usual log-likelihood such as $\ell(b,D,G)$.

Finally, Lemma 4.3 and the representation of $\ell_{\text{p}}(D,G)$ in (4.24) or $\ell_{\text{ap}}(D,G)$ in (4.23) also suggest convenient algorithms for computing the estimators of (b,D,G) and predictor of u through the extended log-likelihood $h(\cdot)$, instead of through the original definitions in Sections 4.1.2–4.1.4. The estimator and predictor $(\hat{b},\hat{u}) = (\hat{b}_{\hat{D},\hat{G}}, \hat{u}_{\hat{D},\hat{G}})$ and the ML estimator $(\hat{D},\hat{G})$ can be jointly determined as

$$(\hat{b},\hat{u},\hat{D},\hat{G}) = \text{argmax}_{b,u,D,G}\left\{h(b,u,D,G) - \frac{1}{2}\log|\Lambda_{D,G}|\right\}. \tag{4.25}$$

The estimator and predictor $(\tilde{b},\tilde{u}) = (\hat{b}_{\tilde{D},\tilde{G}}, \hat{u}_{\tilde{D},\tilde{G}})$ and the REML estimator $(\tilde{D},\tilde{G})$ can be jointly determined as

$$(\tilde{b},\tilde{u},\tilde{D},\tilde{G}) = \text{argmax}_{b,u,D,G}\left\{h(b,u,D,G) - \frac{1}{2}\log|H_{D,G}|\right\}. \tag{4.26}$$

In the case of ML estimation, a block-coordinate descent algorithm alternates between updating (b,u) or (D,G) as follows.

- Pick an initial value (D^0,G^0).
- Compute for $k=1,2,\cdots$,

$$(b^k,u^k) = \text{argmax}_{b,u}\; h(b,u,D^{k-1},G^{k-1}), \tag{4.27a}$$

$$(D^k,G^k) = \text{argmax}_{D,G}\left\{h(b^k,u^k,D,G) - \frac{1}{2}\log|\Lambda_{D,G}|\right\}, \tag{4.27b}$$

  until some termination criterion is met.

In the case of REML estimation, the iterations are

$$(b^k,u^k) = \text{argmax}_{b,u}\; h(b,u,D^{k-1},G^{k-1}), \tag{4.28a}$$

$$(D^k,G^k) = \text{argmax}_{D,G}\left\{h(b^k,u^k,D,G) - \frac{1}{2}\log|H_{D,G}|\right\}, \tag{4.28b}$$

The update (4.27a) or (4.28a) can be solved in a closed form by Lemma 4.3 or by (4.20). The update (4.27b) or (4.28b) does not admit a direct solution, but can be replaced by some partial updates, depending on the parameterization of (D,G).

To illustrate the preceding discussion, suppose that $G=\sigma_\varepsilon^2 I_N$ and $D=\sigma_u^2 I_m$, with parameters σ_ε^2 and σ_u^2. The the extended log-likelihood function (4.17) becomes

$$h(b,u,\sigma_\varepsilon^2,\sigma_u^2)$$

$$= -\frac{\|\mathbf{Y} - \mathbf{X}b - \mathbf{Z}u\|_2^2}{2\sigma_\varepsilon^2} - \frac{\|u\|_2^2}{2\sigma_u^2} - \frac{N}{2}\log(\sigma_\varepsilon^2) - \frac{m}{2}\log(\sigma_u^2) + \text{constant}. \quad (4.29)$$

For ML estimation, the gradient of the objective function in (4.27b), denoted as $\ell_{\mathrm{p}}^k(\sigma_u^2, \sigma_\varepsilon^2)$, can be calculated as (Exercise 4.5)

$$\begin{aligned}
\frac{\partial \ell_{\mathrm{p}}^k}{\partial \sigma_\varepsilon^2} &= \frac{\|\mathbf{Y} - \mathbf{X}b^k - \mathbf{Z}u^k\|_2^2}{2\sigma_\varepsilon^4} - \frac{N}{2\sigma_\varepsilon^2} + \frac{1}{2\sigma_\varepsilon^2}\mathrm{tr}\{(\mathbf{Z}^{\mathrm{T}}\mathbf{Z} + \rho I_m)^{-1}\mathbf{Z}^{\mathrm{T}}\mathbf{Z}\}, \\
\frac{\partial \ell_{\mathrm{p}}^k}{\partial \sigma_u^2} &= \frac{\|u^k\|_2^2}{2\sigma_u^4} - \frac{m}{2\sigma_u^2} + \frac{1}{2\sigma_u^2}\mathrm{tr}\{I_m - (\mathbf{Z}^{\mathrm{T}}\mathbf{Z} + \rho I_m)^{-1}\mathbf{Z}^{\mathrm{T}}\mathbf{Z}\},
\end{aligned} \quad (4.30)$$

where $\rho = \sigma_\varepsilon^2/\sigma_u^2$. Solving the gradient equation, but with ρ fixed at $\rho^{k-1} = (\sigma_\varepsilon^2)^{k-1}/(\sigma_u^2)^{k-1}$, leads to a simple update in place of (4.27b):

$$(\sigma_\varepsilon^2)^k = \frac{\|\mathbf{Y} - \mathbf{X}b^k - \mathbf{Z}u^k\|_2^2}{N - q^{k-1}}, \quad (\sigma_u^2)^k = \frac{\|u^k\|_2^2}{q^{k-1}}, \quad (4.31)$$

where $q^{k-1} = \mathrm{tr}\{(\mathbf{Z}^{\mathrm{T}}\mathbf{Z} + \rho^{k-1} I_m)^{-1}\mathbf{Z}^{\mathrm{T}}\mathbf{Z}\}$. Similarly, for REML estimation, the gradient of the objective function in (4.28b), denoted as $\ell_{\mathrm{ap}}^k(\sigma_u^2, \sigma_\varepsilon^2)$, can be calculated as (Exercise 4.5)

$$\begin{aligned}
\frac{\partial \ell_{\mathrm{ap}}^k}{\partial \sigma_\varepsilon^2} &= \frac{\|\mathbf{Y} - \mathbf{X}b^k - \mathbf{Z}u^k\|_2^2}{2\sigma_\varepsilon^4} - \frac{N}{2\sigma_\varepsilon^2} + \frac{1}{2\sigma_\varepsilon^2}\mathrm{tr}\{(\mathbf{C}^{\mathrm{T}}\mathbf{C} + \rho\,\mathrm{diag}(0_p, I_m))^{-1}\mathbf{C}^{\mathrm{T}}\mathbf{C}\}, \\
\frac{\partial \ell_{\mathrm{ap}}^k}{\partial \sigma_u^2} &= \frac{\|u^k\|_2^2}{2\sigma_u^4} - \frac{m}{2\sigma_u^2} + \frac{1}{2\sigma_u^2}\mathrm{tr}\{I_{p+m} - (\mathbf{C}^{\mathrm{T}}\mathbf{C} + \rho\,\mathrm{diag}(0_p, I_m))^{-1}\mathbf{C}^{\mathrm{T}}\mathbf{C}\},
\end{aligned} \quad (4.32)$$

where $\mathbf{C} = (\mathbf{X}, \mathbf{Z})$ and $\mathrm{diag}(0_p, I_m)$ is the diagonal matrix $\begin{pmatrix} 0_p & \\ & I_m \end{pmatrix}$. Solving the gradient equation, but with ρ fixed at $\rho^{k-1} = (\sigma_\varepsilon^2)^{k-1}/(\sigma_u^2)^{k-1}$, leads to a simple update in place of (4.28b):

$$(\sigma_\varepsilon^2)^k = \frac{\|\mathbf{Y} - \mathbf{X}b^k - \mathbf{Z}u^k\|_2^2}{N - q^{k-1}}, \quad (\sigma_u^2)^k = \frac{\|u^k\|_2^2}{-p + q^{k-1}}, \quad (4.33)$$

where $q^{k-1} = \mathrm{tr}\{(\mathbf{C}^{\mathrm{T}}\mathbf{C} + \rho^{k-1}\,\mathrm{diag}(0_p, I_m))^{-1}\mathbf{C}^{\mathrm{T}}\mathbf{C}\}$. Note that $(\sigma_\varepsilon^2)^k$ in (4.31) or (4.33) is a block-coordinate descent update of σ_ε^2 and, separately, $(\sigma_u^2)^k$ is a block-coordinate descent update of σ_u^2, with ρ fixed at ρ^{k-1}, but they do not make a cycle of block-coordinate descent updates.

4.1.6 Model-based inference

The preceding sections mainly deal with point estimation and prediction. In this section, we discuss model-based inference about the fixed and random effects (b, u), assuming linear mixed model (4.5)–(4.6) is correctly specified with fixed $(\mathbf{X}, \mathbf{Z})$. Model-based inference about variance parameters can also be developed; see for example Searle et al. (1992) and Pinheiro and Bates (2000).

We first discuss inference about (b, u) given (D, G). The following result shows the sampling properties of the point estimator and predictor $(\hat{b}_{D,G}, \hat{u}_{D,G})$.

Lemma 4.4 (Sampling properties of estimator and predictor)**.** *Given (D, G), we have*

$$\begin{pmatrix} \hat{b}_{D,G} - b \\ \hat{u}_{D,G} - u \end{pmatrix} \sim \mathrm{N}(0, H_{D,G}^{-1}), \tag{4.34}$$

where $H_{D,G}$ is defined in (4.19). *If (ε, u) has the mean and variance matrix as in* (4.6) *but may not be multivariate normal, then $(\hat{b}_{D,G} - b, \hat{u}_{D,G} - u)$ has the mean and variance matrix as in* (4.34) *but may not be multivariate normal.*

Proof. From the expression (4.20) for $(\hat{b}_{D,G}, \hat{u}_{D,G})$, direct calculation yields

$$\begin{aligned} \begin{pmatrix} \hat{b}_{D,G} - b \\ \hat{u}_{D,G} - u \end{pmatrix} &= H_{D,G}^{-1} \begin{pmatrix} \mathbf{X}^{\mathrm{T}} G^{-1}(\mathbf{Y} - \mathbf{X}b - \mathbf{Z}u) \\ \mathbf{Z}^{\mathrm{T}} G^{-1}(\mathbf{Y} - \mathbf{X}b - \mathbf{Z}u) - D^{-1}u \end{pmatrix} \\ &= H_{D,G}^{-1} \begin{pmatrix} \mathbf{X}^{\mathrm{T}} G^{-1}\varepsilon \\ \mathbf{Z}^{\mathrm{T}} G^{-1}\varepsilon - D^{-1}u \end{pmatrix}. \end{aligned} \tag{4.35}$$

Then the desired result (4.34) follows from model assumption (4.6). The proof also applies when the multivariate normality in (4.6) is relaxed. □

From the expression (4.21c) for $H_{D,G}^{-1}$, the joint distribution in (4.34) implies the marginal distributions:

$$\hat{b}_{D,G} - b \sim \mathrm{N}(0, (H_{D,G}^{-1})_{11}), \tag{4.36a}$$

$$\hat{u}_{D,G} - u \sim \mathrm{N}(0, (H_{D,G}^{-1})_{22}), \tag{4.36b}$$

where

$$\begin{aligned} (H_{D,G}^{-1})_{11} &= (\mathbf{X}^{\mathrm{T}} V_{D,G}^{-1} \mathbf{X})^{-1}, \\ (H_{D,G}^{-1})_{22} &= \Lambda_{D,G}^{-1} + \Lambda_{D,G}^{-1} \mathbf{Z}^{\mathrm{T}} G^{-1} \mathbf{X} (\mathbf{X}^{\mathrm{T}} V_{D,G}^{-1} \mathbf{X})^{-1} \mathbf{X}^{\mathrm{T}} G^{-1} \mathbf{Z} \Lambda_{D,G}^{-1}. \end{aligned}$$

Note that, because b is fixed, (4.36a) is equivalent to $\hat{b}_{D,G} \sim \mathrm{N}(b, (\mathbf{X}^{\mathrm{T}} V_{D,G}^{-1} \mathbf{X})^{-1})$, as stated earlier in (4.11), whereas, because u is random, (4.36b) does *not* indicate $\hat{u}_{D,G} \sim \mathrm{N}(u, (H_{D,G}^{-1})_{22})$ even conditionally on u. Nevertheless, (4.36b) shows that $E(\hat{u}_{D,G} - u) = 0$ and $\mathrm{var}(\hat{u}_{D,G} - u) = (H_{D,G}^{-1})_{22}$. The latter has an interesting interpretation as follows [Exercise]:

$$\begin{aligned} \mathrm{var}(\hat{u}_{D,G} - u) &= E\{\mathrm{var}(\hat{u}_{D,G} - u | \mathbf{Y})\} + \mathrm{var}\{E(\hat{u}_{D,G} - u | \mathbf{Y})\} \\ &= \mathrm{var}\{u - E(u|\mathbf{Y})\} + \mathrm{var}\{\hat{u}_{D,G} - E(u|\mathbf{Y})\}, \end{aligned} \tag{4.37}$$

where

$$\operatorname{var}\{u - E(u|\mathbf{Y})\} = \Lambda_{D,G}^{-1},$$
$$\operatorname{var}\{\hat{u}_{D,G} - E(u|\mathbf{Y})\} = \Lambda_{D,G}^{-1}\mathbf{Z}^{\mathrm{T}}G^{-1}\mathbf{X}(\mathbf{X}^{\mathrm{T}}V_{D,G}^{-1}\mathbf{X})^{-1}\mathbf{X}^{\mathrm{T}}G^{-1}\mathbf{Z}\Lambda_{D,G}^{-1}.$$

Recall that $\hat{u}_{D,G}$ is defined as $E(u|\mathbf{Y})$ with b replaced by $\hat{b}_{D,G}$. The decomposition in (4.37) makes clear the additional variation caused by estimation of b in $\hat{u}_{D,G}$.

Coincidentally, the variance matrix $H_{D,G}^{-1}$ in (4.34) for $(\hat{b}_{D,G}, \hat{u}_{D,G})$, which are jointly a maximizer of $h(b, u, D, G)$ given (D, G), is precisely the inverse of the (constant) negative Hessian of $h(b, u, D, G)$ in (b, u). This seems to further demonstrate similarity of the extended log-likelihood $h(b, u, D, G)$ to a usual log-likelihood such as $\ell(b, D, G)$, where the variance matrix of $\ddot{b}_{D,G}$ as a maximizer of $\ell(b, D, G)$ is the inverse of the (constant) negative Hessian, $\mathbf{X}^{\mathrm{T}}V_{D,G}\mathbf{X}$, of $\ell(b, D, G)$ in b. However, such a connection is subtle. As explained earlier, (4.34) does *not* indicate $(\hat{b}_{D,G}^{\mathrm{T}}, \hat{u}_{D,G}^{\mathrm{T}})^{\mathrm{T}} \sim \mathrm{N}((b^{\mathrm{T}}, u^{\mathrm{T}})^{\mathrm{T}}, H_{D,G}^{-1})$.

From Lemma 4.4, given (D, G), a prediction interval of confidence size $1-c$ for $x^{\mathrm{T}}b + z^{\mathrm{T}}u$ with fixed covariate vectors (x, z) is

$$(x^{\mathrm{T}}\hat{b}_{D,G} + z^{\mathrm{T}}\hat{u}_{D,G}) \pm z_{c/2}\sqrt{(x^{\mathrm{T}}, z^{\mathrm{T}})H_{D,G}^{-1}(x^{\mathrm{T}}, z^{\mathrm{T}})^{\mathrm{T}}}, \tag{4.38}$$

satisfying

$$P\big(x^{\mathrm{T}}b + z^{\mathrm{T}}u \text{ is contained in the interval (4.38)}\big) = 1 - c.$$

The interval (4.38) is called a prediction interval because $x^{\mathrm{T}}b + z^{\mathrm{T}}u$ is a random quantity.

With unknown (D, G), it seems natural to define the point estimator and predictor for (b, u) as $(\hat{b}_{D,G}, \hat{u}_{D,G})$, by replacing (D, G) with the ML estimator $(\hat{D}, \hat{G})$ or REML estimator $(\tilde{D}, \tilde{G})$. In either case, it can be shown that the additional variation caused by estimation of (D, G) is asymptotically negligible and $(\hat{b}_{\hat{D},\hat{G}} - b, \hat{u}_{\hat{D},\hat{G}} - u)$ or $(\hat{b}_{\tilde{D},\tilde{G}} - b, \hat{u}_{\tilde{D},\tilde{G}} - u)$ is approximately multivariate normal as in Lemma 4.4, provided that the estimator $(\hat{D}, \hat{G})$ or $(\tilde{D}, \tilde{G})$ is consistent for (D, G). This result can be attributed to the following orthogonality:

$$\begin{aligned} &E\left\{\frac{\partial}{\partial \tau}\frac{\partial}{\partial(b,u)}h(b,u,D,G)\right\} \\ &= E\left\{\frac{\partial}{\partial \tau}\begin{pmatrix}\mathbf{X}^{\mathrm{T}}G^{-1}(\mathbf{Y} - \mathbf{X}b - \mathbf{Z}u) \\ \mathbf{Z}^{\mathrm{T}}G^{-1}(\mathbf{Y} - \mathbf{X}b - \mathbf{Z}u) - D^{-1}u\end{pmatrix}\right\} = 0, \end{aligned} \tag{4.39}$$

where τ denotes the variance parameters associated with (D, G). Then the gradient of the right-hand side of (4.35) including $H_{D,G}^{-1}$ with respect to τ also has mean 0, so that the asymptotic behavior of the left-hand side of (4.35) is not affected by estimation of τ. See Section 1.10.4 for a related discussion. Based on the extension of Lemma 4.4, the prediction interval (4.38) with the estimated (D, G) is approximately of size $1 - c$.

4.1.7 Examples

To help understand the general discussion, we provide two examples where the estimators can be explicitly calculated.

Example 4.1. Consider the following random-mean model:

$$\begin{aligned}
&Y_{ij} = b_0 + u_{i0} + \varepsilon_{ij}, \quad i = 1, \ldots, m, j = 1, \ldots, n_i, \\
&(\varepsilon_{ij} : i = 1, \ldots, m, j = 1, \ldots, n_i) \overset{\text{IID}}{\sim} \mathrm{N}(0, \sigma_\varepsilon^2), \\
&(u_{i0} : i = 1, \ldots, m) \overset{\text{IID}}{\sim} \mathrm{N}(0, \sigma_0^2),
\end{aligned}$$

where ε_{ij}'s and u_{i0}'s are independent. This can be put in the general form (4.5) and (4.6) with

$$\mathbf{X} = \mathbf{1}_N, \quad b = b_0, \quad \mathbf{Z} = \begin{pmatrix} \mathbf{1}_{n_1} & & \\ & \ddots & \\ & & \mathbf{1}_{n_m} \end{pmatrix}, \quad u = \begin{pmatrix} u_{10} \\ \vdots \\ u_{m0} \end{pmatrix},$$
$$D = \sigma_0^2 I_m, \quad G = \sigma_\varepsilon^2 I_N,$$

where I_k is the $k \times k$ identity matrix and $\mathbf{1}_k$ is the $k \times 1$ vector of all ones for $k \geq 1$, and $N = \sum_{i=1}^m n_i$. Then $V_{D,G} = \mathbf{Z}D\mathbf{Z}^{\mathrm{T}} + G$ can be calculated as

$$V_{D,G} = \mathrm{diag}(V_1, \ldots, V_m), \text{ with } V_i = \sigma_0^2 \mathbf{1}_{n_i \times n_i} + \sigma_\varepsilon^2 I_{n_i},$$

where $\mathbf{1}_{k \times k}$ is the $k \times k$ matrix of all ones. By using the fact that $(A + cc^{\mathrm{T}})^{-1} = A^{-1} - A^{-1}c(1 + c^{\mathrm{T}}A^{-1}c)^{-1}c^{\mathrm{T}}A^{-1}$ for a matrix A and vector c, it can be shown that

$$V_i^{-1} = \sigma_\epsilon^{-2}(I_{n_i} - \frac{\sigma_0^2}{\sigma_\varepsilon^2 + n_i\sigma_0^2}\mathbf{1}_{n_i \times n_i}).$$

Moreover, it can be shown that $|V_i| = \sigma_\varepsilon^{2(n_i-1)}(\sigma_\varepsilon^2 + n_i\sigma_0^2)$ [Exercise]. To apply (4.8), we calculate

$$\begin{aligned}
\mathbf{X}^{\mathrm{T}} V_{D,G}^{-1}\mathbf{Y} &= \sum_{i=1}^m \mathbf{1}_{n_i}^{\mathrm{T}} V_i^{-1}\mathbf{Y}_i \\
&= \sum_{i=1}^m \sigma_\epsilon^{-2}(1 - \frac{n_i\sigma_0^2}{\sigma_\varepsilon^2 + n_i\sigma_0^2})\mathbf{1}_{n_i}^{\mathrm{T}}\mathbf{Y}_i = \sum_{i=1}^m \frac{n_i\bar{Y}_i}{\sigma_\varepsilon^2 + n_i\sigma_0^2}, \\
\mathbf{X}^{\mathrm{T}} V_{D,G}^{-1}\mathbf{X} &= \sum_{i=1}^m \frac{n_i}{\sigma_\varepsilon^2 + n_i\sigma_0^2},
\end{aligned}$$

where $\mathbf{Y}_i = (Y_{i1}, \ldots, Y_{in_i})^{\mathrm{T}}$ and $\bar{Y}_i = \mathbf{1}_{n_i}^{\mathrm{T}}\mathbf{Y}_i/n_i = \sum_{j=1}^{n_i} Y_{ij}/n_i$, referred to as the ith group mean. Then from (4.8), the ML estimator of b_0 given $(\sigma_0^2, \sigma_\varepsilon^2)$ is

$$\hat{b}_0 = \frac{\sum_{i=1}^m \frac{n_i\bar{Y}_i}{\sigma_\varepsilon^2 + n_i\sigma_0^2}}{\sum_{i=1}^m \frac{n_i}{\sigma_\varepsilon^2 + n_i\sigma_0^2}}, \tag{4.40}$$

which is a weighted average of the group means:

$$\hat{b}_0 = \sum_{i=1}^{m} w_i \bar{Y}_i, \text{ with } w_i \propto \frac{n_i}{\sigma_\varepsilon^2 + n_i\sigma_0^2} \text{ and } \sum_{i=1}^{m} w_i = 1.$$

[The dependency of $\hat{b}_0$ on $(\sigma_0^2, \sigma_\varepsilon^2)$ is suppressed in the notation.] The weight satisfies $w_i \propto \frac{n_i\sigma_0^2}{\sigma_\varepsilon^2+n_i\sigma_0^2} = 1 - \frac{\sigma_\varepsilon^2}{\sigma_\varepsilon^2+n_i\sigma_0^2}$, indicating that the smaller n_i is, the less weight is put on $\bar{Y}_i$.

The point predictor of $u = (u_{10}, \ldots, u_{m0})^{\mathrm{T}}$ given $(\sigma_0^2, \sigma_\varepsilon^2)$ is

$$\hat{u} = D\mathbf{Z}^{\mathrm{T}} V_{D,G}^{-1}(\mathbf{Y} - \mathbf{X}\hat{b}_0) = \begin{pmatrix} \vdots \\ \frac{n_i\sigma_0^2}{\sigma_\varepsilon^2+n_i\sigma_0^2}(\bar{Y}_i - \hat{b}_0) \\ \vdots \end{pmatrix}_{i=1,\ldots,m}.$$

The predictor of u_{i0} can be rewritten as

$$\hat{u}_{i0} = \frac{n_i\sigma_0^2}{\sigma_\varepsilon^2 + n_i\sigma_0^2}(\bar{Y}_i - \hat{b}_0) = (\bar{Y}_i - \hat{b}_0) - \frac{\sigma_\varepsilon^2}{\sigma_\varepsilon^2 + n_i\sigma_0^2}(\bar{Y}_i - \hat{b}_0). \tag{4.41}$$

This represents a linear shrinkage of $\bar{Y}_i - \hat{b}_0$ toward 0. For fixed $\sigma_0^2/\sigma_\varepsilon^2$, the smaller n_i is, the more $\bar{Y}_i - \hat{b}_0$ is shrunk toward 0. For fixed n_i, the smaller $\sigma_0^2/\sigma_\varepsilon^2$ is, the more $\bar{Y}_i - \hat{b}_0$ is shrunk toward 0. As shown later, estimation of $(\sigma_0^2, \sigma_\varepsilon^2)$ and hence $\sigma_0^2/\sigma_\varepsilon^2$ depends on the response data Y_{ij}'s from all groups, not just the ith group. In this way, the amount of shrinkage in the prediction of u_{i0} is determined by pooling information from all groups.

To evaluate the profile log-likelihood (4.9), we calculate

$$\begin{aligned}
(\mathbf{Y} - \mathbf{X}\hat{b}_0)^T V_{D,G}^{-1}(\mathbf{Y} - \mathbf{X}\hat{b}_0) &= \sum_{i=1}^{m}(\mathbf{Y}_i - \hat{b}_0\mathbf{1}_{n_i})^{\mathrm{T}} V_i^{-1}(\mathbf{Y}_i - \hat{b}_0\mathbf{1}_{n_i}) \\
&= \sum_{i=1}^{m} \sigma_\varepsilon^{-2} e_i^{\mathrm{T}}(I_{n_i} - \frac{\sigma_0^2}{\sigma_\varepsilon^2 + n_i\sigma_0^2}\mathbf{1}_{n_i\times n_i})e_i \\
&= \sum_{i=1}^{m} \sigma_\varepsilon^{-2}(n_i\overline{e_i^2} - \frac{\sigma_0^2}{\sigma_\varepsilon^2 + n_i\sigma_0^2}(n_i\bar{e}_i)^2) \\
&= \sum_{i=1}^{m} \sigma_\varepsilon^{-2}(n_i s_i + \frac{n_i\bar{e}_i^2\sigma_\varepsilon^2}{\sigma_\varepsilon^2 + n_i\sigma_0^2}),
\end{aligned}$$

where $e_i = \mathbf{Y}_i - \hat{b}_0\mathbf{1}_{n_i}$, $\bar{e}_i = \sum_{j=1}^{n_i} e_{ij}/n_i$, $\overline{e_i^2} = \sum_{j=1}^{n_i} e_{ij}^2/n_i$, and $s_i = \sum_{j=1}^{n_i}(e_{ij} - \bar{e}_i)^2/n_i$, satisfying $\overline{e_i^2} = s_i + \bar{e}_i^2$. In addition, $\log|V_{D,G}| = \sum_{i=1}^{m}\log|V_i| = \sum_{i=1}^{m}\{(n_i - 1)\log\sigma_\varepsilon^2 + \log(\sigma_\varepsilon^2 + n_i\sigma_0^2)\}$. From (4.9), the profile log-likelihood of $(\sigma_0^2, \sigma_\varepsilon^2)$ is, up to an additive constant,

$$\ell_{\mathrm{p}}(\sigma_0^2, \sigma_\varepsilon^2)$$

$$= -\frac{1}{2}\sum_{i=1}^{m}\{(n_i-1)\log\sigma_\varepsilon^2 + \log(\sigma_\varepsilon^2 + n_i\sigma_0^2)\} - \frac{1}{2}\sum_{i=1}^{m}(\frac{n_i s_i}{\sigma_\varepsilon^2} + \frac{n_i\bar{e}_i^2}{\sigma_\varepsilon^2 + n_i\sigma_0^2}). \tag{4.42}$$

From (4.10), the adjusted profile log-likelihood of $(\sigma_0^2, \sigma_\varepsilon^2)$ is, up to an additive constant,

$$\ell_{\text{ap}}(\sigma_0^2, \sigma_\varepsilon^2) = \ell_{\text{p}}(\sigma_0^2, \sigma_\varepsilon^2) - \frac{1}{2}\log\sum_{i=1}^{m}\frac{n_i}{\sigma_\varepsilon^2 + n_i\sigma_0^2}. \tag{4.43}$$

In general, maximizing the profile log-likelihood or the adjusted profile log-likelihood does not lead to closed-form expressions for the ML estimators $(\hat{\sigma}_0^2, \hat{\sigma}_\varepsilon^2)$ or REML estimators $(\tilde{\sigma}_0^2, \tilde{\sigma}_\varepsilon^2)$.

To gain further understanding, it is instructive to examine the special case where $n_1 = \cdots = n_m = n$. In this case, $\hat{b}_0 = \sum_{i=1}^{m}\bar{Y}_i/m$, $\bar{e}_i = \bar{Y}_i - \hat{b}_0$, and $s_i = \sum_{j=1}^{n}(Y_{ij} - \bar{Y}_i)^2/n$. The profile log-likelihood of $(\sigma_0^2, \sigma_\varepsilon^2)$ reduces to

$$\ell_{\text{p}}(\sigma_0^2, \sigma_\varepsilon^2) = -\frac{m}{2}\{(n-1)\log\sigma_\varepsilon^2 + \log(\sigma_\varepsilon^2 + n\sigma_0^2)\} - \frac{1}{2}\sum_{i=1}^{m}(\frac{ns_i}{\sigma_\varepsilon^2} + \frac{n\bar{e}_i^2}{\sigma_\varepsilon^2 + n\sigma_0^2}),$$

which is separable in $\tau_1 = \sigma_\varepsilon^2$ and $\tau_2 = \sigma_\varepsilon^2 + n\sigma_0^2$. Re-parameterizing ℓ_{p} in terms of (τ_1, τ_2),

$$\ell_{\text{p}}(\tau_1, \tau_2) = -\frac{m}{2}\{(n-1)\log\tau_1 + \log\tau_2\} - \frac{1}{2}\sum_{i=1}^{m}(\frac{ns_i}{\tau_1} + \frac{n\bar{e}_i^2}{\tau_2}),$$

which is separable in (τ_1, τ_2). Setting the derivatives of ℓ_{p} to 0 gives

$$\frac{\partial\ell_{\text{p}}}{\partial\tau_1} = -\frac{m}{2}\frac{n-1}{\tau_1} + \frac{n}{2\tau_1^2}\sum_{i=1}^{m}s_i = 0,$$
$$\frac{\partial\ell_{\text{p}}}{\partial\tau_2} = -\frac{m}{2}\frac{1}{\tau_2} + \frac{n}{2\tau_2^2}\sum_{i=1}^{m}\bar{e}_i^2 = 0.$$

Solving the equations leads to

$$\begin{aligned}\hat{\tau}_1 = \hat{\sigma}_\varepsilon^2 &= \frac{n\sum_{i=1}^{m}s_i}{m(n-1)} = \frac{\sum_{i=1}^{m}\sum_{j=1}^{n}(e_{ij} - \bar{e}_i)^2}{m(n-1)} = \frac{\text{SSW}}{m(n-1)},\\ \hat{\tau}_2 = \widehat{\sigma_\varepsilon^2 + n\sigma_0^2} &= \frac{n\sum_{i=1}^{m}\bar{e}_i^2}{m} = \frac{\text{SSB}}{m},\end{aligned} \tag{4.44}$$

and hence

$$\hat{\sigma}_0^2 = \frac{1}{n}\left(\frac{\text{SSB}}{m} - \frac{\text{SSW}}{m(n-1)}\right),$$

where SSW $= \sum_{i=1}^m \sum_{j=1}^n (e_{ij} - \bar{e}_i)^2$ (called the "sum of squares within"), and SSB $= n\sum_{i=1}^m \bar{e}_i^2$ (called the "sum of squares between"). However, the preceding solution (4.44) ignores the constraint that $\sigma_0^2 \geq 0$ or equivalently $\tau_2 \geq \tau_1$ when maximizing ℓ_{p}. The resulting estimator $\hat{\sigma}_0^2$ takes a negative value if $\frac{\mathrm{SSB}}{m} - \frac{\mathrm{SSW}}{m(n-1)} < 0$. When this occurs, we may reset $\hat{\sigma}_0^2 = 0$ or equivalently $\hat{\tau}_2 = \hat{\tau}_1$, with $\hat{\tau}_1$ and $\hat{\sigma}_\varepsilon^2$ still defined in (4.44).

In the special case of $n_1 = \cdots = n_m = n$, the adjust profile log-likelihood of $(\sigma_0^2, \sigma_\varepsilon^2)$ reduces to, up to an additive constant,

$$\begin{aligned}\ell_{\mathrm{ap}}(\sigma_0^2,\sigma_\varepsilon^2) &= \ell_{\mathrm{p}}(\sigma_0^2,\sigma_\varepsilon^2) + \frac{1}{2}\log(\sigma_\varepsilon^2 + n\sigma_0^2)\\ &= -\frac{m(n-1)}{2}\log\sigma_\varepsilon^2 - \frac{m-1}{2}\log(\sigma_\varepsilon^2 + n\sigma_0^2) - \frac{1}{2}\sum_{i=1}^m\Big(\frac{ns_i}{\sigma_\varepsilon^2} + \frac{n\bar{e}_i^2}{\sigma_\varepsilon^2 + n\sigma_0^2}\Big).\end{aligned}$$

Similar calculation as above (while ignoring the constraint $\sigma_0^2 \geq 0$) yields the following estimators:

$$\begin{aligned}\tilde{\tau}_1 &= \tilde{\sigma}_\varepsilon^2 = \frac{\mathrm{SSW}}{m(n-1)},\\ \tilde{\tau}_2 &= \widetilde{\sigma_\varepsilon^2 + n\sigma_0^2} = \frac{\mathrm{SSB}}{m-1},\end{aligned}$$

and hence

$$\tilde{\sigma}_0^2 = \frac{1}{n}\left(\frac{\mathrm{SSB}}{m-1} - \frac{\mathrm{SSW}}{m(n-1)}\right).$$

Compared with (4.44), the only change is that SSB is divided by $m-1$ (instead of m) as the degrees of freedom associated with SSB. The estimators $(\tilde{\sigma}_0^2, \tilde{\sigma}_\varepsilon^2)$ coincide with the standard analysis-of-variance (ANOVA) estimators, which are known to be unbiased (although $\tilde{\sigma}_0^2$ may be negative). When $\frac{\mathrm{SSB}}{m-1} - \frac{\mathrm{SSW}}{m(n-1)} < 0$. we may reset $\tilde{\sigma}_0^2 = 0$.

Example 4.2. For the random-mean model, the calculation in Example 4.1 follows the approach in Sections 4.1.2–4.1.4, starting with the log-likelihood function of (b, D, G) with u marginalized out. To help understanding, we recalculate the relevant expressions using the extended log-likelihood function discussed in Section 4.1.5.

By the notation in Example 4.1, the extended log-likelihood function is

$$\begin{aligned}h(b_0, u, \sigma_\varepsilon^2, \sigma_0^2) = &-\frac{1}{2\sigma_\varepsilon^2}\sum_{i=1}^m\sum_{j=1}^{n_i}(Y_{ij} - b_0 - u_{i0})^2 - \frac{1}{2\sigma_0^2}\sum_{i=1}^m u_{i0}^2\\ &-\frac{N}{2}\log(\sigma_\varepsilon^2) - \frac{m}{2}\log(\sigma_0^2) + \text{constant}.\end{aligned}$$

The estimator $\hat{b}_0$ and $\hat{u} = (\hat{u}_{10}, \ldots, \hat{u}_{m0})^{\mathrm{T}}$ are jointly determined by solving the equation which sets the gradient of $h(b, u, \sigma_\varepsilon^2, \sigma_0^2)$ to 0:

$$\frac{\partial h}{\partial b_0} = \frac{1}{\sigma_\varepsilon^2} \sum_{i=1}^{m} \sum_{j=1}^{n_i} (Y_{ij} - b_0 - u_{i0}) = 0, \tag{4.45a}$$

$$\frac{\partial h}{\partial u_{i0}} = \frac{1}{\sigma_\varepsilon^2} \sum_{j=1}^{n_i} (Y_{ij} - b_0 - u_{i0}) - \frac{u_{i0}}{\sigma_0^2} = 0. \tag{4.45b}$$

For fixed $\hat{b}_0$, solving (4.45b) for u_{i0} gives

$$\hat{u}_{i0} = \frac{\frac{1}{\sigma_\varepsilon^2} \sum_{j=1}^{n_i} (Y_{ij} - b_0)}{\frac{n_i}{\sigma_\varepsilon^2} + \frac{1}{\sigma_0^2}} = \frac{n_i \sigma_0^2}{\sigma_\varepsilon^2 + n_i \sigma_0^2} (\bar{Y}_i - \hat{b}_0),$$

which is expression (4.41) for $\hat{u}_{i0}$. Summing (4.45b) over i and using (4.45a) and the above expression of $\hat{u}_{i0}$ yields

$$0 = \sum_{i=1}^{m} \hat{u}_{i0} = \sum_{i=1}^{m} \frac{n_i \sigma_0^2}{\sigma_\varepsilon^2 + n_i \sigma_0^2} (\bar{Y}_i - \hat{b}_0),$$

which leads to expression (4.40) for $\hat{b}_0$.

The profile log-likelihood of $(\sigma_0^2, \sigma_\varepsilon^2)$ is (dropping the additive constant)

$$\ell_{\mathrm{p}}(\sigma_0^2, \sigma_\varepsilon^2) = h(\hat{b}_0, \hat{u}, \sigma_0^2, \sigma_\varepsilon^2) - \frac{1}{2} \log |\Lambda_{\sigma_0^2, \sigma_\varepsilon^2}|.$$

With the notation in Example 4.1, the following can be directly calculated using the expressions for $\hat{b}_0$ and $\hat{u}_{i0}$ and $\sum_{j=1}^{n_i} (e_{ij} - \frac{n_i \sigma_0^2}{\sigma_\varepsilon^2 + n_i \sigma_0^2} \bar{e}_i)^2 = \sum_{j=1}^{n_i} (e_{ij} - \bar{e}_i)^2 + n_i (\frac{\sigma_\varepsilon^2}{\sigma_\varepsilon^2 + n_i \sigma_0^2} \bar{e}_i)^2$:

$$\begin{aligned}
& h(\hat{b}_0, \hat{u}, \sigma_\varepsilon^2, \sigma_0^2) \\
&= -\frac{1}{2\sigma_\varepsilon^2} \sum_{i=1}^{m} \sum_{j=1}^{n_i} (Y_{ij} - \hat{b}_0 - \hat{u}_{i0})^2 - \frac{1}{2\sigma_0^2} \sum_{i=1}^{m} \hat{u}_{i0}^2 - \frac{N}{2} \log(\sigma_\varepsilon^2) - \frac{m}{2} \log(\sigma_0^2) \\
&= -\frac{1}{2\sigma_\varepsilon^2} \sum_{i=1}^{m} \sum_{j=1}^{n_i} (e_{ij} - \frac{n_i \sigma_0^2}{\sigma_\varepsilon^2 + n_i \sigma_0^2} \bar{e}_i)^2 - \frac{1}{2\sigma_0^2} \sum_{i=1}^{m} (\frac{n_i \sigma_0^2}{\sigma_\varepsilon^2 + n_i \sigma_0^2} \bar{e}_i)^2 \\
&\quad - \frac{N}{2} \log(\sigma_\varepsilon^2) - \frac{m}{2} \log(\sigma_0^2) \\
&= -\frac{1}{2\sigma_\varepsilon^2} \sum_{i=1}^{m} n_i s_i^2 - \frac{1}{2} \sum_{i=1}^{m} \frac{n_i \bar{e}_i^2}{\sigma_\varepsilon^2 + n_i \sigma_0^2} - \frac{N}{2} \log(\sigma_\varepsilon^2) - \frac{m}{2} \log(\sigma_0^2),
\end{aligned}$$

and

$$\log |\Lambda_{\sigma_0^2, \sigma_\varepsilon^2}| = \log \left| -\frac{\partial^2 h}{\partial u \partial u^{\mathrm{T}}} \right| = \sum_{i=1}^{m} \log \left| -\frac{\partial^2 h}{\partial u_{i0}^2} \right| = \sum_{i=1}^{m} \log(\frac{n_i}{\sigma_\varepsilon^2} + \frac{1}{\sigma_0^2}).$$

Combining the preceding three displays leads to the expression (4.42) for the profile log-likelihood of $(\sigma_0^2, \sigma_\varepsilon^2)$.

The adjusted profile log-likelihood of $(\sigma_0^2, \sigma_\varepsilon^2)$ is (dropping the additive constant)

$$\ell_{\mathrm{ap}}(\sigma_0^2, \sigma_\varepsilon^2) = h(\hat{b}_0, \hat{u}, \sigma_0^2, \sigma_\varepsilon^2) - \frac{1}{2}\log|H_{\sigma_0^2,\sigma_\varepsilon^2}|,$$

where

$$H_{\sigma_0^2,\sigma_\varepsilon^2} = -\begin{pmatrix} \frac{\partial^2 h}{\partial b_0^2} & \frac{\partial^2 h}{\partial b_0 \partial u^{\mathrm{T}}} \\ \frac{\partial^2 h}{\partial u \partial b_0} & \frac{\partial^2 h}{\partial u \partial u^{\mathrm{T}}} \end{pmatrix},$$

and $\frac{\partial^2 h}{\partial b_0 \partial u^{\mathrm{T}}}$ is a $1 \times m$ vector with ith element $\frac{\partial^2 h}{\partial b_0 \partial u_{i0}}$ and $\frac{\partial^2 h}{\partial u \partial u^{\mathrm{T}}}$ is an $m \times m$ diagonal matrix with ith element $\frac{\partial^2 h}{\partial u_{i0}^2}$. The individual second derivatives can be calculated as follows:

$$-\frac{\partial^2 h}{\partial b_0^2} = \frac{N}{\sigma_\varepsilon^2}, \quad -\frac{\partial^2 h}{\partial b_0 \partial u_{i0}} = \frac{n_i}{\sigma_\varepsilon^2}, \quad -\frac{\partial^2 h}{\partial u_{i0}^2} = \frac{n_i}{\sigma_\varepsilon^2} + \frac{1}{\sigma_0^2}.$$

The log determinant can be calculated as

$$\begin{aligned}
&\log|H_{\sigma_0^2,\sigma_\varepsilon^2}| \\
&= \log\left|-\frac{\partial^2 h}{\partial u \partial u^{\mathrm{T}}}\right| + \log\left|(-\frac{\partial^2 h}{\partial b_0^2}) - (-\frac{\partial^2 h}{\partial b_0 \partial u^{\mathrm{T}}})(-\frac{\partial^2 h}{\partial u \partial u^{\mathrm{T}}})^{-1}(-\frac{\partial^2 h}{\partial u \partial b_0})\right| \\
&= \log\left|-\frac{\partial^2 h}{\partial u \partial u^{\mathrm{T}}}\right| + \log\sum_{i=1}^m \left\{\frac{n_i}{\sigma_\varepsilon^2} - (\frac{n_i}{\sigma_\varepsilon^2})^2(\frac{n_i}{\sigma_\varepsilon^2} + \frac{1}{\sigma_0^2})^{-1}\right\} \\
&= \log\left|-\frac{\partial^2 h}{\partial u \partial u^{\mathrm{T}}}\right| + \log\sum_{i=1}^m \frac{n_i}{\sigma_\varepsilon^2 + n_i\sigma_0^2},
\end{aligned}$$

where the first line follows from Exercise 1.4. Combining the preceding calculation with the expression for $\ell_{\mathrm{p}}(\sigma_0^2, \sigma_\varepsilon^2)$ leads to the expression (4.43) for $\ell_{\mathrm{ap}}(\sigma_0^2, \sigma_\varepsilon^2)$.

Example 4.3. Consider the following parallel-line, random-intercept model as in (4.2):

$$\begin{aligned}
&Y_{ij} = b_0 + b_1 X_{ij} + u_{i0} + \varepsilon_{ij}, \quad i = 1, \ldots, m, j = 1, \ldots, n_i, \\
&(\varepsilon_{ij} : i = 1, \ldots, m, j = 1, \ldots, n_i) \overset{\mathrm{IID}}{\sim} \mathrm{N}(0, \sigma_\varepsilon^2), \\
&(u_{i0} : i = 1, \ldots, m) \overset{\mathrm{IID}}{\sim} \mathrm{N}(0, \sigma_0^2),
\end{aligned}$$

where ε_{ij}'s and u_{i0}'s are independent. This can be put in the general form (4.5) and (4.6) with

$$\mathbf{X} = \begin{pmatrix} \mathbf{1}_{n_1} & \mathbf{X}_1 \\ \vdots & \vdots \\ \mathbf{1}_{n_m} & \mathbf{X}_m \end{pmatrix}, \quad b = \begin{pmatrix} b_0 \\ b_1 \end{pmatrix}, \quad \mathbf{Z} = \begin{pmatrix} \mathbf{1}_{n_1} & & \\ & \ddots & \\ & & \mathbf{1}_{n_m} \end{pmatrix}, \quad u = \begin{pmatrix} u_{10} \\ \vdots \\ u_{m0} \end{pmatrix},$$

$D = \sigma_0^2 I_m, \quad G = \sigma_\varepsilon^2 I_N,$

where $\mathbf{X}_i = (X_{i1}, \ldots, X_{in_i})^{\mathrm{T}}$. Then $V_{D,G}$ is the same as in Example 4.1. To apply (4.8), we calculate

$$\begin{aligned}
\mathbf{X}^{\mathrm{T}} V_{D,G}^{-1} \mathbf{Y} &= \sum_{i=1}^{m} \begin{pmatrix} \mathbf{1}_{n_i}^{\mathrm{T}} \\ \mathbf{X}_i^{\mathrm{T}} \end{pmatrix} V_i^{-1} \mathbf{Y}_i \\
&= \sum_{i=1}^{m} \begin{pmatrix} \mathbf{1}_{n_i}^{\mathrm{T}} \\ \mathbf{X}_i^{\mathrm{T}} \end{pmatrix} \sigma_\epsilon^{-2} \left(\mathbf{Y}_i - \frac{n_i \sigma_0^2 \mathbf{1}_{n_i}}{\sigma_\varepsilon^2 + n_i \sigma_0^2} \bar{Y}_i \right) \\
&= \sum_{i=1}^{m} \sigma_\epsilon^{-2} n_i \begin{pmatrix} (1-\rho_i)\bar{Y}_i \\ \overline{X_i Y_i} - \rho_i \bar{X}_i \bar{Y}_i \end{pmatrix}, \\
\mathbf{X}^{\mathrm{T}} V_{D,G}^{-1} \mathbf{X} &= \sum_{i=1}^{m} \begin{pmatrix} \mathbf{1}_{n_i}^{\mathrm{T}} \\ \mathbf{X}_i^{\mathrm{T}} \end{pmatrix} \sigma_\epsilon^{-2} \left((\mathbf{1}_{n_i}, \mathbf{X}_i) - \frac{n_i \sigma_0^2 \mathbf{1}_{n_i}}{\sigma_\varepsilon^2 + n_i \sigma_0^2} (1, \bar{X}_i) \right) \\
&= \sum_{i=1}^{m} \sigma_\epsilon^{-2} n_i \begin{pmatrix} 1-\rho_i & (1-\rho_i)\bar{X}_i \\ (1-\rho_i)\bar{X}_i & \overline{X_i^2} - \rho_i \bar{X}_i^2 \end{pmatrix},
\end{aligned}$$

where $\bar{X}_i = \sum_{j=1}^{n_i} X_{ij}/n_i$, $\overline{X_i^2} = \sum_{j=1}^{n_i} X_{ij}^2/n_i$, $\overline{X_i Y_i} = \sum_{j=1}^{n_i} X_{ij} Y_{ij}/n_i$, and

$$\rho_i = \frac{n_i \sigma_0^2}{\sigma_\varepsilon^2 + n_i \sigma_0^2}.$$

Then from (4.8), the ML estimator of b given $(\sigma_0^2, \sigma_\varepsilon^2)$ is

$$\hat{b} = \left\{ \sum_{i=1}^{m} n_i \begin{pmatrix} 1-\rho_i & (1-\rho_i)\bar{X}_i \\ (1-\rho_i)\bar{X}_i & \overline{X_i^2} - \rho_i \bar{X}_i^2 \end{pmatrix} \right\}^{-1} \left\{ \sum_{i=1}^{m} n_i \begin{pmatrix} (1-\rho_i)\bar{Y}_i \\ \overline{X_i Y_i} - \rho_i \bar{X}_i \bar{Y}_i \end{pmatrix} \right\}.$$

Similarly as in Example 4.1, the prediction of $u = (u_{10}, \ldots, u_{m0})^{\mathrm{T}}$ given $(\sigma_0^2, \sigma_\varepsilon^2)$ is

$$\hat{u} = D\mathbf{Z}^{\mathrm{T}} V_{D,G}^{-1} (\mathbf{Y} - \mathbf{X}\hat{b}) = \begin{pmatrix} \vdots \\ \frac{n_i \sigma_0^2}{\sigma_\varepsilon^2 + n_i \sigma_0^2} (\bar{Y}_i - \hat{b}_0 - \hat{b}_1 X_i) \\ \vdots \end{pmatrix}_{i=1,\ldots,m}.$$

Hence the prediction of u_{i0} is a linear shrinkage of $\bar{Y}_i - \hat{b}_0 - \hat{b}_1 X_i$ toward 0, and the amount of shrinkage depends on n_i and $\sigma_0^2/\sigma_\varepsilon^2$.

In the special case of $n_1 = \cdots = n_m = n$, the estimator reduces to

$$\hat{b} = \begin{pmatrix} m(1-\rho) & (1-\rho)\sum_{i=1}^{m} \bar{X}_i \\ (1-\rho)\sum_{i=1}^{m} \bar{X}_i & \sum_{i=1}^{m} (\overline{X_i^2} - \rho \bar{X}_i^2) \end{pmatrix}^{-1} \begin{pmatrix} (1-\rho)\sum_{i=1}^{m} \bar{Y}_i \\ \sum_{i=1}^{m} (\overline{X_i Y_i} - \rho \bar{X}_i \bar{Y}_i) \end{pmatrix},$$

where $\rho = \frac{n\sigma_0^2}{\sigma_\varepsilon^2 + n\sigma_0^2}$. For further simplification to obtain separate expressions of $\hat{b}_0$ and $\hat{b}_1$, we left-multiply the preceding expression by

$$\Gamma = \begin{pmatrix} 1 & \sum_{i=1}^m \bar{X}_i/m \\ 0 & 1 \end{pmatrix}$$

and calculate the left-hand and right-hand sides respectively:

$$\Gamma\hat{b} = \begin{pmatrix} \hat{b}_0 + (\sum_{i=1}^m \bar{X}_i/m)\hat{b}_1 \\ \hat{b}_1 \end{pmatrix},$$

$$\begin{aligned}
&\Gamma \begin{pmatrix} m(1-\rho) & (1-\rho)\sum_{i=1}^m \bar{X}_i \\ (1-\rho)\sum_{i=1}^m \bar{X}_i & \sum_{i=1}^m (\overline{X_i^2} - \rho\bar{X}_i^2) \end{pmatrix}^{-1} \begin{pmatrix} (1-\rho)\sum_{i=1}^m \bar{Y}_i \\ \sum_{i=1}^m (\overline{X_iY_i} - \rho\bar{X}_i\bar{Y}_i) \end{pmatrix} \\
&= \begin{pmatrix} m(1-\rho) & 0 \\ 0 & \sum_{i=1}^m (\overline{X_i^2} - \rho\bar{X}_i^2) - (1-\rho)(\sum_{i=1}^m \bar{X}_i)^2/m \end{pmatrix}^{-1} \\
&\quad \times \begin{pmatrix} (1-\rho)\sum_{i=1}^m \bar{Y}_i \\ \sum_{i=1}^m (\overline{X_iY_i} - \rho\bar{X}_i\bar{Y}_i) - (1-\rho)\sum_{i=1}^m \bar{Y}_i \sum_{i=1}^m \bar{X}_i/m \end{pmatrix} \\
&= \begin{pmatrix} \sum_{i=1}^m \bar{Y}_i/m \\ \frac{\sum_{i=1}^m (\overline{X_iY_i} - \rho\bar{X}_i\bar{Y}_i) - (1-\rho)\sum_{i=1}^m \bar{Y}_i \sum_{i=1}^m \bar{X}_i/m}{\sum_{i=1}^m (\overline{X_i^2} - \rho\bar{X}_i^2) - (1-\rho)(\sum_{i=1}^m \bar{X}_i)^2/m} \end{pmatrix}.
\end{aligned}$$

Therefore, we obtain

$$\hat{b}_0 = \sum_{i=1}^m \bar{Y}_i/m - (\sum_{i=1}^m \bar{X}_i/m)\hat{b}_1,$$

$$\hat{b}_1 = \frac{\sum_{i=1}^m (\overline{X_iY_i} - \rho\bar{X}_i\bar{Y}_i) - (1-\rho)\sum_{i=1}^m \bar{Y}_i \sum_{i=1}^m \bar{X}_i/m}{\sum_{i=1}^m (\overline{X_i^2} - \rho\bar{X}_i^2) - (1-\rho)(\sum_{i=1}^m \bar{X}_i)^2/m}.$$

The slope estimator has a well-defined limit as $\rho \to 0+$ or equivalently $n\sigma_0^2/\sigma_\varepsilon^2 \to 0$,

$$\lim_{\rho\to 0+} \hat{b}_1 = \frac{\sum_{i=1}^m \overline{X_iY_i} - \sum_{i=1}^m \bar{Y}_i \sum_{i=1}^m \bar{X}_i/m}{\sum_{i=1}^m \overline{X_i^2} - (\sum_{i=1}^m \bar{X}_i)^2/m}, \tag{4.46}$$

or as $\rho \to 1-$ or equivalently $n\sigma_0^2/\sigma_\varepsilon^2 \to \infty$ (provided $n \geq 2$),

$$\lim_{\rho\to 1-} \hat{b}_1 = \frac{\sum_{i=1}^m (\overline{X_iY_i} - \bar{X}_i\bar{Y}_i)}{\sum_{i=1}^m (\overline{X_i^2} - \bar{X}_i^2)}. \tag{4.47}$$

The right-hand side of (4.47) can be shown to be consistent for b_1 provided $n \geq 2$ and $mn \to \infty$, in the setting where $(u_{10}, \ldots, u_{m0})$ are *arbitrarily fixed* and the noises ε_{ij}'s may be non-normal and have unequal variances. The right-hand side of (4.46) can be consistent for b_1 only when $(u_{10}, \ldots, u_{m0})$ are all the same. In fact, the right-hand side of (4.46) is the least-squares estimator of b_1 in the simple linear regression:

$$Y_{ij} = b_0 + b_1 X_{ij} + \varepsilon_{ij},$$

for $i = 1, \ldots, m$ and $j = 1, \ldots, n$.

4.2 Numerical demonstration: Linear mixed models

We demonstrate numerical application of linear mixed models in R using the package `nlme()` (Pinheiro et al., 2020). We leave to readers the exercises of writing self-contained code to implement the statistical methods in Section 4.1, depending on parameterization of variance matrices (D, G), while calling only basic mathematical functions. In the case of $D = \sigma_u^2 I_m$ and $G = \sigma_\varepsilon^2 I_N$, our own R code based on the extended likelihood is presented in Section 4.3.2.

Consider the petrol data from Prater (1956) about the yield of a petroleum refining process with four covariates from 10 samples, as shown below. The response (`Y`) is the yield as a percentage of crude oil to gasoline. The covariates include `SG`, `VP`, and `VP10`, which are constant within sample, and `EP`, which is the end point as desired volatility of gasoline and varies within sample. The crude oil samples are identified by the variable `No`. Analysis of the dataset using linear mixed models is discussed in Venables and Ripley (2002), Section 10.1. The covariates are centered by the sample means, to facilitate the interpretation of intercepts. For our demonstration, we study the association of the response `Y` with the covariate `EP` only. Moreover, we add Gaussian noises to the response `Y`, so that the shrinkage effect of coefficient estimation from mixed modeling becomes more evident (see Figure 4.2). For safe replication, the dataset with noise-added `Y` is saved as an accompanying data file.

```
> petrol <- read.table(file="petrol.data", header=T)
> petrol$No <- as.factor(petrol$No)
>
> # centering the predictors
> petrol[,2:5] <- scale(petrol[,2:5], scale=F)
>
> # adding noise to the response
> set.seed(0)
> petrol$Y <- petrol$Y + rnorm(32, mean=0, sd=5)
>
> petrol[1:5,]
  No    SG       VP   V10          EP        Y
1  A 11.55  4.41875 -51.5 -127.09375 18.51477
2  A 11.55  4.41875 -51.5  -57.09375 20.66883
3  A 11.55  4.41875 -51.5   12.90625 41.34900
4  A 11.55  4.41875 -51.5   74.90625 52.06215
5  B  1.55 -0.68125 -31.5 -114.09375 10.07321
> attach(petrol)
```

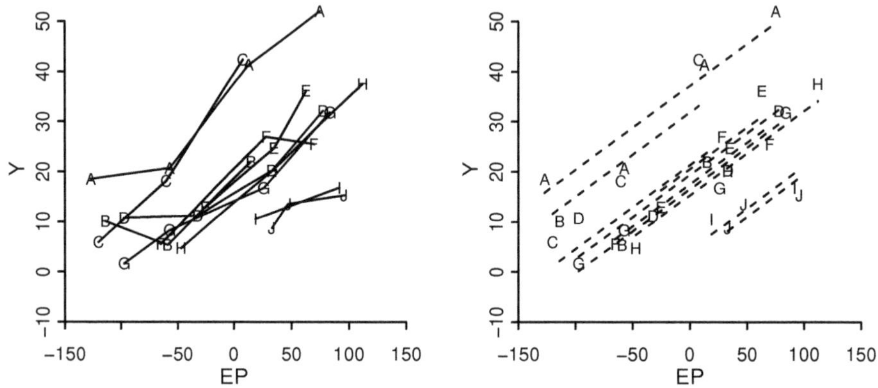

FIGURE 4.1
Scatter plot of Y vs EP in the petrol dataset (left), and fitted lines within oil samples from a fixed-effect, parallel-line model (right). The symbols of the points plotted are the values of No indicating oil samples.

From the scatter plot in Figure 4.1 (left), the response Y appears to be linearly related to the covariate EP within each oil sample, and the associated slopes appear to the same across the oil samples. For fixed-effect analysis, we fit a separate-line regression model and a parallel-line regression model of Y on EP with the grouping variable No, as described in examples (i) and (ii) in Section 4.1.1. The parallel-line model can be expressed as

$$\mathrm{Y}_{ij} = \beta_{i0} + \beta_1 \,\mathrm{EP}_{ij} + \varepsilon_{ij}, \quad i = 1, \ldots, m\,(= 10), j = 1, \ldots, n_i,$$

where i denotes the oil sample (i.e., level of $\mathtt{No}_{ij}$) and j denotes the observation on that sample, and $\varepsilon_{ij} \sim \mathrm{N}(0, \sigma_\varepsilon^2)$. A formal analysis-of-variance test using the built-in function anova() for comparing fitted models in R confirms the adequacy of the parallel-line model with p-value 0.28. The fitted lines are shown in Figure 4.1 (right).

```
> pet1.lm <- lm(Y ~ No *EP, data=petrol)
> pet2.lm <- lm(Y ~ No +EP, data=petrol)
>
> anova(pet2.lm, pet1.lm)
......
  Res.Df    RSS Df Sum of Sq      F Pr(>F)
1     21 544.96
2     12 263.74  9    281.23 1.4218 0.2795
```

For mixed-effect analysis, we apply the function lme() in the package nlme to fit a parallel-line, random-intercept model of Y on EP with the grouping factor No. See the accompanying R code file for fitting of other mixed models. As discussed in examples (iv) in Section 4.1.1 and in Example 4.3 in Section 4.1.7, the random-intercept model states that

$$\mathtt{Y}_{ij} = b_0 + b_1\,\mathrm{EP}_{ij} + u_{i0} + \varepsilon_{ij}, \quad i = 1, \ldots, m\,(= 10), j = 1, \ldots, n_i,$$

where i denotes the oil sample (i.e., level of $\mathtt{No}_{ij}$) and j denotes the observation on that sample, $\varepsilon_{ij} \sim \mathrm{N}(0, \sigma_\varepsilon^2)$ and $u_{i0} \sim \mathrm{N}(0, \sigma_0^2)$ independently. From the output below with REML estimation, **summary(pet2.lme)** gives the following estimates

$$\hat{b}_0 = 18.77, \quad \hat{b}_1 = 0.16, \quad \hat{\sigma}_0 = 9.91, \quad \hat{\sigma}_\varepsilon = 5.12.$$

Moreover, **pet2.lme$coef** gives the fixed-effect estimates $(\hat{b}_0, \hat{b}_1)$ as above and the random-effect estimates (or predictions) corresponding to the 10 oil samples:

$$(\hat{u}_{10}, \hat{u}_{20}, \ldots, \hat{u}_{m0}) = (17.11, 2.01, \ldots, -14.4).$$

Finally, **pet2.lme$fitted** gives the fitted and predicted values at different levels of grouping by columns. The first column corresponds to the population level with fixed effects only, $\hat{b}_0 + \hat{b}_1\mathrm{EP}_{ij}$, and the second column corresponds to the level of oil samples with both fixed and random effects, $\hat{b}_0 + \hat{b}_1\mathrm{EP}_{ij} + \hat{u}_{i0}$.

```
> library(nlme)
> # default is REML
> pet2.lme <- lme(Y ~ EP, random = ~1 | No, data=petrol)
> summary(pet2.lme)
Linear mixed-effects model fit by REML
......
Random effects:
 Formula: ~1 | No
        (Intercept) Residual
StdDev:    9.908272 5.116323

Fixed effects: Y ~ EP
                Value Std.Error DF   t-value p-value
(Intercept) 18.771808  3.266044 21  5.747568       0
EP           0.160875  0.015315 21 10.504681       0
 Correlation:
   (Intr)
EP -0.013
......
Number of Observations: 32
```

```
Number of Groups: 10
>
> pet2.lme$coef
$fixed
(Intercept)          EP
 18.7718079   0.1608751


$random
$random$No
  (Intercept)
A  17.1122730
B   2.0139112
C  11.6086293
D   0.5170764
E   1.7735135
F  -1.0861136
G  -2.2019988
H  -2.9616359
I -12.3744389
J -14.4012171
>
> pet2.lme$fitted
        fixed          No
1  -1.6744170 15.4378569
2   9.5868428 26.6991168
3  20.8481027 37.9603766
4  30.8223615 47.9346354
5   0.4169598  2.4308710
......
28 21.8133536  9.4389147
29 33.5572389 21.1828000
30 24.0656056  9.6643884
31 26.3178575 11.9166404
32 34.2007395 19.7995223
```

Figure 4.2 (left) shows the estimated intercepts, $(\hat{\beta}_{10}, \ldots, \hat{\beta}_{m0})$ from the fixed-effect, parallel-line model `pet2.lm` and $(\hat{b}_0 + \hat{u}_{10}, \ldots, \hat{b}_0 + \hat{u}_{m0})$ from the mixed-effect, parallel-line model `pet2.lme`. The latter estimates can be related to the former estimates through shrinkage toward some center value. The amount of shrinkage is small because $\hat{\sigma}_0^2/\hat{\sigma}_\varepsilon^2 = (9.91/5.12)^2$ is relatively large. Figure 4.2 (right) shows the fitted values at the population level (which

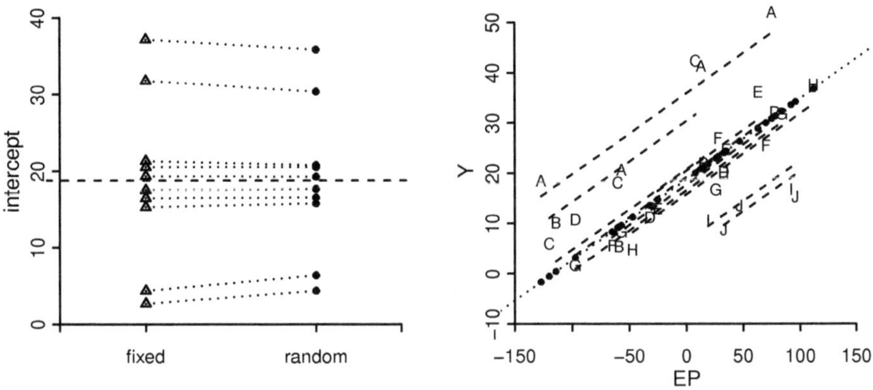

FIGURE 4.2
Plot of the estimated intercepts from the fixed-effect, parallel-line model and from the mixed-effect, parallel-line model, labeled as "fixed" or "random" (left) and plot of the fitted values at the population level (• and dotted line) and at the oil-sample level (dashed lines) from the mixed-effect, parallel-line model (right).

form a straight line in EP) and at the oil-sample level (which form a straight line in EP within each oil sample). The fitted lines within oil samples are also shrunk toward some center (less spread out), compared with those in Figure 4.1 (right).

The default estimation method in the function `lme()` is REML. Alternatively, ML estimation can be obtained by setting the argument `method="ML"`. From the output below, there are small differences between the REML and ML estimates of fixed effects and variance components. For the original data without adding noise to Y, the difference between REML and ML estimation is more appreciable as noted in Venables and Ripley (2002).

```
> pet2.ML.lme <- lme(Y ~ EP, random=~1 | No, method="ML", data=petrol)
> summary(pet2.ML.lme)
Linear mixed-effects model fit by maximum likelihood
......
Random effects:
 Formula: ~1 | No
        (Intercept) Residual
StdDev:    9.338972 5.006102

Fixed effects: Y ~ EP
                Value Std.Error DF  t-value p-value
(Intercept) 18.779327  3.189123 21 5.888555       0
```

```
EP          0.160412  0.015458 21 10.377066      0
 Correlation:
   (Intr)
EP -0.014
......
Number of Observations: 32
Number of Groups: 10
```

4.3 Applications to penalized and smoothing splines

There is an interesting connection which can be exploited between ridge-penalized and smoothing splines (Chapter 3) and linear mixed models, to facilitate selection of the tuning parameters when fitting penalized and smoothing splines.

4.3.1 Algebraic connection

From Section 3.4, the finite-dimensional representation of smoothing splines is

$$f(x;\theta) = b_0 + b_1 x + \sum_{i=1}^{n} c_i K_1(X_i, x),$$

where $\theta = (b^{\mathrm{T}}, c^{\mathrm{T}})^{\mathrm{T}}$, $b = (b_0, b_1)^{\mathrm{T}}$ and $c = (c_1, \ldots, c_n)^{\mathrm{T}}$. For fixed $\lambda > 0$, the coefficient vector θ is estimated as $\hat{\theta}_\lambda = (b_\lambda^{\mathrm{T}}, c_\lambda^{\mathrm{T}})^{\mathrm{T}}$ in (3.36), i.e.,

$$\hat{\theta}_\lambda = \operatorname{argmin}_{\theta=(b^{\mathrm{T}},c^{\mathrm{T}})^{\mathrm{T}}} \left\{ \frac{1}{n} \|\mathbf{Y} - \mathbf{K}_0 b - \mathbf{K}_1 c\|_2^2 + \lambda c^{\mathrm{T}} \mathbf{K}_1 c \right\},$$

where

$$\mathbf{Y} = \begin{pmatrix} Y_1 \\ \vdots \\ Y_n \end{pmatrix}, \quad \mathbf{K}_0 = \begin{pmatrix} 1 & X_1 \\ \vdots & \vdots \\ 1 & X_n \end{pmatrix}, \quad \mathbf{K}_1 = \begin{pmatrix} K_1(X_1, X_1) & \ldots & K_1(X_1, X_n) \\ \vdots & \vdots & \vdots \\ K_1(X_n, X_1) & \ldots & K_1(X_n, X_n) \end{pmatrix}.$$

To connect with smoothing splines, consider a linear mixed model:

$$\mathbf{Y} = \underbrace{\mathbf{K}_0 b}_{\mathbf{X}b} + \underbrace{\mathbf{K}_1 c}_{\mathbf{Z}u} + \varepsilon, \tag{4.48}$$

where $\mathbf{Y}$, $\mathbf{K}_0$, $\mathbf{K}_1$ are defined above as in (3.36), and

$$\begin{pmatrix} c \\ \varepsilon \end{pmatrix} \sim \mathrm{N}\left(0, \begin{pmatrix} \sigma_0^2 \mathbf{K}_1^{-1} & 0 \\ 0 & \sigma_\varepsilon^2 I_N \end{pmatrix}\right),$$

i.e., $D = \sigma_0^2 \mathbf{K}_1^{-1}$ and $G = \sigma_\varepsilon^2 I_N$. By Lemma 4.3, given $\tau = (\sigma_0^2, \sigma_\varepsilon^2)$, the estimator and predictor $(\hat{b}_\tau, \hat{c}_\tau)$ are defined as jointly a minimizer of the negative extended log-likelihood function, which up to an additive constant reduces to

$$\begin{aligned} &\frac{1}{2}(\mathbf{Y} - \mathbf{X}b - \mathbf{Z}u)^{\mathrm{T}} G^{-1} (\mathbf{Y} - \mathbf{X}b - \mathbf{Z}u) + \frac{1}{2} u^{\mathrm{T}} D^{-1} u \\ &= \frac{1}{2\sigma_\varepsilon^2} \|\mathbf{Y} - \mathbf{K}_0 b - \mathbf{K}_1 c\|_2^2 + \frac{1}{2\sigma_0^2} c^{\mathrm{T}} \mathbf{K}_1 c \\ &= \frac{1}{2\sigma_\varepsilon^2} \left(\|\mathbf{Y} - \mathbf{K}_0 b - \mathbf{K}_1 c\|_2^2 + \frac{\sigma_\varepsilon^2}{\sigma_0^2} c^{\mathrm{T}} \mathbf{K}_1 c \right). \end{aligned}$$

Therefore, $(\hat{b}_\tau, \hat{c}_\tau)$ coincide with $(\hat{b}_\lambda, \hat{c}_\lambda)$ in (3.36) for the smoothing spline with the tuning parameter $\lambda = \frac{1}{n}\frac{\sigma_\varepsilon^2}{\sigma_0^2}$. By this connection, an estimator of $\frac{1}{n}\frac{\sigma_\varepsilon^2}{\sigma_0^2}$ derived from the ML or REML estimators of $(\sigma_0^2, \sigma_\varepsilon^2)$ in the linear mixed model (4.48) can be used as an "automated" choice of λ for the smoothing spline. [Question: Can this choice be shown to yield the standard convergence rate for smoothing splines?]

For ease of implementation, the linear mixed model (4.48) can be reparameterized such that the variance matrix for the random-coefficient vector is proportional to an identity matrix. In fact, let $\tilde{c} = Qc$, where $Q = \mathbf{K}_1^{1/2}$ is the matrix square root of $\mathbf{K}_1$. Then $\mathbf{K}_1 c = Q\tilde{c}$ and $\tilde{c} \sim \mathrm{N}(0, \sigma_0^2 I_n)$. Hence (4.48) can equivalently stated as

$$\mathbf{Y} = \underbrace{\mathbf{K}_0 b}_{\mathbf{X}b} + \underbrace{Q\tilde{c}}_{\mathbf{Z}u} + \varepsilon, \tag{4.49}$$

where

$$\begin{pmatrix} \tilde{c} \\ \varepsilon \end{pmatrix} \sim \mathrm{N}\left(0, \begin{pmatrix} \sigma_0^2 I_n & 0 \\ 0 & \sigma_\varepsilon^2 I_N \end{pmatrix}\right),$$

i.e., $D = \sigma_0^2 I_n$ and $G = \sigma_\varepsilon^2 I_N$. This linear mixed model can be directly fitted by using the R package `nlme`, to implement the smoothing spline with an automated choice of λ.

A similar application can also be exploited for implementing ridge penalized splines as discussed in Section 3.1. A cubic spline model with k knots can be stated as

$$Y_i = b_0 + b_1 Z_i + b_1 Z_i^2 + b_3 Z_i^3 + \sum_{j=1}^{k} c_j \{(Z_i - \xi_j)_+\}^3 + \varepsilon_i, \quad i = 1, \dots, n,$$

where $b = (b_0, \dots, b_3)^{\mathrm{T}}$ and $c = (c_1, \dots, c_k)^{\mathrm{T}}$ are $4 + k$ unknown coefficients. For fixed $\lambda > 0$, the ridge estimator of (b, c) is

$$(\hat{b}_\lambda, \hat{c}_\lambda) = \mathrm{argmin}_{(b,c)} \left\{ \frac{1}{n} \|\mathbf{Y} - \mathbf{X}b - \mathbf{Z}c\|_2^2 + \lambda \|c\|_2^2 \right\},$$

where

$$\mathbf{Y} = \begin{pmatrix} Y_1 \\ \vdots \\ Y_n \end{pmatrix}, \quad \mathbf{X} = \begin{pmatrix} 1 & Z_1 & Z_2^2 & Z_1^3 \\ \vdots & \vdots & \vdots & \vdots \\ 1 & Z_n & Z_n^2 & Z_n^3 \end{pmatrix},$$

$$\mathbf{Z} = \begin{pmatrix} \{(Z_1 - \xi_1)_+\}^3 & \dots & \{(Z_1 - \xi_k)_+\}^3 \\ \vdots & \vdots & \vdots \\ \{(Z_n - \xi_1)_+\}^3 & \dots & \{(Z_n - \xi_k)_+\}^3 \end{pmatrix}.$$

To capture the ridge penalized spline, consider a linear mixed model:

$$\mathbf{Y} = \underbrace{\mathbf{X}b}_{\mathbf{X}b} + \underbrace{\mathbf{Z}c}_{\mathbf{Z}u} + \varepsilon, \tag{4.50}$$

where

$$\begin{pmatrix} c \\ \varepsilon \end{pmatrix} \sim \mathrm{N}\left(0, \begin{pmatrix} \sigma_0^2 I_k & 0 \\ 0 & \sigma_\varepsilon^2 I_n \end{pmatrix}\right),$$

i.e., $D = \sigma_0^2 I_k$ and $G = \sigma_\varepsilon^2 I_n$. By Lemma 4.3, given $\tau = (\sigma_0^2, \sigma_\varepsilon^2)$, the estimator and predictor $(\hat{b}_\tau, \hat{c}_\tau)$ are defined as

$$\begin{aligned}(\hat{b}_\tau, \hat{c}_\tau) &= \mathrm{argmin}_{(b,c)} \left\{ \frac{1}{2\sigma_\varepsilon^2} \|\mathbf{Y} - \mathbf{X}b - \mathbf{Z}c\|_2^2 + \frac{1}{2\sigma_0^2} \|c\|_2^2 \right\} \\ &= \mathrm{argmin}_{(b,c)} \left\{ \|\mathbf{Y} - \mathbf{X}b - \mathbf{Z}c\|_2^2 + \frac{\sigma_\varepsilon^2}{\sigma_0^2} \|c\|_2^2 \right\},\end{aligned}$$

which coincides with $(\hat{b}_\lambda, \hat{c}_\lambda)$ with the choice $\lambda = \frac{1}{n}\frac{\sigma_\varepsilon^2}{\sigma_0^2}$. Therefore an estimator of $\frac{1}{n}\frac{\sigma_\varepsilon^2}{\sigma_0^2}$ in the linear mixed model (4.50) can be used as an automated choice of the ridge parameter λ.

4.3.2 Numerical demonstration

We demonstrate application of linear mixed modeling to select tuning parameters when fitting ridge-penalized and smoothing splines. We return to the ozone dataset as used in Sections 3.2 and 3.5 and study univariate smoothing of `ozone` against the Dagget pressure gradient (`dpg`).

```
> data <- read.table(file="LAozone.data", header=T, sep=",")
> attach(data)
```

First, we fit ridge-penalized cubic splines with 9 knots at the $10\%, 20\%, \dots, 90\%$ quantiles of the covariate data `dpg`, using the functions `tr.pow()` and `pen.spline0()` from Section 3.2 and `lme()` from the package `nlme`. From the output below, the basis matrix `X1` at the data points (excluding the constant) is of row dimension $n = 330$ and column dimension $p = 3 + k = 12$ for $k = 9$ knots. The first 3 columns `X1[,1:3]` together with the constant form the matrix $\mathbf{X}$ and the remaining columns `X1[,4:12]` form the matrix $\mathbf{Z}$. Then `lme()` is applied with proper specification of fixed and random effects to fit linear mixed model (4.50). From `summary(ozone.lme)`, we obtain the following estimates for variance components:

$$\hat{\sigma}_0 = \texttt{117.2968}, \quad \hat{\sigma}_\epsilon = \texttt{7.369685}.$$

Then the tuning parameter `lam` (corresponding to $n\lambda$) is selected as $\frac{\hat{\sigma}_\epsilon^2}{\hat{\sigma}_0^2}$, and `pen.spline0()` is applied to fit a penalized cubic spline. The fitted values from `lme()` including the predicted random-effects (i.e., second column of `ozone.lme$fitted`) and those from `pen.spline0()` can be verified to coincide with each other.

```
> knots <- quantile(dpg, seq(.1,.9,.1))
> ends <- c(-70, 110)
> X1 <- tr.pow(dpg, degree=3, knots=knots, ends=ends)  #at data points
> dim(X1)
[1] 330  12
>
> const <- rep(1, length(ozone))
> ozone.lme <- lme(ozone ~ X1[,1:3],
+                   random = list(const=pdIdent(~X1[,4:12]-1)))  #REML
> summary(ozone.lme)
Linear mixed-effects model fit by REML
......
Random effects:
 Formula: ~X1[, 4:12] - 1 | const
 Structure: Multiple of an Identity
        X1[, 4:12]1 X1[, 4:12]2 X1[, 4:12]3 X1[, 4:12]4 X1[, 4:12]5
StdDev:    117.2968    117.2968    117.2968    117.2968    117.2968
        X1[, 4:12]6 X1[, 4:12]7 X1[, 4:12]8 X1[, 4:12]9 Residual
StdDev:    117.2968    117.2968    117.2968    117.2968 7.369685

Fixed effects: ozone ~ X1[, 1:3]
                 Value Std.Error  DF    t-value p-value
(Intercept)    3.22136   3.09842 326  1.0396790  0.2993
X1[, 1:3]1    -3.95732  33.87034 326 -0.1168373  0.9071
X1[, 1:3]2   148.60305 124.95438 326  1.1892584  0.2352
X1[, 1:3]3  -194.82253 160.38525 326 -1.2147160  0.2254
......
```

```
Number of Observations: 330
Number of Groups: 1

> lam <- (7.369685 / 117.2968)^2
> out <- pen.spline0(ozone, X1, degree=3, lam=lam)
> out$df
[1] 4.741521
>
> cbind(ozone.lme$fitted, out$fit)
          fixed      const
1     8.4756166  8.4763528  8.4763527
2     8.6635209  8.6645649  8.6645648
3    13.8847762 14.7481447 14.7481448
4    13.9748592 14.3400224 14.3400225
5     6.9947433  6.9947433  6.9947431
......
326  11.9962835 12.0346486 12.0346486
327  10.5069092 10.5166082 10.5166082
328   6.4628934  6.4628934  6.4628932
329  14.0395253 14.5194517 14.5194518
330  13.5580882 13.7405112 13.7405113
```

Figure 4.3 shows the penalized cubic spline with the tuning parameter λ selected from mixed modeling in the (2,2) plot and, for comparison, penalized cubic splines obtained with several fixed choices of the tuning parameter. The following R code illustrates how the (2,2) plot can be produced. Each fitted spline function is evaluated at the grid points `x`, not just the data points `dpg`, and then the function values are plotted.

```
> out4 <- pen.spline0(ozone, X1, degree=3, lam=2.9)
> out4$df
[1] 4.002353
>
> x <- seq(-80,120,2)
> bX1 <- tr.pow(x, degree=3, knots=knots, ends=ends)  #at grid points
>
> plot(dpg, ozone, xlim=c(-80,120), ylim=c(-5,40))
> lines(x, cbind(1,bX1)%*%out4$coef, col="blue")
> abline( v=knots, lty=2)
>
> lines(x, cbind(1,bX1)%*%out$coef, col="red", lty=2)
```

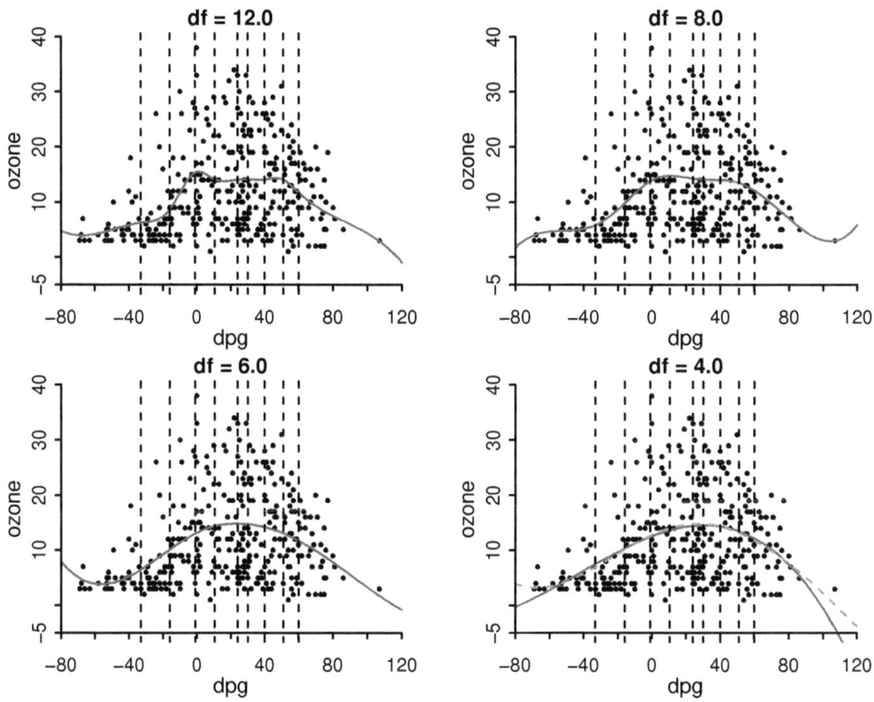

FIGURE 4.3
Plots of ridge-penalized cubic splines for `ozone` against `dpg` in the ozone dataset. For fixed choices of tuning parameter (solid lines in the four plots), the degrees of freedom are stated in the plot titles. For the selected tuning parameter from mixed modeling (dashed line in the (2,2) plot), the degrees of freedom is 4.74. Vertical lines are placed at the nine knots which are quantiles of `dpg`.

Next, we fit smoothing cubic splines using the functions `kernel.smsp()` and `sm.spline()` from Section 3.5 and `lme()` from the package `nlme`. From the output below, the kernel matrix `K` at the data points is of dimension 330×330, corresponding to the matrix $\mathbf{K}_1$ in Section 4.3.1. The matrix square root of $\mathbf{K}_1$ is computed as `rK`, using an eigenvalue decomposition of `K`. Then `lme()` is applied with proper specification of fixed and random effects to fit linear mixed model (4.49), re-parameterized from model (4.48). From `summary(ozone.lme)`, we obtain the following estimates for variance components:

$$\hat{\sigma}_0 = \texttt{71.94386}, \quad \hat{\sigma}_\epsilon = \texttt{7.350853}.$$

Then the tuning parameter `lam` (corresponding to $n\lambda$) is selected as $\frac{\hat{\sigma}_\epsilon^2}{\hat{\sigma}_0^2}$, and `sm.spline()` is applied to fit a smoothing cubic spline. The fitted values from `lme()` including the predicted random-effects (i.e., second column of `ozone.lme$fitted`) and those from `pen.spline0()` can be verified to coincide with each other.

```
> ends <- c(-70, 110)
> K <- kernel.smsp(dpg, dpg, ends=ends)[,-1]
> dim(K)
[1] 330 330
> eK <- eigen(K, symm=T)
> rK <- eK$vec %*% diag( sqrt(ifelse(eK$val <1e-16, 0, eK$val)) ) %*%
+       t(eK$vec)
>
> const <- rep(1, length(ozone))
> ozone.lme <- lme(ozone ~ dpg, random = list(const=pdIdent(~rK-1)))

> summary(ozone.lme)
Linear mixed-effects model fit by REML
 Data: NULL
      AIC      BIC    logLik
  2273.01 2288.182 -1132.505

Random effects:
 Formula: ~rK - 1 | const
 Structure: Multiple of an Identity
             rK1      rK2      rK3      rK4      rK5      rK6
StdDev: 71.94386 71.94386 71.94386 71.94386 71.94386 71.94386
......
           rK325    rK326    rK327    rK328    rK329    rK330
StdDev: 71.94386 71.94386 71.94386 71.94386 71.94386 71.94386
        Residual
StdDev: 7.350853

Fixed effects: ozone ~ dpg
               Value Std.Error  DF   t-value p-value
(Intercept) 9.389194  8.501414 328 1.1044273  0.2702
dpg         0.097046  0.146445 328 0.6626776  0.5080
 Correlation:
    (Intr)
dpg 0.977
......
Number of Observations: 330
```

```
Number of Groups: 1

> lam <- (7.350853 / 71.94386)^2
> out <- sm.spline(ozone, dpg, lam=lam, ends)
> out$df
[1] 5.231804
>
> cbind(ozone.lme$fitted, out$fit)
        fixed     const
1    6.963052  8.202975  8.202975
2    7.060097  8.399376  8.399376
3   11.815337 14.505232 14.505232
4   10.844880 14.422191 14.422191
5    6.186686  6.808542  6.808542
......
326  8.903966 12.432307 12.432307
327  8.030555 10.542545 10.542545
328  5.895549  6.362874  6.362874
329 11.136017 14.486199 14.486199
330 10.165560 14.093876 14.093876
```

Figure 4.4 shows the cubic smoothing spline with the tuning parameter λ selected from mixed modeling in the (2,2) plot and, for comparison, cubic smoothing splines obtained with several fixed choices of the tuning parameter. The following R code illustrates how the (2,2) plot can be produced. Each fitted spline function is evaluated at the grid points `x`, not just the data points `dpg`, and then the function values are plotted.

```
> out4 <- sm.spline(ozone, dpg, lam=3.7e-2, ends)
> out4$df
[1] 3.997657
>
> x <- seq(-80,120,2)
> bX <- kernel.smsp(x, dpg, ends=ends)
>
> plot(dpg, ozone, xlim=c(-80,120), ylim=c(-5,40))
> lines(x, cbind(1,bX)%*%c(out4$cL, out4$cK), col="blue")
>
> lines(x, cbind(1,bX)%*%c(out$cL, out$cK), col="red", lty=2)
```

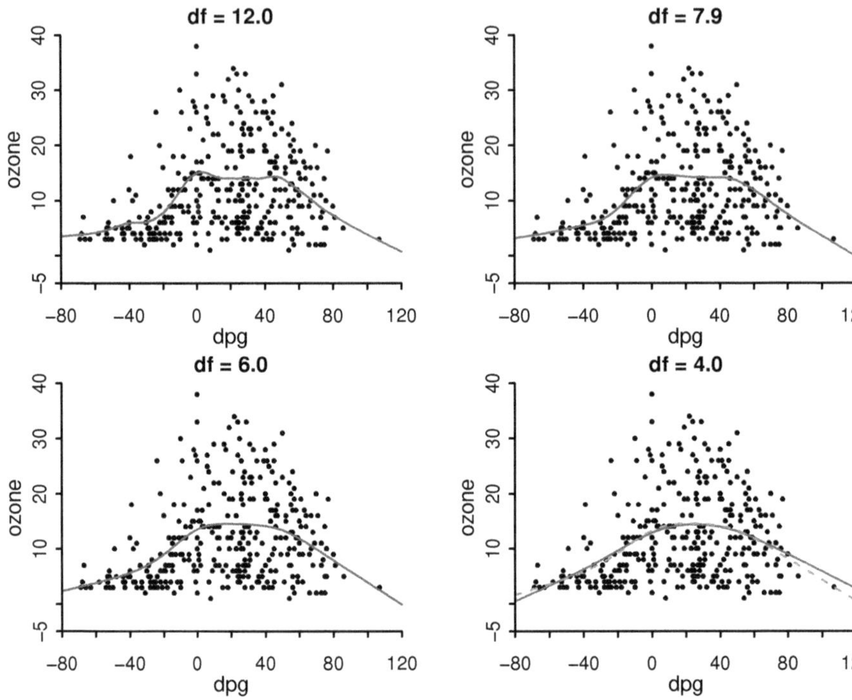

FIGURE 4.4
Plots of cubic smoothing splines for `ozone` against `dpg` in the ozone dataset. For fixed choices of tuning parameter (solid lines in the four plots), the degrees of freedom are stated in the plot titles. For the selected tuning parameter from mixed modeling (dashed line in the (2,2) plot), the degrees of freedom is 5.23.

Finally, we write the following function `my.lme()` to implement the algorithms based on the extended likelihood for fitting linear mixed models with two variance parameters $(\sigma_u^2, \sigma_\varepsilon^2)$ as discussed in Section 4.1.5. The algorithm for ML or REML estimation is defined by (4.27a) and (4.31) or by (4.28a) and (4.33) respectively. To save space, the lines for ML estimation are omitted and can be found in the accompanying R code file. The function `my.lme()` returns a list including the estimated fixed effects (`b`), the predicted random effects (`u`), the estimated variance parameters (`sigs`), the degrees of freedom (`df`) which is the value of q^k after the final iteration, convergence status (`conv`), number of iterations executed (`iter`), and the objective values along the iterations (`obj`) which are defined by the negative extended log-likelihood $h(\cdot)$ together with the adjustment term, i.e., $\frac{1}{2}\log|H_{D,G}|$ for REML estimation.

```
my.lme <- function(y, x, z, init=NULL, niter=100, eps=1e-9,
                   method="REML") {
  x <- cbind(1,x)
  xz <- cbind(x,z)
  n <- dim(x)[1]
  p <- dim(x)[2]
  m <- dim(z)[2]

  if (is.null(init))
    init <- c(1,1)

  sig.e <- init[1]
  sig.u <- init[2]
  lam <- (sig.e/sig.u)^2

  A <- diag(p+m)
  A[1:p, 1:p] <- 0

  # algorithm based on extended log-likelihood
  obj <- rep(NA, niter)

  conv <- 0
  i <- 1
  while (i< niter & conv<=0) {

    coef <- solve( t(xz)%*%xz/n+lam/n*A, t(xz)%*%y/n )
    b <- coef[1:p]
    u <- coef[p+1:m]

    if (method=="ML") {
      ......
    } else {  # REML
      df <- sum(diag( solve(t(xz)%*%xz/n+lam/n*A, t(xz)%*%xz/n) ))

      sig.e <- sqrt( sum((y-xz%*%coef)^2) /(n-df) )
      sig.u <- sqrt( sum(u^2)/(df-p) )
      lam <- (sig.e/sig.u)^2

      obj[i] <- sum((y-xz%*%coef)^2)/(2*sig.e^2)+sum(u^2)/(2*sig.u^2)+
                 (n-p-m)/2*log(sig.e^2) + m/2*log(sig.u^2) +
                 sum(log(eigen(t(xz)%*%xz+lam*A)$val))/2
```

```
      obj[i] <- obj[i]/n
    }

    if (i>1) {
      if (obj[i-1]-obj[i] < eps)
        conv <- 1
    }
    i <- i+1
  }

  list(b=b, u=u, sigs=c(sig.e,sig.u), df=df,
       conv=conv, iter=i, obj=obj[1:(i-1)])
}
```

We apply the function `my.lme()` to fit the linear mixed model associated with ridge-penalized splines as discussed earlier. From the output below, the estimated fixed effects (`b`) and variance parameters (`sigs`) match, up to several leading digits, those from `summary(ozone.lme)` obtained using `lme()`. Moreover, for REML estimation, the degrees of freedom (`df`) from `my.lme()` is 4.74, the same as that (`out$df`) from fitting a ridge-penalized spline with the tuning parameter `lam` based on the estimated $(\sigma_u^2, \hat{\sigma}_\varepsilon^2)$.

```
> my.out <- my.lme(ozone, x=X1[,1:3], z=X1[,4:12])
> my.out
$b
[1]    3.221079   -3.952846  148.585623 -194.802977

$u
[1]  8.078663 33.408815 56.268766 53.037718 36.117244 29.342306 21.836185
[8] 16.370617 11.357056

$sigs
[1]   7.369686 117.291900

$df
[1] 4.741406
......
```

Similarly, we apply the function `my.lme()` to fit the linear mixed model associated with smoothing splines as discussed in Section 4.3. The estimated fixed effects (`b`) and variance parameters (`sigs`) agree with those from `summary(ozone.lme)` obtained using `lme()`, and the degrees of freedom (`df`),

5.23, coincides with that from fitting a smoothing spline with the tuning parameter `lam` implied from the estimated $(\sigma_u^2, \hat{\sigma}_\varepsilon^2)$.

```
> my.out <- my.lme(ozone, x=dpg, z=rK)
> my.out
$b
[1] 9.38979598 0.09705751
......
$sigs
[1]  7.350856 71.940425

$df
[1] 5.231471
......
```

4.4 Generalized lincar mixed models

4.4.1 Data and model

Linear mixed models are designed for continuous, Gaussian-like response data. To accommodate other types of response data (for example, binary responses), generalized linear mixed models can be formulated by introducing random effects into generalized linear regression.

Each of Examples (i)–(iv) with grouped data in Section 4.1 can be extended such that the response variables Y_{ij} are, for instance, binary (taking value 0 or 1). For illustration, we examine the extension of Example (iv), continued in Example 4.3. Consider the following parallel-line, random-intercept logistic model:

$$\begin{aligned}
&Y_{ij}|u_i \sim \text{Bernoulli}(p_{ij}),\\
&\quad\text{with } p_{ij} = \text{expit}(b_0 + b_1 X_{ij} + u_{i0}), \quad i = 1, \dots, m, j = 1, \dots, n_i,\\
&(Y_{ij} : j = 1, \dots, n_i) \text{ given } u_i \text{ are independent of each other and}\\
&\quad\text{are independent of } (u_l : 1 \le l \le m, l \ne i), \quad i = 1, \dots, m,\\
&(u_{i0} : i = 1, \dots, m) \overset{\text{IID}}{\sim} \text{N}(0, \sigma_0^2),
\end{aligned} \tag{4.51}$$

where the covariates X_{ij}'s are scalar and fixed. As in Example 4.3, let

$$\mathbf{X} = \begin{pmatrix} \mathbf{1}_{n_1} & \mathbf{X}_1 \\ \vdots & \vdots \\ \mathbf{1}_{n_m} & \mathbf{X}_m \end{pmatrix}, \quad b = \begin{pmatrix} b_0 \\ b_1 \end{pmatrix}, \quad \mathbf{Z} = \begin{pmatrix} \mathbf{1}_{n_1} & & \\ & \ddots & \\ & & \mathbf{1}_{n_m} \end{pmatrix}, \quad u = \begin{pmatrix} u_{10} \\ \vdots \\ u_{m0} \end{pmatrix},$$

where $\mathbf{X}_i = (X_{i1}, \ldots, X_{in_i})^{\mathrm{T}}$. The logistic mixed model above can be stated as

$$\begin{aligned} \mathbf{Y}|u &\sim \text{Bernoulli}(\text{expit}(\mathbf{X}b + \mathbf{Z}u)), \\ u &\sim \text{N}(0, \sigma_0^2 I_m), \end{aligned}$$

where expit$(\cdot)$ is applied componentwise and Bernoulli$(\cdot)$ denotes the distribution of independent Bernoulli variables with respective probabilities.

In general, a generalized linear mixed model can be formulated as follows. Let $\mathbf{X} = (X_1, \ldots, X_N)^{\mathrm{T}}$ and $\mathbf{Z} = (Z_1, \ldots, Z_N)^{\mathrm{T}}$ be $N \times p$ and $N \times m$ covariate matrices, being fixed (or conditioned on), and $\mathbf{Y}$ be an $N \times 1$ response vector. [Note that this notation differs from Example 4.3 mentioned above.]

- $\mathbf{Y}$ given u satisfies the mean restriction

$$E(\mathbf{Y}|u) = g^{-}(\mathbf{X}b + \mathbf{Z}u),$$

  where b is a $p \times 1$ vector of fixed coefficients, u is an $m \times 1$ vector of random coefficients, $g^{-}(\cdot)$ is applied componentwise.

- $\mathbf{Y}$ given u is a vector of independent and identically distributed variables from an exponential family in the form

$$f(y; \theta, \phi) = \exp\left\{\frac{y\theta - a(\theta)}{\phi} + c(y, \phi)\right\},$$

  where θ and $\phi > 0$ are two scalar parameters and $a(\cdot)$ and $c(\cdot)$ are *known*, differentiable functions. By (2.6), the vector of canonical parameters, $\boldsymbol{\theta} = (\theta_1, \ldots, \theta_N)^{\mathrm{T}}$, is related to the vector of linear predictors $\boldsymbol{\eta} = \mathbf{X}b + \mathbf{Z}u$ as

$$\boldsymbol{\theta} = (a')^{-}\{g^{-}(\boldsymbol{\eta})\},$$

  where $(a')^{-}(\cdot)$ and $g^{-}(\cdot)$ are applied componentwise. For simplicity, the dispersion parameters are assumed to be equal, $\phi_i = \phi$ for $i = 1, \ldots, N$.

- $u \sim \text{N}(0, D)$, where the variance matrix D is parameterized by some variance parameters τ (which is suppressed in the notation).

Although the generalization seems conceptually straightforward, estimation and prediction in generalized linear mixed models are computationally more demanding than in linear mixed models as discussed below.

4.4.2 Estimation and prediction: Overview

We discuss likelihood-based methods for estimation and prediction. From the modeling assumptions, the joint density of $(\mathbf{Y}, u)$ is

$$f(\mathbf{y}, u; b, \phi, D)$$

$$= \exp\left\{\frac{\mathbf{y}^{\mathrm{T}}\boldsymbol{\theta} - \mathbf{1}_N^{\mathrm{T}}a(\boldsymbol{\theta})}{\phi} + \mathbf{1}_N^{\mathrm{T}}c(\mathbf{y},\phi)\right\}|2\pi D|^{-1/2}\exp\left(-\frac{u^{\mathrm{T}}D^{-1}u}{2}\right),$$

where $\mathbf{1}_N$ is an $N \times 1$ vector of ones. The marginal density of $\mathbf{Y}$ is

$$f(\mathbf{y}; b, \phi, D) = \int f(\mathbf{y}, u; b, \phi, D)\,\mathrm{d}u,$$

which is analytically intractable due to m-dimensional integration. Hence the log-likelihood function, $\ell(b, \phi, D) = \log f(\mathbf{Y}; b, \phi, D)$, is also analytically intractable. The ML estimators of (b, ϕ, D) are defined as

$$(\hat{b}, \hat{\phi}, \hat{D}) = \mathrm{argmax}_{(b,\phi,D)}\,\ell(b, \phi, D).$$

Specialized methods are needed to compute these estimators. There are three broad classes: (i) Laplace approximation and variations, (ii) Monte Carlo extensions of the Expectation-Maximization algorithm, and (ii) stochastic approximation. For the latter two methods, sampling is required from the conditional distribution $f(u|\mathbf{Y}; b, \phi, D)$ described below. We discuss Laplace approximation based methods in Section 4.4.3. For a discussion of sampling-based methods, see for example McCulloch et al. (2008) and Ruppert et al. (2003).

A point predictor of u can be defined as the mean or the mode of the conditional density $f(u|\mathbf{Y}; \hat{b}, \hat{\phi}, \hat{D})$, where

$$f(u|\mathbf{Y}; b, \phi, D) = \frac{f(\mathbf{Y}, u; b, \phi, D)}{f(\mathbf{Y}; b, \phi, D)} \propto f(\mathbf{Y}, u; b, \phi, D),$$

i.e., the conditional density admits an unnormalized density $f(\mathbf{Y}, u; b, \phi, D)$ in a closed form, but the normalizing constant $f(\mathbf{Y}; b, \phi, D)$ is analytically intractable. Hence standard optimization algorithms can be used to compute the mode of $f(u|\mathbf{Y}; b, \phi, D)$, but Monte Carlo integration is often required to compute the mean of $f(u|\mathbf{Y}; b, \phi, D)$.

Similarly as in Section 4.3, there is also a useful connection between ridge-penalized splines and smoothing splines for generalized smoothing (Section 3.8) and generalized linear mixed models. See, for example, Ruppert et al. (2003) for more information.

4.4.3 Estimation and prediction: Laplace approximation

We discuss approximate likelihood estimation based on Laplace approximation and other accompanying approximations. The log-likelihood function of (b, ϕ, D) can be expressed as $\ell(b, \phi, D) = \log \int \mathrm{e}^{h(b,u,\phi,D)}\,\mathrm{d}u$, where

$$\begin{aligned} h(b, u, \phi, D) &= \log f(\mathbf{Y}, u; b, \phi, D), \\ &= \frac{\mathbf{Y}^{\mathrm{T}}\boldsymbol{\theta} - \mathbf{1}_N^{\mathrm{T}}a(\boldsymbol{\theta})}{\phi} + \mathbf{1}_N^{\mathrm{T}}c(\mathbf{Y},\phi) - \frac{u^{\mathrm{T}}D^{-1}u}{2} - \frac{1}{2}\log|2\pi D| \quad (4.52) \end{aligned}$$

is the log joint density of $(\mathbf{Y}, u)$ and called the extended log-likelihood function of (b, u, ϕ, D). For simplicity, assume that a canonical link $g(\cdot)$ is used, satisfying $g = (a')^-$ as in (2.12). Then $\boldsymbol{\theta} = \boldsymbol{\eta} = \mathbf{X}b + \mathbf{Z}u$. As used frequently later, we denote

$$H(b,u,\phi,D) = \begin{pmatrix} H_{bb} & H_{bu} \\ H_{ub} & H_{uu} \end{pmatrix} (b,u,\phi,D) = -\begin{pmatrix} \frac{\partial^2 h}{\partial b \partial b^{\mathrm{T}}} & \frac{\partial^2 h}{\partial u \partial u^{\mathrm{T}}} \\ \frac{\partial^2 h}{\partial u \partial b^{\mathrm{T}}} & \frac{\partial^2 h}{\partial u \partial u^{\mathrm{T}}} \end{pmatrix} (b,u,\phi,D),$$

i.e., $H(b, u, \phi, D)$ is the negative Hessian of $h(b, u, \phi, D)$ with respect to (b, u). By similar calculation as in (2.21),

$$H(b,u,\phi,D) = \begin{pmatrix} \mathbf{X}^{\mathrm{T}} W_{b,u,\phi} \mathbf{X} & \mathbf{X}^{\mathrm{T}} W_{b,u,\phi} \mathbf{Z} \\ \mathbf{Z}^{\mathrm{T}} W_{b,u,\phi} \mathbf{X} & \mathbf{Z}^{\mathrm{T}} W_{b,u,\phi} \mathbf{Z} + D^{-1} \end{pmatrix},$$

where $W_{b,u,\phi}$ is the $N \times N$ diagonal matrix with ith diagonal element $\frac{1}{\phi} a''(X_i^{\mathrm{T}} b + Z_i^{\mathrm{T}} u)$ depending on (b, u, ϕ). The matrix $H(b, u, \phi, D)$ mirrors $H_{D,G}$ in Section 4.1.5, but depends on both coefficient parameters (b, u) and variance parameters (ϕ, D).

To approximate the integral $\int e^{h(b,u,\phi,D)} \, du$, let $\hat{u}_{b,\phi,D} = \operatorname{argmax}_u \; h(b, u, \phi, D)$ and consider a Taylor expansion of $h(\cdot)$ about $\hat{u}_{b,\phi,D}$:

$$h(b,u,\phi,D) \approx h(b, \hat{u}_{b,\phi,D}, \phi, D) - \frac{1}{2}(u - \hat{u}_{b,\phi,D})^{\mathrm{T}} \Lambda_{b,\phi,D} (u - \hat{u}_{b,\phi,D}),$$

where $\Lambda_{b,\phi,D} = H_{uu}(b, \hat{u}_{b,\phi,D}, \phi, D)$. Substituting the approximation into $\int e^{h(b,u,\phi,D)} \, du$ yields

$$\begin{aligned} H \int e^{h(b,u,\phi,D)} \, du &\approx e^{h(b,\hat{u}_{b,\phi,D},\phi,D)} \int e^{-\frac{1}{2}(u-\hat{u}_{b,\phi,D})^{\mathrm{T}} \Lambda_{b,\phi,D} (u-\hat{u}_{b,\phi,D})} \, du \\ &= e^{h(b,\hat{u}_{b,\phi,D},\phi,D)} \cdot \left| 2\pi \Lambda_{b,\phi,D}^{-1} \right|^{1/2}, \end{aligned}$$

where the second line uses $\int e^{-\frac{1}{2} u^{\mathrm{T}} \Sigma^{-1} u} \, du = |2\pi\Sigma|^{1/2}$ because the integration of a multivariate normal density is 1. This is known as *Laplace approximation*, which is satisfactory when the integrand decays fast to zero away for u from $\hat{u}_{b,\phi,D}$. Accordingly, $\ell(b, \phi, D) = \log \int e^{h(b,u,\phi,D)} \, du$ is approximated as

$$\begin{aligned} \ell(b,\phi,D) &\approx h(b, \hat{u}_{b,\phi,D}, \phi, D) + \log \int e^{-\frac{1}{2}(u-\hat{u}_{b,\phi,D})^{\mathrm{T}} \Lambda_{b,\phi,D} (u-\hat{u}_{b,\phi,D})} \, du \\ &= \underbrace{h(b, \hat{u}_{b,\phi,D}, \phi, D) - \frac{1}{2} \log \left| \frac{\Lambda_{b,\phi,D}}{2\pi} \right|}_{\ell_{\mathrm{appr}}(b,\phi,D)}. \end{aligned}$$

Note that $h(b, \hat{u}_{b,\phi,D}, \phi, D)$ depends on b through both arguments b and $\hat{u}_{b,\phi,D}$, and $\log |\Lambda_{b,\phi,D}|$ also depends on b. In principle, approximate ML estimators

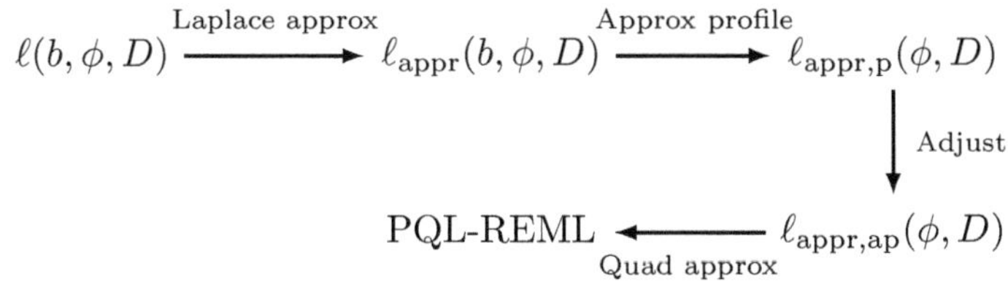

FIGURE 4.5
Approximations involved from the exact log-likelihood $\ell(b, \phi, D)$ to PQL with REML estimation (PQL-REML).

of (b, ϕ, D) can be defined by maximizing $\ell_{\text{appr}}(b, \phi, D)$, but the optimization seems computationally complex.

Alternatively, we ignore the dependency of $\log|\Lambda_{b,\phi,D}|$ on b, and define an approximate ML estimator of b given (ϕ, D) as

$$\hat{b}_{\phi,D} = \operatorname{argmax}_b \, h(b, \hat{u}_{b,\phi,D}, \phi, D).$$

and an approximate profile log-likelihood of (ϕ, D) as

$$\begin{aligned} \ell_{\text{appr,p}}(\phi, D) &= \ell_{\text{appr}}(\hat{b}_{\phi,D}, \phi, D) \\ &= h(\hat{b}_{\phi,D}, \hat{u}_{\phi,D}, \phi, D) - \frac{1}{2}\log\left|\frac{\Lambda_{\hat{b}_{\phi,D},\phi,D}}{2\pi}\right| \\ &= \left\{\max_b \, h(b, \hat{u}_{b,\phi,D}, \phi, D)\right\} - \frac{1}{2}\log\left|\frac{\Lambda_{\hat{b}_{\phi,D},\phi,D}}{2\pi}\right|, \end{aligned} \tag{4.53}$$

where $\hat{u}_{\phi,D} = \hat{u}_{\hat{b}_{\phi,D},\phi,D}$, and hence $(\hat{b}_{\phi,D}, \hat{u}_{\phi,D}) = \operatorname{argmax}_{b,u} \, h(b, u, \phi, D)$ jointly and $\Lambda_{\hat{b}_{\phi,D},\phi,D} = H_{uu}(\hat{b}_{\phi,D}, \hat{u}_{\phi,D}, \phi, D)$. Note that $\ell_{\text{appr,p}}(\phi, D) \neq \max_b \, \ell_{\text{appr}}(b, \phi, D)$, in spite of the subscript $_{\text{p}}$ indicating "profile". The resulting approximate ML estimator of (ϕ, D) is

$$(\hat{\phi}, \hat{D}) = \operatorname{argmax}_{\phi,D} \, \ell_{\text{appr,p}}(\phi, D),$$

and the approximate ML estimator of b is $\hat{b}_{\hat{\phi},\hat{D}}$, i.e., $\hat{b}_{\phi,D}$ with $(\phi, D) = (\hat{\phi}, \hat{D})$. Therefore, there are two approximations involved in the preceding derivation: Laplace approximation and the ignorance of the dependency of $\log|\Lambda_{b,\phi,D}|$ on b. See Figure 4.5.

For a potential improvement, the approximate profile log-likelihood of (ϕ, D) can be adjusted by analogy with the adjusted profile likelihood of variance parameters in linear mixed models (Section 4.1.3). Ignoring the log determinant term in (4.53), $\ell_{\text{appr,p}}(\phi, D)$ is obtained by maximizing $h(b, \hat{u}_{b,\phi,D}, \phi, D)$ over b, and hence can be adjusted by further subtracting half of the log determinant of the negative Hessian of $h(b, \hat{u}_{b,\phi,D}, \phi, D)$ in

b evaluated at the mode $\hat{b}_{\phi,D}$. This yields the approximate adjusted profile log-likelihood of (ϕ, D):

$$\begin{aligned}
&\ell_{\text{appr,ap}}(\phi, D) \\
&= \left\{ h(b, \hat{u}_{b,\phi,D}, \phi, D) - \frac{1}{2} \log\det\left(-\frac{\partial^2 h(b, \hat{u}_{b,\phi,D}, \phi, D)}{\partial b \partial b^{\mathrm{T}}} \right) \right\} \Big|_{b=\hat{b}_{\phi,D}} \\
&\quad - \frac{1}{2} \log \left| \frac{\Lambda_{\hat{b}_{\phi,D},\phi,D}}{2\pi} \right|. \qquad (4.54)
\end{aligned}$$

The resulting estimator of (ϕ, D), called an approximate REML estimator, is

$$(\tilde{\phi}, \tilde{D}) = \operatorname{argmax}_{\phi,D}\, \ell_{\text{appr,p}}(\phi, D),$$

and the approximate REML estimator of b is $\hat{b}_{\tilde{\phi},\tilde{D}}$, i.e., $\hat{b}_{\phi,D}$ with $(\phi, D) = (\tilde{\phi}, \tilde{D})$.

Remarkably, there is a simplification for the expression (4.54), in spite of the dependency of $h(b, \hat{u}_{b,\phi,D}, \phi, D)$ on b through both arguments b and $\hat{u}_{b,\phi,D}$. It can be shown that (Exercise 4.12)

$$\begin{aligned}
&- \frac{\partial^2 h(b, \hat{u}_{b,\phi,D}, \phi, D)}{\partial b \partial b^{\mathrm{T}}} \Big|_{b=\hat{b}_{\phi,D}} \\
&= (H_{bb} - H_{bu} H_{uu}^{-1} H_{ub})(b, u, \phi, D) \Big|_{(b,u)=(\hat{b}_{\phi,D}, \hat{u}_{\phi,D})}.
\end{aligned}$$

In addition, note that $\Lambda_{\hat{b}_{\phi,D},\phi,D} = H_{uu}(\hat{b}_{\phi,D}, \hat{u}_{\phi,D}, \phi, D)$. Then the approximate adjusted profile log-likelihood of (ϕ, D) in (4.54) is up to an additive constant,

$$\begin{aligned}
&\ell_{\text{appr,ap}}(\phi, D) \\
&= \left\{ h(b, u, \phi, D) - \frac{1}{2} \log |H_{bb} - H_{bu} H_{uu}^{-1} H_{ub}| - \frac{1}{2} \log |H_{uu}| \right\} \Big|_{(b,u)=(\hat{b}_{\phi,D}, \hat{u}_{\phi,D})} \\
&= \left\{ h(b, u, \phi, D) - \frac{1}{2} \log |H| \right\} \Big|_{(b,u)=(\hat{b}_{\phi,D}, \hat{u}_{\phi,D})}, \qquad (4.55)
\end{aligned}$$

where $H = H(b, u, \phi, D)$, and the third line follows because $|H| = |H_{uu}| \cdot |H_{bb} - H_{bu} H_{uu}^{-1} H_{ub}|$, evaluated at $(b, u) = (\hat{b}_{\phi,D}, \hat{u}_{\phi,D})$ (Exercise 4.12). Therefore, $\ell_{\text{appr,ap}}(\phi, D)$ is obtained by maximizing the extended log-likelihood $h(\cdot)$ over (b, u) and then subtracting $\frac{1}{2} \log |H|$, half of the log determinant of the negative Hessian of $h(\cdot)$ in (b, u) at the mode $(\hat{b}_{\phi,D}, \hat{u}_{\phi,D})$. This is similar to (4.24) for the adjusted profile log-likelihood of (D, G) in linear mixed models, with a difference that $h(\cdot)$ is quadratic in (b, u) and hence has a constant Hessian. See Exercise 4.13 for another (more direct) derivation of $\ell_{\text{appr,ap}}(\phi, D)$ in (4.55) by applying the Laplace approximation to the integrated likelihood with a flat prior on b.

To summarize, the estimator and predictor $(\hat{b}, \hat{u}) = (\hat{b}_{\hat{\phi},\hat{D}}, \hat{u}_{\hat{\phi},\hat{D}})$ and the approximate ML estimator $(\hat{\phi}, \hat{D})$ can be jointly determined from an alternating optimization:

$$\begin{aligned} (\hat{b}, \hat{u}) &= \text{argmax}_{b,u}\, h(b, u, \hat{\phi}, \hat{D}), \\ (\hat{\phi}, \hat{D}) &= \text{argmax}_{\phi,D} \left\{ h(\hat{b}, \hat{u}, \phi, D) - \frac{1}{2} \log |H_{uu}(\hat{b}, \hat{u}, \phi, D)| \right\}. \end{aligned} \tag{4.56}$$

The estimator and predictor $(\tilde{b}, \tilde{u}) = (\hat{b}_{\tilde{\phi},\tilde{D}}, \hat{u}_{\tilde{\phi},\tilde{D}})$ and the approximate REML estimator $(\tilde{\phi}, \tilde{D})$ can be jointly determined from an alternating optimization:

$$\begin{aligned} (\tilde{b}, \tilde{u}) &= \text{argmax}_{b,u}\, h(b, u, \tilde{\phi}, \tilde{D}), \\ (\tilde{\phi}, \tilde{D}) &= \text{argmax}_{\phi,D} \left\{ h(\tilde{b}, \tilde{u}, \phi, D) - \frac{1}{2} \log |H(\tilde{b}, \tilde{u}, \phi, D)| \right\}. \end{aligned} \tag{4.57}$$

In general, each alternating optimization problem cannot be represented as a single (joint) optimization problem such as (4.25) or (4.25) in linear mixed models. Nevertheless, the preceding definitions lead to alternating updating algorithms which resemble the block-coordinate updates (4.27a)–(4.27b) and (4.28a)–(4.28b) in linear mixed models.

In the case of ML estimation, an alternating update algorithm for solving (4.56) proceeds as follows in updating (b, u) or (ϕ, D).

- Pick an initial value (ϕ^0, D^0).
- Compute for $k = 1, 2, \cdots,$

$$(b^k, u^k) = \text{argmax}_{b,u}\, h(b, u, \phi^{k-1}, D^{k-1}), \tag{4.58a}$$

$$(\phi^k, D^k) = \text{argmax}_{\phi,D} \left\{ h(b^k, u^k, \phi, D) - \frac{1}{2} \log |H_{uu}(b^k, u^k, \phi, D)| \right\}, \tag{4.58b}$$

  until some termination criterion is met.

In the case of REML estimation, the iterations for solving (4.57) are

$$(b^k, u^k) = \text{argmax}_{b,u}\, h(b, u, \phi^{k-1}, D^{k-1}), \tag{4.59a}$$

$$(\phi^k, D^k) = \text{argmax}_{\phi,D} \left\{ h(b^k, u^k, \phi, D) - \frac{1}{2} \log |H(b^k, u^k, \phi, D)| \right\}, \tag{4.59b}$$

The update (4.58a) or (4.59a) can be solved by fitting ridge-penalized generalized linear regression with the initial values of (b, u) set to (b^{k-1}, u^{k-1}) or, for computational efficiency, replaced by a single Newton–Raphson update. The update (4.58b) or (4.59b) can be either fully solved or replaced by some partial updates.

A further (third) approximation can be made by relating the updates (4.58a)–(4.58b) and (4.59a)–(4.59b) to fitting linear mixed models. For simplicity, we discuss only the REML updates with a Newton–Raphson update in (4.59a), and similar ideas can be applied to the ML updates. As discussed in Sections 2.2 and 2.5, a Newon–Raphson update in (4.59a) for fitting ridge-penalized generalized linear regression can be expressed as solving a ridge-penalized weighted least-squares problem. For each i, the ith working response is

$$\tilde{Y}_i(b,u) = X_i^{\mathrm{T}} b + Z_i^{\mathrm{T}} u + \frac{Y_i - \mu_i(b,u)}{V_i(b,u)},$$

where $\mu_i(b,u) = a'(X_i^{\mathrm{T}} b + Z_i^{\mathrm{T}} u) = g^{-}(X_i^{\mathrm{T}} b + Z_i^{\mathrm{T}} u)$ and $V_i(b,u) = a''(X_i^{\mathrm{T}} b + Z_i^{\mathrm{T}} u)$. Let $\tilde{\mathbf{Y}}(b,u)$ be the working response vector with the ith element $\tilde{Y}_i(b,u)$, and $W_{b,u,\phi}$ be the diagonal weight matrix as before. Then the quadratic approximation by the Taylor expansion of $h(b,u,\phi,D)$ about (b^{k-1},u^{k-1}) with fixed (ϕ,D) is

$$\begin{aligned} h_Q(b,u,\phi,D;b^{k-1},u^{k-1}) = &-\frac{1}{2}\left\| W^{1/2}_{b^{k-1},u^{k-1},\phi}(\tilde{\mathbf{Y}}(b^{k-1},u^{k-1}) - \mathbf{X}b - \mathbf{Z}u)\right\|_2^2 \\ &- \frac{u^{\mathrm{T}} D^{-1} u}{2} - \frac{1}{2}\log|2\pi D| + \text{constant}. \end{aligned} \tag{4.60}$$

Performing a Newton–Raphson update in (4.59a) (which is equivalent to replacing $h(b,u,\phi,D)$ by $h_Q(b,u,\phi,D;b^{k-1},u^{k-1})$ in (4.59a)) and further replacing $h(b,u,\phi,D)$ by $h_Q(b,u,\phi,D;b^k,u^k)$ in (4.59b) yields the iterations:

$$\begin{aligned} (b^k,u^k) &= \mathrm{argmax}_{b,u}\ h_Q(b,u,\phi^{k-1},D^{k-1};b^{k-1},u^{k-1}), \\ (\phi^k,D^k) &= \mathrm{argmax}_{\phi,D}\left\{h_Q(b^k,u^k,\phi,D;b^k,u^k) - \frac{1}{2}\log|H(b^k,u^k,\phi,D)|\right\}. \end{aligned}$$

Then it can be shown that $(b^{k+1},u^{k+1},\phi^k,D^k)$ are jointly determined as

$$\begin{aligned} &(b^{k+1},u^{k+1},\phi^k,D^k) \\ &= \mathrm{argmax}_{b,u,\phi,D}\left\{h_Q(b,u,\phi,D;b^k,u^k) - \frac{1}{2}\log|H(b^k,u^k,\phi,D)|\right\}, \end{aligned} \tag{4.61}$$

and $H(b^k,u^k,\phi,D)$ is the (constant) negative Hessian of the quadratic function $h_Q(b,u,\phi,D;b^k,u^k)$ in (b,u). In other words, $(b^{k+1},u^{k+1},\phi^k,D^k)$ corresponds to exactly REML estimation and prediction (4.26) in the "working" linear mixed model at (b^k,u^k):

$$\tilde{\mathbf{Y}}(b^k,u^k) = \mathbf{X}b + \mathbf{Z}u + \varepsilon,$$

with the variance matrices D and $G = W^{-1}_{b^k,u^k,\phi}$ for u and ε respectively. The approximate method of iteratively fitting linear mixed models in (4.61) is commonly called the penalized quasi-likelihood (PQL) method with REML estimation, and can be derived from various perspectives for fitting generalized linear mixed models (Schall, 1991; Breslow & Clayton, 1993; Wolfinger & O'Connell, 1993; McGilchrist, 1994). See Figure 4.5 for a summary of approximations involved.

For nonlinear mixed models, the PQL-REML estimators and predictor, $(\tilde{b}_Q, \tilde{u}_Q, \tilde{\phi}_Q, \tilde{D}_Q)$, defined by the limit of $(b^{k+1}, u^{k+1}, \phi^k, D^k)$ in (4.61), in general differ from the estimators and predictor $(\tilde{b}, \tilde{u}, \tilde{\phi}, \tilde{D})$ defined by the limit of (b^k, u^k, ϕ^k, D^k) in (4.59a)–(4.59b) (which satisfy (4.57)). [Question: Can $(\tilde{b}_Q, \tilde{u}_Q, \tilde{\phi}_Q, \tilde{D}_Q)$ be formally characterized by optimizing an objective function?] However, when ϕ is known and ϕ^k is dropped from the iterations as in fitting logistic mixed models, the estimators and predictor $(\tilde{b}_Q, \tilde{u}_Q, \tilde{D}_Q)$ become the same as $(\tilde{b}, \tilde{u}, \tilde{D})$, because replacing $h(b, u, \phi, D)$ by $h_Q(b, u, \phi, D; b^k, u^k)$ in (4.59b) does not alter the optimization over D.

For illustration, consider the logistic mixed model:

$$\begin{aligned} \mathbf{Y}|u &\sim \text{Bernoulli}(\text{expit}(\mathbf{X}b + \mathbf{Z}u)), \\ u &\sim \text{N}(0, \sigma_u^2 I_m). \end{aligned} \tag{4.62}$$

The extended log-likelihood in (4.52) is, with $\phi = 1$ (known),

$$\begin{aligned} &h(b, u, \sigma_u^2) \\ &= \mathbf{Y}^{\mathrm{T}}(\mathbf{X}b + \mathbf{Z}u) - \mathbf{1}_N^{\mathrm{T}} \log(1 + \mathrm{e}^{\mathbf{X}b+\mathbf{Z}u}) - \frac{\|u\|_2^2}{2\sigma_u^2} - \frac{m}{2}\log(\sigma_u^2) + \text{constant}. \end{aligned} \tag{4.63}$$

The weight matrix $W_{b,u}$ is diagonal with ith diagonal element $W(b, u)_{ii} = \text{expit}(X_i^{\mathrm{T}} b + Z_i^{\mathrm{T}} u)(1 - \text{expit}(X_i^{\mathrm{T}} b + Z_i^{\mathrm{T}} u))$. From the general expression (4.60) with $\phi = 1$, the quadratic approximation to (4.63) anchored at (b^{k-1}, u^{k-1}) is

$$\begin{aligned} h_Q(b, u, \sigma_u^2; b^{k-1}, u^{k-1}) = &-\frac{1}{2}\left\| W_{b^{k-1},u^{k-1}}^{1/2}(\tilde{\mathbf{Y}}(b^{k-1}, u^{k-1}) - \mathbf{X}b - \mathbf{Z}u)\right\|_2^2 \\ &- \frac{\|u\|_2^2}{2\sigma_u^2} - \frac{m}{2}\log(\sigma_u^2) + \text{constant}, \end{aligned} \tag{4.64}$$

where $\tilde{\mathbf{Y}}(b, u)$ is the working response vector with ith element

$$\tilde{Y}_i(b, u) = X_i^{\mathrm{T}} b + Z_i^{\mathrm{T}} u + \frac{Y_i - \text{expit}(X_i^{\mathrm{T}} b + Z_i^{\mathrm{T}} u)}{W(b, u)_{ii}}.$$

To obtain (b^k, u^k), a Newon–Raphson update in (4.58a) or (4.59a) is equivalent to optimizing (4.64) over (b, u), i.e., solving a ridge-penalized weighted least squares problem. Algebraically, (4.64) is the extended log-likelihood in the "working" linear mixed model at (b^k, u^k) is

$$\tilde{\mathbf{Y}}(b^k, u^k) = \mathbf{X}b + \mathbf{Z}u + \varepsilon,$$

with the variance matrices $D = \sigma_u^2 I_m$ and $G = W_{b^k,u^k}^{-1}$ for u and ε respectively.

For ML estimation, the gradient of the objective function in (4.58b), denoted as $\ell^k_{\text{appr,p}}(\sigma_u^2)$, can be calculated as (Exercise 4.14)

$$\frac{\partial \ell^k_{\text{appr,p}}}{\partial \sigma_u^2} = \frac{\|u^k\|_2^2}{2\sigma_u^4} - \frac{m}{2\sigma_u^2} + \frac{1}{2\sigma_u^2}\text{tr}\{I_m - (\mathbf{Z}^{\text{T}}W^k\mathbf{Z} + \sigma_u^{-2}I_m)^{-1}\mathbf{Z}^{\text{T}}W^k\mathbf{Z}\}. \tag{4.65}$$

where $W^k = W_{b^k,u^k}$. Solving the gradient equation, but with σ_u^2 inside tr(·) replaced by $(\sigma_u^2)^{k-1}$, leads to a simple, partial update:

$$(\sigma_u^2)^k = \frac{\|u^k\|_2^2}{q^{k-1}}, \tag{4.66}$$

where $q^{k-1} = \text{tr}\{(\mathbf{Z}^{\text{T}}W^k\mathbf{Z} + (\sigma_u^2)^{k-1}I_m)^{-1}\mathbf{Z}^{\text{T}}W^k\mathbf{Z}\}$. Similarly, for REML estimation, the gradient of the objective function in (4.59b) or (4.61), denoted as $\ell^k_{\text{appr,ap}}(\sigma_u^2)$, can be calculated as (Exercise 4.14)

$$\begin{aligned}&\frac{\partial \ell^k_{\text{appr,ap}}}{\partial \sigma_u^2}\\ &= \frac{\|u^k\|_2^2}{2\sigma_u^4} - \frac{m}{2\sigma_u^2} + \frac{1}{2\sigma_u^2}\text{tr}\{I_{p+m} - (\mathbf{C}^{\text{T}}W^k\mathbf{C} + \sigma_u^{-2}\text{diag}(0_p, I_m))^{-1}\mathbf{C}^{\text{T}}W^k\mathbf{C}\},\end{aligned} \tag{4.67}$$

where $\mathbf{C} = (\mathbf{X}, \mathbf{Z})$. Solving the gradient equation, but with σ_u^2 inside tr(·) replaced by $(\sigma_u^2)^{k-1}$, leads to a simple, partial update:

$$(\sigma_u^2)^k = \frac{\|u^k\|_2^2}{-p + q^{k-1}}, \tag{4.68}$$

where $q^{k-1} = \text{tr}\{(\mathbf{C}^{\text{T}}W^k\mathbf{C} + (\sigma_u^2)^{k-1}\text{diag}(0_p, I_m))^{-1}\mathbf{C}^{\text{T}}W^k\mathbf{C}\}$.

The quadratic approximation (4.64) to the extended log-likelihood can be seemingly expanded by introducing a free dispersion parameter ϕ:

$$\begin{aligned}&h_Q(b, u, \phi, \sigma_u^2; b^{k-1}, u^{k-1})\\ &= -\frac{1}{2}\left\|(\phi^{-1}W_{b^{k-1},u^{k-1}})^{1/2}(\tilde{\mathbf{Y}}(b^{k-1}, u^{k-1}) - \mathbf{X}b - \mathbf{Z}u)\right\|_2^2\\ &\quad - \frac{N}{2}\log(\phi) - \frac{\|u\|_2^2}{2\sigma_u^2} - \frac{m}{2}\log(\sigma_u^2) + \text{constant},\end{aligned} \tag{4.69}$$

which corresponds to the extended log-likelihood in the "working" linear mixed model

$$\tilde{\mathbf{Y}}(b^k, u^k) = \mathbf{X}b + \mathbf{Z}u + \varepsilon,$$

with the variance matrices $D = \sigma_u^2 I_m$ and $G = \phi W^{-1}_{b^k,u^k}$ for u and ε respectively. Substituting (4.69) into the iteration (4.61) leads to the PQL-REML method for fitting the logistic mixed model with an additional dispersion parameter ϕ, which can be thought to account for over- or under-dispersion (McCullagh & Nelder, 1989). The objective function (4.69) is a weighted version of the extended log-likelihood function (4.29) in a linear mixed model with variance parameters $(\sigma_\varepsilon^2, \sigma_u^2)$. As a result, similar partial updates for (ϕ, σ_u^2) can be derived as (4.31) or (4.33) for ML or REML estimation respectively.

4.4.4 Approximate model-based inference

The approximate methods in Section 4.4.3 deal with point estimation and prediction. The associated approximations can also be exploited to conduct model-based inference. Assume that the generalized linear mixed model is correctly specified with fixed $(\mathbf{X}, \mathbf{Z})$. As in Section 4.4.3, assume that a canonical link is used.

We first discuss approximate inference about (b, u) given (ϕ, D). Although several estimators of (ϕ, D) are presented in Section 4.4.3, the point estimator and predictor of (b, u) given (ϕ, D) are always defined as

$$(\hat{b}_{\phi,D}, \hat{u}_{\phi,D}) = \text{argmax}_{b,u}\, h(b, u, \phi, D),$$

where (b, u) above denotes possible values to be optimized over. By mimicking asymptotic properties of M-estimation (e.g. Tan, 2025), we expect that

$$\begin{aligned}
\begin{pmatrix} \hat{b}_{\phi,D} - b \\ \hat{u}_{\phi,D} - u \end{pmatrix} &\approx E^{-1}\{H(b, u, \phi, D)\} \frac{\partial}{\partial(b, u)} h(b, u, \phi, D) \\
&= E^{-1}\{H(b, u, \phi, D)\} \begin{pmatrix} \frac{1}{\phi}\mathbf{X}^{\mathrm{T}}(\mathbf{Y} - a'(\mathbf{X}b + \mathbf{Z}u)) \\ \frac{1}{\phi}\mathbf{Z}^{\mathrm{T}}(\mathbf{Y} - a'(\mathbf{X}b + \mathbf{Z}u)) - D^{-1}u \end{pmatrix} \qquad (4.70\text{a}) \\
&= E^{-1}\{H(b, u, \phi, D)\} \begin{pmatrix} \frac{1}{\phi}\mathbf{X}^{\mathrm{T}}\boldsymbol{\varepsilon} \\ \frac{1}{\phi}\mathbf{Z}^{\mathrm{T}}\boldsymbol{\varepsilon} - D^{-1}u \end{pmatrix} \\
&\overset{\text{approx}}{\sim} \mathrm{N}(0, E^{-1}\{H(b, u, \phi, D)\}), \qquad (4.70\text{b})
\end{aligned}$$

where $\boldsymbol{\varepsilon} = \mathbf{Y} - a'(\mathbf{X}b + \mathbf{Z}u)$, and (b, u) above denotes the underlying fixed and random effects. The last line is expected from the central limit theorem and the expressions

$$E\begin{pmatrix} \frac{1}{\phi}\mathbf{X}^{\mathrm{T}}\boldsymbol{\varepsilon} \\ \frac{1}{\phi}\mathbf{Z}^{\mathrm{T}}\boldsymbol{\varepsilon} - D^{-1}u \end{pmatrix} = 0, \qquad (4.71\text{a})$$

$$\text{var}\begin{pmatrix} \frac{1}{\phi}\mathbf{X}^{\mathrm{T}}\boldsymbol{\varepsilon} \\ \frac{1}{\phi}\mathbf{Z}^{\mathrm{T}}\boldsymbol{\varepsilon} - D^{-1}u \end{pmatrix} = E\{H(b, u, \phi, D)\}. \qquad (4.71\text{b})$$

The expressions above can be verified using the model assumption that $(Y_1, \ldots, Y_n)$ given u are independent and identically distributed, and Y_i has mean $a'(X_i^{\mathrm{T}}b + Z_i^{\mathrm{T}}u)$ and variance $\phi a''(X_i^{\mathrm{T}}b + Z_i^{\mathrm{T}}u)$ (Exercise 4.15).

From the approximate distribution (4.70b) and estimation of the variance matrix $E^{-1}\{H(b,u,\phi,D)\})$ by $H^{-1}(\hat{b}_{\phi,D},\hat{u}_{\phi,D},\phi,D)$, an approximate prediction interval of size $1-c$ for $x^{\mathrm{T}}b+z^{\mathrm{T}}u$ with fixed covariate vectors (x,z) is

$$(x^{\mathrm{T}}\hat{b}_{\phi,D}+z^{\mathrm{T}}\hat{u}_{\phi,D}) \pm z_{c/2}\sqrt{(x^{\mathrm{T}},z^{\mathrm{T}})H^{-1}(\hat{b}_{\phi,D},\hat{u}_{\phi,D},\phi,D)(x^{\mathrm{T}},z^{\mathrm{T}})^{\mathrm{T}}}, \quad (4.72)$$

which is expected to satisfy

$$P\big(x^{\mathrm{T}}b+z^{\mathrm{T}}u \text{ is contained in the interval (4.72)}\big) \approx 1-c.$$

The prediction interval (4.72) is algebraically the same as the prediction interval (4.38) in the "working" linear mixed model at $(\hat{b}_{\phi,D},\hat{u}_{\phi,D})$:

$$\tilde{\mathbf{Y}}(\hat{b}_{\phi,D},\hat{u}_{\phi,D}) = \mathbf{X}b+\mathbf{Z}u+\varepsilon,$$

with the variance matrices D and $G=W^{-1}_{\hat{b}_{\phi,D},\hat{u}_{\phi,D},\phi}$ for u and ε respectively. Applying the inverse link function $g^{-}(\cdot)=a'(\cdot)$ to the two end points in (4.72) yields a prediction interval for the expected response with covariate vectors (x,z).

With unknown (ϕ,D), the point estimator and predictor of (b,u) are defined as $(\hat{b}_{\phi,D},\hat{u}_{\phi,D})$ by replacing (ϕ,D) with the approximate ML estimator $(\hat{\phi},\hat{D})$, REML estimator $(\tilde{\phi},\tilde{D})$, or further approximate ones such as $(\tilde{\phi}_Q,\tilde{D}_Q)$. In each case, provided the estimator of (ϕ,D) is consistent, the additional variation caused by estimation of (ϕ,D) is expected to be negligible due to the following orthogonality:

$$\begin{aligned} &E\left\{\frac{\partial}{\partial\tau}\frac{\partial}{\partial(b,u)}h(b,u,\phi,D)\right\} \\ &= E\left\{\frac{\partial}{\partial\tau}\begin{pmatrix}\frac{1}{\phi}\mathbf{X}^{\mathrm{T}}(\mathbf{Y}-a'(\mathbf{X}b+\mathbf{Z}u)) \\ \frac{1}{\phi}\mathbf{Z}^{\mathrm{T}}(\mathbf{Y}-a'(\mathbf{X}b+\mathbf{Z}u))-D^{-1}u\end{pmatrix}\right\}=0,\end{aligned}$$

where τ denotes the variance parameters associated with (ϕ,D). This is similar to the discussion about the orthogonality (4.39) in linear mixed models. As a result, the prediction interval (4.72) with the estimated (ϕ,D) is expected to be also approximately of size $1-c$.

4.5 Numerical demonstration: Logistic mixed models

We demonstrate numerical implementation of the penalized quasi-likelihood (PQL) method for fitting logistic mixed models in R. Among existing R packages, the function `glmmPQL()` in the package `MASS` (Ripley et al., 2020) implements PQL with default estimation of the dispersion ϕ, whereas `glmer()` in the package `lme4` (Bates et al., 2020) uses adaptive Gauss–Hermite approximation to the log-likelihood of fixed effects and variance parameters (which can be more accurate than Laplace approximation). We provide self-contained code to directly fit logistic mixed model (4.62) with known dispersion $\phi = 1$ as discussed in Section 4.4.3. In the accompanying R code file, we also provide R code to fit logistic mixed model (4.62) with estimation of ϕ, which produces results numerically matching those from `glmmPQL()` in our example.

Consider the parallel-line, random-intercept logistic model (4.51) with the following specification for $i = 1, \ldots, m\,(= 10), j = 1, \ldots, n\,(= 15)$:

$$\begin{aligned} P(Y_{ij} = 1|u_i) &= \text{expit}(b_0 + b_1 X_{ij} + u_{i0}), \\ X_{ij} &\sim \text{N}(j/n, 0.1^2), \\ u_{i0} &\sim \text{N}(0, \sigma_0^2). \end{aligned} \tag{4.73}$$

The model is modified from the simulation study in McCulloch (1997), where $X_{ij} = j/n$. We add Gaussian noises to covariates X_{ij}, to avoid the situation where $(X_{ij} : j = 1, \ldots, n)$ are the same across different groups $i = 1, \ldots, m$, which tends to yield few distinct intercept estimates in subsequent analysis (see Exercise 4.16). We simulate data from the model with $(b_0, b_1) = (0, 2)$ and $\sigma_0^2 = 0.5$, as shown in the R output below. For safe replication, the data frame `sim.data` is also saved as an accompanying data file.

```
> set.seed(1)
>
> m <- 10
> n <- 15
>
> xmat <- matrix(rep((1:n)/n,m) +
+           rnorm(m*n,0,0.1), ncol=m,nrow=n) #randomly perturbed
> xmat <- apply(xmat,2,sort)
> x <- c(xmat) # x arranged as x11,x21,...,x12,x22,...
>
> u <- rnorm(m,0,sqrt(0.5))    # true value of sig.u = sqrt(0.5)
> b <- c(0,2)       # true value of b = (0,2)
> eta <- b[1]+b[2]*x + rep(u,each=n)
>
> y <- rbinom(m*n,1,prob = expit(eta))    # true value of sig.e = 1
> ymat <- matrix(y,ncol=m,nrow=n)
> ymat
      [,1] [,2] [,3] [,4] [,5] [,6] [,7] [,8] [,9] [,10]
 [1,]    0    1    1    1    0    0    0    0    0     1
 [2,]    1    1    1    0    0    1    1    1    0     1
 [3,]    1    1    0    1    1    0    1    0    0     0
```

```
 [4,]    0    1    1    1    1    0    1    0    1     1
 [5,]    0    0    0    1    0    0    1    0    0     1
 [6,]    0    0    1    1    0    1    1    0    0     1
 [7,]    0    1    0    0    1    1    1    1    0     1
 [8,]    1    1    1    0    0    0    1    1    0     1
 [9,]    1    1    0    1    1    0    1    1    1     1
[10,]    1    1    1    1    0    0    1    1    1     1
[11,]    0    1    1    1    1    0    1    0    0     0
[12,]    1    1    1    1    1    1    1    1    1     1
[13,]    1    1    0    0    0    0    1    1    1     1
[14,]    1    1    1    1    1    1    1    1    1     1
[15,]    1    1    1    1    1    0    1    0    1     1
>
> zlab <- as.factor( rep(1:m,each=n) )
> z <- model.matrix(~ -1+zlab)
>
> sim.data <- data.frame(y, x, zlab, u=u[zlab])
```

For fixed-effect analysis, we fit a parallel-line logistic regression model of `y` on `x` with the grouping variable `zlab`, i.e., for $i = 1, \ldots, m\,(= 10), j = 1, \ldots, n\,(= 15)$,

$$P(Y_{ij} = 1) = \text{expit}(\beta_{i0} + \beta_1 X_{ij}).$$

From `summary(sim.glm)` below, the intercepts $\beta_{10}, \beta_{20}, \ldots, \beta_{m0}$ associated with the $m = 10$ groups are estimated to be $-0.7274, 0.8263, \ldots, 1.0008$, and the common slope β_1 is estimated to be 2.1559. Figure 4.6 shows the fitted values at the linear-predictor scale (`sim.eta`) and at the probability scale (`sim.fitted`).

```
> sim.glm <- glm(y ~ -1+x +zlab, data=sim.data,
+             family=binomial(link="logit"))  #removing the intercept
> summary(sim.glm)
......
Coefficients:
       Estimate Std. Error z value Pr(>|z|)
x       2.15589    0.70776   3.046  0.00232 **
zlab1  -0.72740    0.66007  -1.102  0.27046
zlab2   0.82634    0.83237   0.993  0.32083
zlab3  -0.40909    0.66431  -0.616  0.53802
zlab4  -0.08129    0.69054  -0.118  0.90629
zlab5  -1.04416    0.66175  -1.578  0.11459
zlab6  -1.92465    0.71591  -2.688  0.00718 **
```

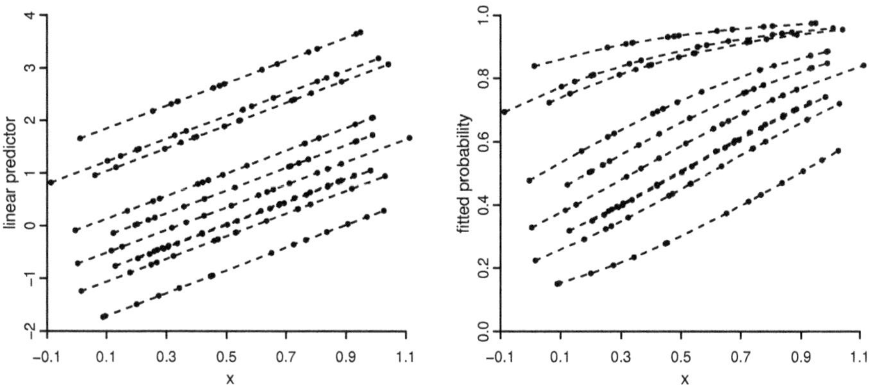

FIGURE 4.6
Fitted values against covariates within groups from a fixed-effect, parallel-line logistic model, at the linear-predictor scale (left) and at the probability scale (right).

```
zlab7    1.62998    1.08482    1.503  0.13296
zlab8   -1.06502    0.66194   -1.609  0.10763
zlab9   -1.27604    0.65870   -1.937  0.05272 .
zlab10   1.00825    0.81299    1.240  0.21491
......
>
> sim.eta <- as.vector(cbind(x,z)%*%sim.glm$coef)
> sim.fitted <- expit(sim.eta)
```

We write the following function `logit.mixed()` to implement the PQL method for fitting logistic mixed model (4.62) with known dispersion $\phi = 1$ as discussed in Section 4.4.3. The ML or REML update for the variance parameter is defined by (4.66) or by (4.68) respectively. It is helpful to note that `logit.mixed()` is similar to the function `my.lme()` for fitting linear mixed models in Section 4.3.2. The function `logit.mixed()` returns a list including the estimated fixed effects (`b`), the predicted random effects (`u`), the estimated variance parameter (`sig.u`), the degrees of freedom (`df`) which is the value of q^k after the final iteration, convergence status (`conv`), number of iterations executed (`iter`), and the L_∞ norm of the difference vector between the successive updates of fixed effects before termination.

```
logit.mixed <- function(y, x, z, init=NULL, niter=100, eps=1e-9,
                        method="REML") {
  x <- cbind(1,x)
  xz <- cbind(x,z)
```

```
n <- dim(x)[1]
p <- dim(x)[2]
m <- dim(z)[2]

if (is.null(init))
  init <- c(1, rep(0,p+m))

sig.u <- init[1]
coef <- init[-1]

lam <- (1/sig.u)^2
eta <- as.vector(xz%*%coef)
w <- expit(eta)*(1-expit(eta))
y.til <- eta + (y-expit(eta))/w

A <- diag(p+m)
A[1:p, 1:p] <- 0

#
conv <- 0
i <- 1
while (i< niter & conv<=0) {
  coef.old <- coef
  coef <- solve( t(xz)%*%(w*xz)/n+lam/n*A, t(xz)%*%(w*y.til)/n )
  b <- coef[1:p]
  u <- coef[p+1:m]

  eta <- as.vector(xz%*%coef)
  w <- expit(eta)*(1-expit(eta))
  y.til <- eta + (y-expit(eta))/w

  if (method=="ML") {
    df <- sum(diag(solve(t(z)%*%(w*z)/n+lam/n*diag(m), t(z)%*%(w*z)/n)))
    sig.u <- sqrt( sum(u^2)/df )
    lam <- (1/sig.u)^2
  } else {  # REML
    df <- sum(diag(solve(t(xz)%*%(w*xz)/n+lam/n*A, t(xz)%*%(w*xz)/n)))
    sig.u <- sqrt( sum(u^2)/(df-p) )
    lam <- (1/sig.u)^2
  }
```

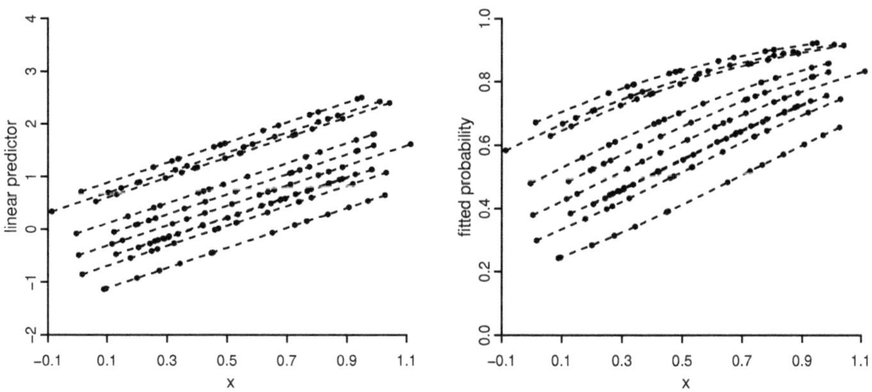

FIGURE 4.7
Fitted values against covariates within groups from a mixed-effect, parallel-line logistic model, at the linear-predictor scale (left) and at the probability scale (right).

```
    diff <- max(abs(coef-coef.old))
    if (diff < eps)
      conv <- 1
    i <- i+1
  }

  list(b=b, u=u, sig.u=sig.u, df=df, conv=conv, iter=i, diff=diff)
}
```

We apply the function `glm.mixed()` to fit the parallel-line, random-intercept logistic model to the simulated data. From the output below, ML estimation give the following estimates

$$\hat{b}_0 = -0.29, \quad \hat{b}_1 = 1.88, \quad \hat{\sigma}_0 = 0.74,$$

whereas REML estimation give the following estimates

$$\hat{b}_0 = -0.29, \quad \hat{b}_1 = 1.90, \quad \hat{\sigma}_0 = 0.82.$$

Figure 4.7 shows the within-group, REML fitted (more precisely, predicted) values at the linear-predictor scale (`my.eta`) and at the probability scale (`my.fitted`). These fitted lines and curves are shrunk toward some center (less spread out), compared with those in Figure 4.6 from fixed-effect analysis.

```
> logit.mixed(y, x, z, method="ML")
$b
```

```
[1] -0.2865842  1.8798833


$u
 [1] -0.193421297  0.652812923  0.007037265  0.204284807 -0.399011294
 [6] -0.947920017  0.907083387 -0.412654124 -0.554936430  0.736724781


$sig.u
[1] 0.7416778
......
$diff
[1] 5.840926e-10


>
> my.out <- logit.mixed(y, x, z, method="REML")
> my.out
$b
[1] -0.2904899  1.9024582


$u
 [1] -0.211386076  0.702120757  0.002714238  0.214379099 -0.430210828
 [6] -1.015056107  0.984149497 -0.444731391 -0.595454204  0.793475015


$sig.u
[1] 0.816128
......
$diff
[1] 8.628773e-10


>
> my.eta <- cbind(1,x)%*%my.out$b + my.out$u[zlab]
> my.fitted <- expit(my.eta)
```

Figure 4.8 (left) directly shows how the estimated intercepts $(\hat{\beta}_{10}, \ldots, \hat{\beta}_{m0})$ from the fixed-effect, parallel-line logistic model are shrunk toward $(\hat{b}_0 + \hat{u}_{10}, \ldots, \hat{b}_0 + \hat{u}_{m0})$ based on REML estimation from the mixed-effect, parallel-line logistic model. As the data are simulated from a known distribution, Figure 4.8 (right) plots the two sets of estimated intercepts against the underlying ("true") intercepts $(b_0 + u_{10}, \ldots, b_0 + u_{m0})$. The mean squared error from the mixed-effect analysis is much smaller than that from the fixed-effect analysis.

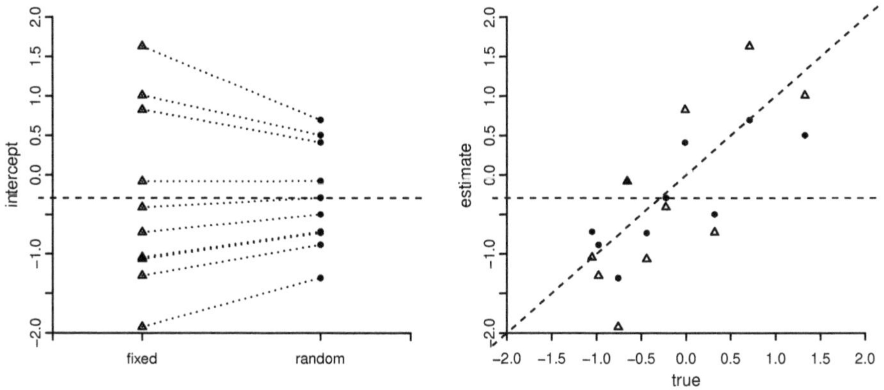

FIGURE 4.8
Plot of the estimated intercepts from the fixed-effect, parallel-line logistic model and from the mixed-effect, parallel-line logistic model, labeled as "fixed" or "random" (left) and plot of the estimated intercepts (fixed △ and random •) against the true intercepts (right).

```
> inter.mat <- cbind(sim.glm$coef[1+1:10],
+                    my.out$b[1] + my.out$u)
>
> mean((inter.mat[,1] -(b[1]+u))^2)
[1] 0.4949392
> mean((inter.mat[,2] -(b[1]+u))^2)
[1] 0.2368501
```

4.6 Appendices

4.6.1 Shrinkage estimation of normal means

As mentioned in Sectinon 4.1.1, estimation and prediction in mixed models can be related to empirical Bayes estimation. For a random-mean model, both are connected with James–Stein shrinkage estimation of normal means. We describe these connections and the results of James and Stein (1961), which are ground-breaking and can be seen to signify potential advantages of mixed-model analysis, empirical Bayes estimation, and related techniques.

Consider the random-mean model with $b_0 = 0$ fixed in Example 4.1, i.e.,

$$\begin{aligned} &Y_{ij} = u_i + \varepsilon_{ij}, \quad i = 1, \ldots, m, j = 1, \ldots, n_i, \\ &(\varepsilon_{ij} : i = 1, \ldots, m, j = 1, \ldots, n_i) \overset{\text{IID}}{\sim} \mathrm{N}(0, \sigma_\varepsilon^2), \\ &(u_i : i = 1, \ldots, m) \overset{\text{IID}}{\sim} \mathrm{N}(0, \sigma_u^2), \end{aligned} \tag{4.74}$$

where ε_{ij}'s and u_i's are independent. Assume further that $n_1 = \cdots = n_m = n\,(\geq 2)$. From Example 4.1, given $(\sigma_u^2, \sigma_\varepsilon^2)$, the (optimal) prediction of u_i is

$$\hat{u}_i(\sigma_u^2, \sigma_\varepsilon^2) = \frac{n\sigma_u^2}{\sigma_\varepsilon^2 + n\sigma_u^2}\bar{Y}_i = \left(1 - \frac{\sigma_\varepsilon^2}{\sigma_\varepsilon^2 + n\sigma_u^2}\right)\bar{Y}_i, \quad i = 1, \ldots, m.$$

The prediction MSE of $\hat{u}_i(\sigma_u^2, \sigma_\varepsilon^2)$ for u_i is, by Lemma 4.2,

$$E[\{\hat{u}_i(\sigma_u^2, \sigma_\varepsilon^2) - u_i\}^2] = \frac{\sigma_u^2\sigma_\varepsilon^2}{\sigma_\varepsilon^2 + n\sigma_u^2}, \tag{4.75}$$

which, for any $\sigma_u^2, \sigma_\varepsilon^2 > 0$, is smaller than the MSE of the group mean $\bar{Y}_i$, $E\{(\bar{Y}_i - u_i)^2\} = \frac{\sigma_\varepsilon^2}{n}$. The ML estimators of $(\sigma_u^2, \sigma_\varepsilon^2)$, with the constraint $\sigma_u^2 \geq 0$ ignored, are determined by

$$\hat{\sigma}_\varepsilon^2 = \frac{\mathrm{SSW}}{m(n-1)} = \frac{\sum_{i=1}^m \sum_{j=1}^n (Y_{ij} - \bar{Y}_i)^2}{m(n-1)}, \tag{4.76a}$$

$$\widehat{\sigma_\varepsilon^2 + n\sigma_u^2} = \frac{\mathrm{SSB}}{m} = \frac{n\sum_{i=1}^m \bar{Y}_i^2}{m}, \tag{4.76b}$$

which are unbiased estimators of σ_ε^2 and $\sigma_\varepsilon^2 + n\sigma_u^2$ respectively under model (4.74). The prediction of u_i with estimated $(\sigma_u^2, \sigma_\varepsilon^2)$ is

$$\hat{u}_i = \left(1 - \frac{m(\hat{\sigma}_\varepsilon^2/n)}{\sum_{s=1}^m \bar{Y}_s^2}\right)\bar{Y}_i, \quad i = 1, \ldots, m.$$

Similarly as in Example 4.1, this represents a linear shrinkage of the group mean $\bar{Y}_i$. The shrinkage coefficient, however, depends on all group means $(\bar{Y}_1, \ldots, \bar{Y}_m)$ as well as the estimated noise variance $\hat{\sigma}_\varepsilon^2$. For large m, by the law of large numbers, the prediction MSE of $\hat{u}_i$ for u_i is expected to be close to that of $\hat{u}_i(\sigma_u^2, \sigma_\varepsilon^2)$ in (4.75).

In the case of known σ_ε^2, the random-mean model (4.74) can be interpreted as a Bayesian model: independently over $i = 1, \ldots, n$,

$$\text{data:} \quad (Y_{ij} : j = 1, \ldots, n_i) \mid u \overset{\text{IID}}{\sim} \mathrm{N}(u_i, \sigma_\varepsilon^2), \tag{4.77a}$$

$$\text{prior:} \quad u_i \sim \mathrm{N}(0, \sigma_u^2), \tag{4.77b}$$

where $u = (u_1, \ldots, u_m)^{\mathrm{T}}$ are data parameters, and σ_u^2 is a prior parameter. Then the posterior mean (or Bayes rule) of u_i is easily shown to be $\hat{u}_i(\sigma_u^2, \sigma_\varepsilon^2)$. With $\sigma_\varepsilon^2 + n\sigma_u^2$ estimated by (4.76b), the estimated Bayes rule of u_i is

$$\hat{u}_i(\sigma_\varepsilon^2) = \left(1 - \frac{m(\sigma_\varepsilon^2/n)}{\sum_{s=1}^m \bar{Y}_s^2}\right)\bar{Y}_i, \quad i = 1, \ldots, m,$$

hence called an empirical Bayes estimator of u_i. For large m, by the law of large numbers, the prediction MSE of $\hat{u}_i(\sigma_\varepsilon^2)$ for u_i, or equivalently the Bayes risk of $\hat{u}_i(\sigma_\varepsilon^2)$, is expected to be close to that of $\hat{u}_i(\sigma_u^2, \sigma_\varepsilon^2)$ in (4.75).

The class of James–Stein estimators of u is $\delta_c = (\delta_{1,c}, \dots, \delta_{m,c})^{\mathrm{T}}$ with

$$\delta_{i,c} = \left(1 - \frac{c(\hat{\sigma}_\varepsilon^2/n)}{\sum_{s=1}^m \bar{Y}_s^2}\right) \bar{Y}_i, \quad i = 1, \dots, m,$$

where $c \geq 0$ is a constant. The estimator $\hat{u} = (\hat{u}_1, \dots, \hat{u}_m)^{\mathrm{T}}$ above corresponds to δ_c with the choice $c = m$. In the case of known σ_ε^2, the class of James–Stein estimators is $\delta_c(\sigma_\varepsilon^2) = (\delta_{1,c}(\sigma_\varepsilon^2), \dots, \delta_{m,c}(\sigma_\varepsilon^2))^{\mathrm{T}}$ with

$$\delta_{i,c}(\sigma_\varepsilon^2) = \left(1 - \frac{c(\sigma_\varepsilon^2/n)}{\sum_{s=1}^m \bar{Y}_s^2}\right) \bar{Y}_i, \quad i = 1, \dots, m.$$

The empirical Bayes estimator $\hat{u}(\sigma_\varepsilon^2) = (\hat{u}_1(\sigma_\varepsilon^2), \dots, \hat{u}_m(\sigma_\varepsilon^2))^{\mathrm{T}}$ above corresponds to $\delta_c(\sigma_\varepsilon^2)$ with the choice $c = m$.

James and Stein (1961) studied the (frequentist) risk of $\delta_c(\sigma_\varepsilon^2)$ with known σ_ε^2 and that of δ_c with unknown σ_ε^2 under the data model (4.77a) only, regardless of (4.77b). The risk of an estimator $\delta = (\delta_1, \dots, \delta_m)^{\mathrm{T}}$ for u is defined as

$$R(\delta, u) = \sum_{i=1}^m E\{(\delta_i - u_i)^2 | u\},$$

where conditioning on u indicates the expectation being taken under model (4.77a), and the dependency of $R(\delta, u)$ is suppressed in the notation. For example, the risk of group means $\bar{Y}_{1:m} = (\bar{Y}_1, \dots, \bar{Y}_m)^T$ is $R(\hat{Y}_{1:m}, u) = m(\sigma_\varepsilon^2/n)$, independent of u. The findings of James and Stein (1961) can be summarized as follows, based on the sampling properties that under model (4.77a),

$$\begin{aligned}
&\bar{Y}_i \sim \mathrm{N}(u_i, \frac{\sigma_\varepsilon^2}{n}) \quad \text{independently over } i = 1, \dots, m, \\
&m(n-1)\hat{\sigma}_\varepsilon^2 \sim \sigma_\varepsilon^2 \chi^2_{m(n-1)}, \\
&(\bar{Y}_1, \dots, \bar{Y}_m) \text{ and } \hat{\sigma}_\varepsilon^2 \text{ are independent.}
\end{aligned}$$

See Fourdrinier et al. (2018), Chapter 2, for an accessible exposition of James–Stein estimators.

Proposition 4.1 (Risk of James–Stein estimator). *Consider the data model (4.77a) with $m \geq 3$. Let $\sigma^2 = \sigma_\varepsilon^2/n$ and $k = m(n-1)$.*

(i) With known σ_ε^2, the risk of $\delta_c(\sigma_\varepsilon^2)$ is

$$R(\delta_c(\sigma_\varepsilon^2), u) = m\sigma^2 + \sigma^4 E\left\{\frac{c(c - 2(m-2))}{\sum_{s=1}^m \bar{Y}_s^2} | u\right\}.$$

Hence $\delta_c(\sigma_\varepsilon^2)$ dominates $\bar{Y}_{1:m}$, $R(\delta_c(\sigma_\varepsilon^2), u) < m\sigma^2$, if $0 < c < 2(m-2)$. The optimal choice of c in minimizing $R(\delta_c(\sigma_\varepsilon^2), u)$ uniformly over $u \in \mathbb{R}^m$ is $c = m - 2$.

(ii) With unknown σ_ε^2, the risk of δ_c is

$$R(\delta_c, u) = m\sigma^2 + \sigma^4 E\left\{ \frac{c(\frac{k+2}{k}c - 2(m-2))}{\sum_{s=1}^m \bar{Y}_s^2} \Big| u \right\}.$$

Hence δ_c dominates $\bar{Y}_{1:m}$, $R(\delta_c, u) < m\sigma^2$, if $0 < c < \frac{k}{k+2}2(m-2)$. The optimal choice of c in minimizing $R(\delta_c, u)$ uniformly over $u \in \mathbb{R}^m$ is $c = \frac{k}{k+2}(m-2)$.

Proposition 4.1 shows that for suitable c, the James–Stein estimator δ_c or $\delta_c(\sigma_\varepsilon^2)$ achieves a smaller risk than the group means $\bar{Y}_{1:m}$, which are the ML estimators of u under model (4.77a), uniformly over $u \in \mathbb{R}^m$ provided $m \geq 3$. In particular, with known σ_ε^2, the empirical Bayes estimator $\hat{u}(\sigma_\varepsilon^2)$ achieves a smaller risk than $\bar{Y}_{1:m}$ provided $m > 4$. Such improvement over ML estimation was surprising at the time of James and Stein (1961).

Proposition 4.1 can also be used to quantify the Bayes risk of the James–Stein estimators under (4.77a)–(4.77b). Given a prior π on u, the Bayes risk of δ is $E(\delta, \pi) = E\{R(\delta, u)\} = \sum_{i=1}^m E\{(\delta_i - u_i)^2\}$, namely the sum of prediction MSEs of δ_i for u_i over $i = 1, \ldots, m$ under mixed model (4.74). Under Bayesian model (4.77a)–(4.77b), it can be shown that $E(\frac{m-2}{\sum_{s=1}^m \bar{Y}_s^2}) = \frac{1}{\sigma_u^2 + \sigma_\varepsilon^2/n}$ and hence from Proposition 4.1 (i) with known σ_ε^2,

$$\begin{aligned} R(\delta_{m-2}(\sigma_\varepsilon^2), \pi) &= m\frac{\sigma_u^2\sigma_\varepsilon^2}{\sigma_\varepsilon^2 + n\sigma_u^2} + 2\frac{\sigma_\varepsilon^4/n}{\sigma_\varepsilon^2 + n\sigma_u^2} \\ &= R(\hat{u}(\sigma_u^2, \sigma_\varepsilon^2), \pi) + 2\frac{\sigma_\varepsilon^4/n}{\sigma_\varepsilon^2 + n\sigma_u^2}, \end{aligned}$$

where the second line follows from the expression (4.75). Therefore, the Bayes risk of $\delta_{m-2}(\sigma_\varepsilon^2)$ or the aggregate prediction MSE over components of u is close that of the Bayes rule $\hat{u}(\sigma_u^2, \sigma_\varepsilon^2)$, which is linear in m, by a difference independent of m.

The preceding discussion assumes equal group sizes, $n_1 = \cdots = n_m$. All group means $\bar{Y}_i$ have equal variances and are shrunk by the same coefficient. For unequal group sizes, there are subtle issues about how group means are shrunk differently (i.e., by different coefficients) depending on their variances. See Morris and Lysy (2012) and Tan (2015) for further information.

Exercises

Ex 4.1. For $n \geq 2$, let $Y_1, \ldots, Y_n$ be IID from $\mathrm{N}(\mu, \sigma^2)$, with unknown $\theta = (\mu, \sigma^2)$.

(i) Compute the conditional density, $p(y_2, \ldots, y_n | \bar{y}; \theta)$, of $(Y_2, \ldots, Y_n)$ given $\bar{Y} = \bar{y}$, where $\bar{y} = n^{-1} \sum_{i=1}^n y_i$. Show that $p(y_2, \ldots, y_n | \bar{y}; \theta)$ depends on σ^2 only, and hence can be denoted as $p(y_2, \ldots, y_n | \bar{y}; \sigma^2)$.

(ii) Show that the conditional likelihood $p(Y_2, \ldots, Y_n | \bar{Y}; \sigma^2)$ as a function of σ^2 is maximized at $\tilde{\sigma}^2 = (n-1)^{-1} \sum_{i=1}^n (Y_i - \bar{Y})^2$.

Ex 4.2. In the context of Section 4.1.3 (restricted ML estimation), show that

$$
\begin{aligned}
&-\frac{1}{2}(\mathbf{y} - \mathbf{X}b)^{\mathrm{T}} V_{D,G}^{-1} (\mathbf{y} - \mathbf{X}b) + \frac{1}{2}(\hat{b} - b)^{\mathrm{T}} (\mathbf{X}^{\mathrm{T}} V_{D,G}^{-1} \mathbf{X})(\hat{b} - b) \\
&= -\frac{1}{2}(\mathbf{y} - \mathbf{X}\hat{b})^{\mathrm{T}} V_{D,G}^{-1} (\mathbf{y} - \mathbf{X}\hat{b}),
\end{aligned}
$$

provided $\hat{b} = (\mathbf{X}^{\mathrm{T}} V_{D,G}^{-1} \mathbf{X})^{-1} (\mathbf{X}^{\mathrm{T}} V_{D,G}^{-1} \mathbf{y})$.

Ex 4.3.

(i) For linear mixed model (4.5)–(4.6), the integrated log-likelihood function of (D, G) with a flat prior on b, $f(b) \propto 1$, is defined as

$$
\ell_{\text{int}}(D, G) = \log \int \mathrm{e}^{\ell(b, D, G)} \, \mathrm{d}b = \log \int f(\mathbf{Y}; b, D, G) \, \mathrm{d}b,
$$

where $\ell(b, D, G) = \log f(\mathbf{Y}; b, D, G)$ as in (4.7). Directly evaluate the integral above and show that $\ell_{\text{int}}(D, G) = \ell_{\text{ap}}(D, G)$, the adjusted profile log-likelihood of (D, G) in (4.10). In contrast, the profile log-likelihood of (D, G) is obtained by maximization as $\max_b \ell(b, D, G)$.

(ii) The integrated log-likelihood function of (D, G) can also be expressed in terms of the extended log-likelihood $h(\cdot)$ as

$$
\ell_{\text{int}}(D, G) = \log \int \mathrm{e}^{h(b, u, D, G)} \, \mathrm{d}b\mathrm{d}u = \log \int f(\mathbf{Y}, u; b, D, G) \, \mathrm{d}b\mathrm{d}u,
$$

where $h(b, u, D, G) = \log f(\mathbf{Y}, u; b, D, G)$ as in (4.17). Directly evaluate the integral above and show that this yields the expression (4.24) for the adjusted profile log-likelihood $\ell_{\text{ap}}(D, G)$ in terms of the extended log-likelihood $h(\cdot)$.

Ex 4.4. In Section 4.1.5 (extended likelihood), Lemma 4.3 is proved using the method of substitution: first solving (4.18b) for u given b and then solving (4.18a) for b.

(i) Explain whether and, if yes, how the result can be proved by first solving (4.18a) for b given u and then solving (4.18b) for u.

(ii) Provide an alternative proof of Lemma 4.3 by calculating the right-hand side of (4.20).

Ex 4.5.

(i) Verify the gradient expressions stated in (4.30) and (4.32).

[Hint: Use the fact, known as Jacob's formula, that for an invertible matrix $A(\theta)$ parameterized by θ, $\frac{\partial}{\partial\theta}\log|A(\theta)| = \mathrm{tr}\{A(\theta)^{-1}\frac{\partial}{\partial\theta}A(\theta)\}$.]

(ii) Show that the expressions in (4.30) with (b^k, u^k) replaced by $(\hat{b}_{D,G}, \hat{u}_{D,G})$ give the gradients $\frac{\partial \ell_{\mathrm{p}}}{\partial \sigma_\varepsilon^2}$ and $\frac{\partial \ell_{\mathrm{p}}}{\partial \sigma_u^2}$, and those in (4.32) with (b^k, u^k) replaced by $(\hat{b}_{D,G}, \hat{u}_{D,G})$ give the gradients $\frac{\partial \ell_{\mathrm{ap}}}{\partial \sigma_\varepsilon^2}$ and $\frac{\partial \ell_{\mathrm{ap}}}{\partial \sigma_u^2}$, where $\ell_{\mathrm{p}} = \ell_{\mathrm{p}}(\sigma_u^2, \sigma_\varepsilon^2)$ and $\ell_{\mathrm{ap}} = \ell_{\mathrm{ap}}(\sigma_u^2, \sigma_\varepsilon^2)$.

Ex 4.6. In the context of Section 4.1.6 (model-based inference), show that $\mathrm{var}(\hat{u}_{D,G} - u)$ can be decomposed as in (4.37).

Ex 4.7. In the context of Example 4.1 (random-mean model), show that

$$V_i^{-1} = \sigma_\epsilon^{-2}(I_{n_i} - \frac{\sigma_0^2}{\sigma_\varepsilon^2 + n_i\sigma_0^2}\mathbf{1}_{n_i \times n_i}),$$
$$|V_i| = \sigma_\varepsilon^{2(n_i-1)}(\sigma_\varepsilon^2 + n_i\sigma_0^2).$$

Ex 4.8. Consider Example 4.1 (random-mean model).

(i) Calculate the gradients $(\frac{\partial \ell_{\mathrm{p}}}{\partial \sigma_\varepsilon^2}, \frac{\partial \ell_{\mathrm{p}}}{\partial \sigma_0^2})$ and $(\frac{\partial \ell_{\mathrm{ap}}}{\partial \sigma_\varepsilon^2}, \frac{\partial \ell_{\mathrm{ap}}}{\partial \sigma_0^2})$ using the explicit expressions of ℓ_{p} and ℓ_{ap} shown in Example 4.1, and verify that the gradient expressions obtained are the same as described in Exercise 4.5 (ii).

(ii) Simplify the expressions of updates (4.27a) and (4.31), and study the behavior of the algorithm defined by (4.27a) and (4.31) in the special case $n_1 = \cdots = n_m = n$. In particular, what does the algorithm converge to, if any, when $\hat{\sigma}_0^2 < 0$?

(iii) Simplify the expressions of updates (4.28a) and (4.33), and study the behavior of the algorithm defined by (4.28a) and (4.33) in the special case $n_1 = \cdots = n_m = n$. In particular, what does the algorithm converge to, if any, when $\tilde{\sigma}_0^2 < 0$?

Ex 4.9. Consider the parallel-line, random-intercept model in Example 4.3:

$$Y_{ij} = b_0 + b_1 X_{ij} + u_{i0} + \varepsilon_{ij}, \quad i = 1, \ldots, m, j = 1, \ldots, n_i,$$

where (b_0, b_1) are fixed coefficients, and $u = (u_{10}, \ldots, u_{m0})^{\mathrm{T}}$ and $\epsilon = (\epsilon_{11}, \ldots, \epsilon_{1,n_1}, \ldots, \epsilon_{m1}, \ldots, \epsilon_{m,n_m})^{\mathrm{T}}$ are random with the joint distribution,

$$\begin{pmatrix} u \\ \epsilon \end{pmatrix} \sim \mathrm{N}\left(0, \begin{pmatrix} \sigma_u^2 I_m & 0 \\ 0 & \sigma_\epsilon^2 I_N \end{pmatrix}\right).$$

For fixed $(\sigma_u^2, \sigma_\epsilon^2)$, the ML estimator of (b_0, b_1) and the prediction of u are calculated in Example 4.3.

(i) Derive the profile log-likelihood function and the score equation for ML estimation of $(\sigma_u^2, \sigma_\epsilon^2)$. Provide your answer in the simplest form possible.

(ii) Derive the *adjusted* profile log-likelihood function and the estimating equation for $(\sigma_u^2, \sigma_\epsilon^2)$. Provide your answer in the simplest form possible.

(iii) In the special case where all the group sizes are equal, i.e., $n_i = n$ for $i = 1, \ldots, m$, simplify your answers in (i) and (ii) as much as possible.

Ex 4.10. For the dataset `petrol.data` with noise added to the response `Y`, the parallel-line, random-intercept model as in Exercise 4.9 is fitted using `lme()` in Section 4.2.

(i) Provide a contour plot, using `contour()` in R, of the profile log-likelihood function for $(\sigma_u^2, \sigma_\epsilon^2)$ possibly in the log scale, and verify that the contour has a maximum at the ML estimates of $(\sigma_u^2, \sigma_\epsilon^2)$ from `lme()` as reported in Section 4.2.

(ii) Provide a contour plot of the *adjusted* profile log-likelihood function for $(\sigma_u^2, \sigma_\epsilon^2)$ possibly in the log scale, and verify that the contour has a maximum at the REML estimates of $(\sigma_u^2, \sigma_\epsilon^2)$ from `lme()` as reported in Section 4.2.

(iii) Write your own R program to compute the estimates of (b_0, b_1) and predicted values of $(u_{01}, \ldots, u_{0m})$, given the REML estimates of $(\sigma_u^2, \sigma_\epsilon^2)$. Show that you can replicate the corresponding results in Section 4.2.

(iv) Write your own R program to compute the ML and REML estimates of $(\sigma_u^2, \sigma_\epsilon^2)$, using the algorithms based on the extended likelihood in Section 4.1.5.

Ex 4.11. For the dataset `LAozone.data`, a ridge-penalized cubic spline is fitted with the tuning parameter λ selected through fitting linear mixed model (4.50) in Section 4.3.2.

(i) Provide a contour plot of the *adjusted* profile log-likelihood function for $(\sigma_u^2, \sigma_\epsilon^2)$ possibly in the log scale, and verify the REML estimates of $(\sigma_u^2, \sigma_\epsilon^2)$ from `lme()` in Section 4.3.2. You do not need to compute the REML estimates of $(\sigma_u^2, \sigma_\epsilon^2)$.

(ii) Find the estimate of the tuning parameter λ, based on the leave-one-out cross validation. See equation (3.48) in Section 3.9.1. Plot the criterion function for choosing λ, and also plot the fitted curve to the data, along with that based on REML estimates of $(\sigma_u^2, \sigma_\epsilon^2)$.

(iii) Find the estimate of the tuning parameter λ based on the generalized cross validation (GCV). See equation (3.51) in Section 3.9.1. Plot the criterion function for choosing λ, and also plot the fitted curve to the data, along with that based on REML estimates of $(\sigma_u^2, \sigma_\epsilon^2)$.

Ex 4.12. Let $g(b, u)$ be a twice-differentiable function such as $-h(b, u, \phi, D)$ with fixed (ϕ, D) in (4.52). In addition, let

$$H(b,u) = \begin{pmatrix} H_{bb} & H_{bu} \\ H_{ub} & H_{uu} \end{pmatrix}(b,u) = \begin{pmatrix} \frac{\partial^2 g}{\partial b \partial b^{\mathrm{T}}} & \frac{\partial^2 g}{\partial u \partial u^{\mathrm{T}}} \\ \frac{\partial^2 g}{\partial u \partial b^{\mathrm{T}}} & \frac{\partial^2 g}{\partial u \partial u^{\mathrm{T}}} \end{pmatrix}(b,u).$$

(i) For each b, let $\hat{u}_b = \operatorname{argmax}_u g(b, u)$. Show that for any b,

$$\frac{\partial g(b, \hat{u}_b)}{\partial b} = \frac{\partial g(b,u)}{\partial b}\bigg|_{u=\hat{u}_b},$$

$$\frac{\partial^2 g(b, \hat{u}_b)}{\partial b \partial b^{\mathrm{T}}} = \left\{ \frac{\partial^2 g(b,u)}{\partial b \partial b^{\mathrm{T}}} \quad \left(\frac{\partial g(b,u)}{\partial b \partial u^{\mathrm{T}}}\right)\left(\frac{\partial g(b,u)}{\partial u \partial u^{\mathrm{T}}}\right)^{-1} \left(\frac{\partial g(b,u)}{\partial u \partial b^{\mathrm{T}}}\right) \right\}\bigg|_{u=\hat{u}_b}$$

$$= (H_{bb} - H_{bu} H_{uu}^{-1} H_{ub})(b,u)\bigg|_{u=\hat{u}_b},$$

provided that $H_{uu}(b, \hat{u}_b)$ is nonsingular.

(ii) Show that for any b,

$$|H(b, \hat{u}_b)| = |H_{uu}(b, \hat{u}_b)| \cdot \left| \frac{\partial^2 g(b, \hat{u}_b)}{\partial b \partial b^{\mathrm{T}}} \right|,$$

provided that $H_{uu}(b, \hat{u}_b)$ is nonsingular.

[Hint: For any (b, u),

$$|H(b,u)| = |H_{uu}(b,u)| \cdot |(H_{bb} - H_{bu} H_{uu}^{-1} H_{ub})(b,u)|.$$

See a general matrix result in Exercise 1.4.]

Ex 4.13. For a generalized linear mixed model, the integrated log-likelihood function of (ϕ, D) with a flat prior on b, $f(b) \propto 1$, is defined as

$$\ell_{\text{int}}(\phi, D) = \log \int e^{h(b,u,\phi,D)}\, db du = \log \int f(\mathbf{Y}, u; b, \phi, D)\, db du,$$

where $h(b, u, \phi, D) = \log f(\mathbf{Y}, u; b, \phi, D)$ as in (4.52). Apply the Laplace approximation directly to the integral above and show that this yields the approximate adjusted profile log-likelihood $\ell_{\text{appr,ap}}(\phi, D)$ in (4.55).

Ex 4.14. Verify the gradient expressions stated in (4.65) and (4.67).

Ex 4.15. Verify the mean and variance stated in (4.71a)–(4.71b).

Ex 4.16. Consider a parallel-line, random-intercept logistic model as in (4.73) but with $X_{ij} = j/n$. Simulate data from the model with $(b_0, b_1) = (0, 2)$ and $\sigma_0^2 = 0.5$ as follows.

```
> set.seed(1)
>
> m <- 10
> n <- 15
>
> xmat <- matrix(rep((1:n)/n,m), ncol=m,nrow=n)
> x <- c(xmat) # x arranged as x11,x21,...,x12,x22,...
>
> u <- rnorm(m,0,sqrt(0.5))    # true value of sig.u = sqrt(0.5)
> b <- c(0,2)        # true value of b = (0,2)
> eta <- b[1]+b[2]*x + rep(u,each=n)
>
> y <- rbinom(m*n,1,prob = expit(eta))   # true value of sig.e = 1
> ymat <- matrix(y,ncol=m,nrow=n)
> ymat
      [,1] [,2] [,3] [,4] [,5] [,6] [,7] [,8] [,9] [,10]
 [1,]    1    0    0    1    1    1    0    1    0     0
 [2,]    0    0    1    1    0    0    0    1    1     0
 [3,]    1    1    0    1    1    0    1    1    1     1
 [4,]    1    0    0    1    1    1    1    1    1     1
 [5,]    1    1    1    0    0    0    1    1    1     1
 [6,]    1    0    1    1    1    0    1    1    1     1
 [7,]    1    1    1    1    1    1    1    1    1     0
 [8,]    1    0    1    1    1    1    1    1    1     1
 [9,]    0    1    0    1    1    0    1    1    1     0
[10,]    1    1    1    1    1    1    1    0    0     0
[11,]    1    1    0    1    1    1    0    1    1     1
[12,]    1    1    1    1    1    1    1    1    1     1
[13,]    1    1    1    1    1    1    1    1    1     1
[14,]    1    1    1    1    1    0    1    0    1     1
[15,]    0    1    1    0    1    1    1    1    1     1
>
> zlab <- as.factor( rep(1:m,each=n) )
> z <- model.matrix(~ -1+zlab)
>
> sim.data <- data.frame(y, x, zlab, u=u[zlab])
```

(i) Fit a fixed-effect, parallel-line logistic model as below. Explain why the estimated intercepts for groups $j = 1, 7$ are the same, those for groups $j = 2, 3, 10$ are the same, and those for groups $j = 4, 5, 8, 9$ are the same, as shown in `summary(sim.glm)`.

(ii) Fit a mixed-effect, parallel-line logistic model as below. Explain why the predicted intercepts for groups $j = 1, 7$ are the same, those for groups $j = 2, 3, 10$ are the same, and those for groups $j = 4, 5, 8, 9$ are the same, as shown in `my.out$u`.

(iii) Study algebraically whether the property observed in (ii) about the prediction of u_{i0}'s also holds when exact maximum likelihood estimation is used.

```
> sim.glm <- glm(y ~ -1+x +zlab, data=sim.data,
+                   family=binomial(link="logit"))
> summary(sim.glm)
......
Coefficients:
       Estimate Std. Error z value Pr(>|z|)
x        2.0415     0.7477   2.730  0.00632 **
zlab1    0.3989     0.7384   0.540  0.58903
zlab2   -0.3391     0.6714  -0.505  0.61355
zlab3   -0.3391     0.6714  -0.505  0.61355
zlab4    0.9066     0.8339   1.087  0.27694
zlab5    0.9066     0.8339   1.087  0.27694
zlab6   -0.6493     0.6637  -0.978  0.32790
zlab7    0.3989     0.7384   0.540  0.58903
zlab8    0.9066     0.8339   1.087  0.27694
zlab9    0.9066     0.8339   1.087  0.27694
zlab10  -0.3391     0.6714  -0.505  0.61355
......
>
> my.out <- logit.mixed(y, x, z, method="REML", niter=1000)
> my.out
$b
[1] 0.2473345 1.9161903

$u
 [1]  1.035559e-08 -3.106678e-08 -3.106678e-08  3.106678e-08
 [5]  3.106678e-08 -5.177797e-08  1.035559e-08  3.106678e-08
 [9]  3.106678e-08 -3.106678e-08

$sig.u
[1] 0.0001425456
```

```
......
$iter
[1] 670

$diff
[1] 9.987847e-10
```

Bibliography

Bates, D. M., Mächler, M., Bolker, B., Walker, S., et al. (2020). *lme4: Linear Mixed-Effects Models using 'Eigen' and S4* [R package version 1.1-37]. (cit. on p. 251)

Beck, A., & Teboulle, M. (2009). A fast iterative shrinkage-thresholding algorithm for linear inverse problems. *SIAM Journal on Imaging Sciences*, 2, 183–202. (cit. on p. 41)

Bohning, D., & Lindsay, B. G. (1988). Monotonicity of quadratic approximation algorithms. *Annals of Institute of Statistical Mathematics*, 40, 641–663. (cit. on p. 44)

Breiman, L., & Friedman, J. H. (1985). Estimating optimal transformations for multiple regression and correlation (with discussion). *Journal of American Statistical Association*, 80, 580–598. (cit. on p. 138)

Breiman, L., Friedman, J. H., Olshen, R. A., & Stone, C. J. (1984). *Classification and Regression Trees*. Chapman & Hall. (cit. on pp. 59, 123)

Breslow, N. E., & Clayton, D. G. (1993). Approximate inference in generalized linear mixed models. *Journal of American Statistical Association*, 88, 9–25. (cit. on p. 247)

Buja, A., Berk, R., Brown, L., George, E., Pitkin, E., Traskin, M., Zhao, L., & Zhang, K. (2019). Models as approximations I: Consequences illustrated with linear regression. *Statistical Science*, 34, 523–544. (cit. on p. 15)

Buja, A., Brown, L., Kuchibhotla, A. K., Berk, R., George, E., & Zhao, L. (2019). Models as approximations II: A model-free theory of parametric regression. *Statistical Science*, 34, 545–565. (cit. on p. 15)

Casella, G., & Berger, R. L. (2001). *Statistical Inference (2nd edition)*. Cengage Learning. (cit. on pp. x, 8, 11, 65)

Chen, H. Y. (2007). A semiparametric odds ratio model for measuring association. *Biometrics*, 63, 413–421. (cit. on p. 33)

Cox, D. R., & Reid, N. (1987). Parameter orthogonality and approximate conditional inference (with discussion). *Journal of Royal Statistical Society, Ser. B*, 49, 1–39. (cit. on p. 205)

Draper, N. R., & Smith, H. (1998). *Applied Regression Analysis (3rd edition)*. Wiley. (cit. on pp. 1, 24)

Efron, B., & Morris, C. (1973). Stein's estimation rule and its competitors — An empirical Bayes approach. *Journal of the American Statistical Association*, 68, 117–130. (cit. on p. 203)

Fourdrinier, D., Strawderman, W. E., & Wells, M. T. (2018). *Shrinkage Estimation*. Springer. (cit. on p. 259)

Friedman, J. H., Hastie, T., Hoefling, H., & Tibshirani, R. (2007). Pathwise coordinate optimization. *Annals of Applied Statistics*, 2, 302–332. (cit. on pp. 41, 118)

Friedman, J. H., Hastie, T., Tibshirani, R., Narasimhan, B., Tay, K., Simon, N., & Qian, J. (2021). *glmnet: Lasso and Elastic-Net Regularized Generalized Linear Models* [R package version 4.1-2]. (cit. on pp. 41, 46, 118, 119)

Fu, W. (1998). Penalized regressions: The bridge versus the lasso. *Journal of Computational and Graphical Statistics*, 7, 397–416. (cit. on p. 41)

Gelman, A., & Hill, J. (2006). *Data Analysis Using Regression and Multilevel/Hierarchical Models*. Cambridge University Press. (cit. on p. 199)

Ghosh, S., & Tan, Z. (2022). Doubly robust semiparametric inference using regularized calibrated estimation with high-dimensional data. *Bernoulli*, 28, 1675–1703. (cit. on pp. 33, 66, 69)

Gu, C. (2013). *Smoothing Spline ANOVA Models (2nd edition)*. Springer. (cit. on pp. 131, 155, 169)

Hastie, T., Tibshirani, R., & Friedman, J. H. (2009). *Elements of Statistical Learning (2nd edition)*. Springer. (cit. on pp. 17, 24, 39, 59, 112, 113, 123, 131, 145, 192)

Henderson, C. R. (1953). Estimation of variance and covanance components. *Biometrics*, 9, 226–252. (cit. on p. 208)

Henderson, C. R., Kempthorne, O., Searle, S. R., & Von Krosigk, C. N. (1959). Estimation of environmental and genetic trends from records subject to culling. *Biometrics*, 15, 192–218. (cit. on p. 208)

Hiriart-Urruty, J.-B., & Lemaréchal, C. (2001). *Fundamentals of Convex Analysis*. Springer. (cit. on pp. 73, 74)

Huber, P. J. (1967). The behavior of maximum likelihood estimates under nonstandard conditions. *Proceedings of the Fifth Berkeley Symposium on Mathematical Statistics and Probability*, 5, 221–233. (cit. on pp. 15, 104)

James, W., & Stein, C. (1961). Estimation with quadratic loss. *Proceedings of the Fourth Berkeley Symposium on Mathematical Statistics and Probability*, 1, 361–379. (cit. on pp. 203, 257, 259, 260)

Javanmard, A., & Montanari, A. (2014). Confidence intervals and hypothesis testing for high-dimensional regression. *Journal of Machine Learning Research*, 15, 2869–2909. (cit. on pp. 20, 26)

Kang, J. D., & Schafer, J. L. (2007). Demystifying double robustness: A comparison of alternative strategies for estimating a population mean from incomplete data (with discussion). *Statistical Science*, 22, 523–539. (cit. on p. 33)

Kim, J. K., & Haziza, D. (2014). Doubly robust inference with missing data in survey sampling. *Statistica Sinica*, 24, 375–394. (cit. on p. 33)

Lee, Y., Nelder, Y. A., & Pawitan, Y. (2017). *Generalized Linear Models with Random Effects: Unified Analysis via H-likelihood (2nd edition)*. Chapman & Hall. (cit. on pp. 199, 205, 208)

Liang, L., & Tan, Z. (2024). On ridge estimation in high-dimensional rotationally sparse linear regression. `https://arxiv.org/abs/2405.00974` (cit. on p. 85)

Liang, L., & Tan, Z. (2025). Debiased prediction inference with non-sparse loadings in misspecified high-dimensional regression models. `https://arxiv.org/abs/2507.10944` (cit. on p. 87)

Mallows, C. L. (1973). Some comments on Cp. *Technometrics*, 15, 661–675. (cit. on p. 189)

Mammen, E., & van de Geer, S. (1997). Locally adaptive regression splines. *Annals of Statistics*, 25, 387–413. (cit. on pp. 131, 178, 179)

Manski, C. F. (1988). *Analog Estimation Methods in Econometrics.* Chapman & Hall. (cit. on pp. 15, 104)

McCullagh, P., & Nelder, J. A. (1989). *Generalized Linear Models (2nd edition)*. Chapman & Hall. (cit. on pp. 88, 249)

McCulloch, C. E. (1997). Maximum likelihood algorithms for generalized linear mixed models. *Journal of the American Statistical Association*, 92, 162–170. (cit. on p. 251)

McCulloch, C. E., Searle, S. R., & Neuhaus, J. M. (2008). *Generalized, Linear, and Mixed Models (2nd edition)*. Wiley. (cit. on pp. 199, 241)

McGilchrist, C. A. (1994). Estimation in generalized mixed models. *Journal of the Royal Statistical Society, Ser. B*, 56, 61–69. (cit. on p. 247)

Morris, C. N., & Lysy, M. (2012). Shrinkage estimation in multilevel normal models. *Statistical Science*, 27, 115–134. (cit. on p. 260)

Osborne, M., Presnell, B., & Turlach, B. (2000). A new approach to variable selection in least squares problem. *IMA Journal of Numerical Analysis*, 20, 389–404. (cit. on p. 41)

Parikh, N., & Boyd, S. (2014). Proximal Algorithms. *Foundations and Trends in Optimization*, 1, 123–231. (cit. on p. 41)

Pinheiro, J. C., & Bates, D. M. (2000). *Mixed-Effect Models in S and S-PLUS.* Springer. (cit. on pp. 199, 213)

Pinheiro, J. C., Bates, D. M., & R Core Team. (2020). *nlme: Linear and Nonlinear Mixed Effects Models* [R package version 3.1-149]. (cit. on p. 223)

Prater, N. H. (1956). Estimate gasoline yields from crudes. *Petroleum Refiner*, 35, 236–238. (cit. on p. 223)

R Core Team. (2020). *R: A Language and Environment for Statistical Computing* [R version 4.0.3]. R Foundation for Statistical Computing. (cit. on pp. ix, x)

Ripley, B. D., Venables, W. N., Bates, D. M., Hornik, K., Gebhardt, A., & Firth, D. (2020). *MASS: Support Functions and Datasets for Venables and Ripley's MASS* [R package version 7.3-53]. (cit. on pp. 127, 250)

Robins, J. M., Mark, S. D., & Newey, W. K. (1992). Estimating exposure effects by modelling the expectation of exposure conditional on confounders. *Biometrics*, 48, 479–495. (cit. on p. 33)

Robins, J. M., & Rotnitzky, A. (2001). Comment on "Inference for semiparametric models: Some questions and an answer" by Bickel and Kwon. *Statistica Sinica*, 11, 920–936. (cit. on pp. 29, 33)

Robins, J. M., Rotnitzky, A., & Zhao, L. P. (1994). Estimation of regression coefficients when some regressors are not always observed. *Journal of American Statistical Association*, 89, 846–866. (cit. on p. 33)

Robins, J. M., Tchetgen-Tchetgen, E., Li, L., & van der Vaart, A. W. (2009). Semiparametric minimax rates. *Electronic Journal of Statistics*, 3, 1305–1321. (cit. on p. 71)

Robinson, G. K. (1991). That BLUP is a good thing: The estimation of random effects. *Statistical Science*, 6, 15–31. (cit. on p. 207)

Rousseauw, J., du Plessis, J., Benade, A., Jordaan, P., Kotze, J., Jooste, P., & Ferreira, J. (1983). Coronary risk factor screening in three rural communities. *South African Medical Journal*, 64, 430–436. (cit. on p. 112)

Ruppert, D., Wand, M. P., & Carroll, R. J. (2003). *Semiparametric Regression.* Cambridge University Press. (cit. on pp. 131, 199, 241)

Schall, R. (1991). Estimation in generalized linear models with random effects. *Biometrika*, 78, 719–727. (cit. on p. 247)

Scharfstein, D. O., Rotnitzky, A., & Robins, J. M. (1999). Adjusting for nonignorable drop-out using semiparametric nonresponse models (with discussion). *Journal of American Statistical Association*, 94, 1096–1120. (cit. on p. 33)

Searle, S. R. (1971). *Linear Models.* Wiley. (cit. on p. 1)

Searle, S. R., Casella, G., & McCulloch, C. E. (1992). *Variance Components*. Wiley. (cit. on pp. 199, 213)

Severini, T. A., & Staniswalis, J. G. (1994). Quasi-likelihood estimation in semiparametric models. *Journal of American Statistical Association*, 89, 501–511. (cit. on p. 26)

Speckman, P. (1988). Kernel smoothing in partial linear models. *Journal of Royal Statistical Society, Ser. B*, 50, 413–436. (cit. on p. 26)

Stamey, T., Kabalin, J., McNeal, J., Johnstone, I., Freiha, F., Redwine, E., & Yang, N. (1989). Prostate specific antigen in the diagnosis and treatment of adenocarcinoma of the prostate II radical prostatectomy treated patients. *Journal of Urology*, 16, 1076–1083. (cit. on p. 17)

Tan, Z. (2007). Understanding OR, PS, and DR, Comment on "Demystifying double robustness: A comparison of alternative strategies for estimating a population mean from incomplete data" by Kang and Schafer. *Statistical Science*, 22, 560–568. (cit. on p. 33)

Tan, Z. (2010). Bounded, efficient, and doubly robust estimation with inverse weighting. *Biometrika*, 97, 661–682. (cit. on p. 33)

Tan, Z. (2015). Improved minimax estimation of a multivariate normal mean under heteroscedasticity. *Bernoulli*, 21, 574–603. (cit. on p. 260)

Tan, Z. (2019). On doubly robust estimation for logistic partially linear models. *Statistics and Probability Letters*, 155, 108577. (cit. on p. 33)

Tan, Z. (2020a). Model-assisted inference for treatment effects using regularized calibrated estimation with high-dimensional data. *Annals of Statistics*, 48, 811–837. (cit. on pp. 33, 66, 69)

Tan, Z. (2020b). Regularized calibrated estimation of propensity scores with model misspecification and high-dimensional data. *Biometrika*, 107, 137–158. (cit. on p. 118)

Tan, Z. (2025). *A Course in Large-sample and High-dimensional Theory*. Book draft. (cit. on pp. x, 20, 26, 31, 103, 249)

Tan, Z., & Sun, B. (2020). *RCAL: Regularized Calibrated Estimation* [R package version 2.0]. (cit. on pp. 41, 51, 118, 119)

Tchetgen Tchetgen, E. J., Robins, J. M., & Rotnitzky, A. (2010). On doubly robust estimation in a semiparametric odds ratio model. *Biometrika*, 97, 171–180. (cit. on p. 33)

Tseng, P. (2001). Convergence of a block coordinate descent method for nondifferentiable minimization. *Journal of Optimization Theory and Application*, 109, 475–494. (cit. on p. 41)

Tsiatis, A. A. (2006). *Semiparametric Theory and Missing Data*. Springer. (cit. on p. 29) van de Geer, S., Bühlmann, P., Ritov, Y., & Dezeure, R. (2014). On asymptotically optimal confidence regions and tests for high-dimensional models. *Annals of Statistics*, 42, 1166–1202. (cit. on pp. 20, 26)

van der Vaart, A. W. (1998). *Asymptotic Statistics*. Cambridge University Press. (cit. on p. 104)

Venables, W. N., & Ripley, B. D. (2002). *Modern Applied Statistics with S (4th edition)*. Springer. (cit. on pp. 127, 223, 227)

Vermeulen, K., & Vansteelandt, S. (2015). Bias-reduced doubly robust estimation. *Journal of American Statistical Association*, 110, 1024–1036. (cit. on p. 33)

Wahba, G. (1990). *Spline Models for Observational Data*. SIAM. (cit. on pp. 131, 169)

Weisberg, S. (2014). *Applied Linear Regression (4th edition)*. Wiley. (cit. on p. 1)

White, H. (1980). A Heteroskedasticity-Consistent Covariance Matrix Estimator and a Direct Test for Heteroskedasticity. *Econometrica*, 48, 817–838. (cit. on p. 65)

White, H. (1982). Maximum likelihood estimation of misspecified models. *Econometrica*, 50, 1–25. (cit. on pp. 15, 104)

Wolfinger, R., & O'Connell, M. (1993). Generalized linear mixed models a pseudo-likelihood approach. *Journal of statistical Computation and Simulation*, 48, 233–243. (cit. on p. 247)

Wu, T. T., & Lange, K. (2010). The MM alternative to EM. *Statistical Science*, 25, 492–505. (cit. on p. 42)

Zhang, C.-H., & Zhang, S. S. (2014). Confidence intervals for low dimensional parameters in high dimensional linear models. *Journal of the Royal Statistical Society, Ser. B*, 76, 217–242. (cit. on pp. 20, 26)

Index

For Product Safety Concerns and Information please contact our EU
representative GPSR@taylorandfrancis.com
Taylor & Francis Verlag GmbH, Kaufingerstraße 24, 80331 München, Germany

www.ingramcontent.com/pod-product-compliance
Ingram Content Group UK Ltd.
Pitfield, Milton Keynes, MK11 3LW, UK
UKHW022220190926
475702UK00003B/458

* 9 7 8 1 0 4 1 1 5 2 7 7 4 *